HOUSING

The Essential Foundatic

Housing: The Essential Foundations provides a comprehensive, multi-disciplinary introduction to housing studies. Introducing the many diverse aspects of housing within a single volume, this topical book is essential reading for students embarking on degree and diploma courses in housing, surveying, town planning and other related subjects. Professionals within these fields will also find the book valuable as a source of up-to-date information and data.

Uniquely multi-disciplinary and including a wealth of illustrations and examples, *Housing: The Essential Foundations* focuses on key topics which include:

- housing policy and finance prior to and after Thatcherism
- equal opportunities and housing organisations
- town planning and housing development
- housing design and development
- environmental health and housing
- property and housing law
- management, policy-making and politics
- future policy issues under the new Labour government after 1997

Housing, often the largest item of personal expenditure, is humankind's most essential need after nourishment. Examining ways to satisfy this need, whether through an adequate injection of public or private investment or through mixed funding schemes, the authors stress the importance of housing market activity that accords with good planning practice, legislation, democratic decision-making, economy and efficiency.

Paul Balchin is Reader in Urban Economics and **Maureen Rhoden** is Course Director and Senior Lecturer in Housing Studies, both at the University of Greenwich.

HOUSING

The Essential Foundations

Edited by Paul Balchin and Maureen Rhoden

London and New York

First published 1998
by Routledge
11 New Fetter Lane, London EC4P 4EE

Simultaneously published in the USA and Canada
by Routledge
29 West 35th Street, New York, NY 10001

Typeset in Garamond by J&L Composition Ltd, Filey, North Yorkshire

Printed and bound in Great Britain by the Bath Press

British Library Cataloguing in Publication Data
A catalogue record for this book is available from the British Library

Library of Congress Cataloging in Publication Data
Housing: the essential foundations/edited by Paul Balchin and Maureen Rhoden.
p. cm.
Simultaneously published in the USA and Canada
Includes bibliographical references and index.
1. Housing—Great Britain. 2. Housing policy—Great Britain.
I. Balchin, Paul N. II. Rhoden, Maureen, 1959– .
HD7333.A3H6975 1998
363.5′0941—dc21 97–28274

ISBN 0–415–16007–3
0–415–16008–1 (pbk)

CONTENTS

PLATES

FIGURES

TABLES

NOTES ON CONTRIBUTORS

Paul Balchin Reader in Urban Economics, School of Land and Construction Management, University of Greenwich, London. Member of the European Network for Housing Research.

Gregory Bull Senior Lecturer in Urban Economics, School of Land and Construction Management, University of Greenwich, London. Member of the European Network for Housing Research.

Pauline Forrester Senior Lecturer in Environmental Health, School of Environmental Sciences, University of Greenwich, London. Member of the European Network for Housing Research.

David Isaac Associate Head and Professor of Real Estate Management, School of Land and Construction Management, University of Greenwich, London.

R. Shean McConnell Associate Senior Lecturer in Town Planning and Management, School of Land and Construction Management, University of Greenwich, London. Formerly, Head of Department of Town Planning, the Polytechnic of the South Bank.

John O'Leary Senior Lecturer in Planning, School of Land and Construction Management, University of Greenwich, London.

Mark Pawlowski Reader in Property Law, School of Law, University of Greenwich, London. Visiting Lecturer in Land Registration, University College, London.

Maureen Rhoden Senior Lecturer in Housing, School of Land and Construction Management, University of Greenwich, London. Member of the European Network for Housing Research.

Jane Weldon Senior Lecturer in Construction Technology, School of Land and Construction Management, University of Greenwich, London.

INTRODUCTION

In the United Kingdom, as in most developed economies, housing is a major and often the largest item in personal expenditure. It is also an important determinant of people's life chances and, next to agriculture, housing normally constitutes the largest single land use. Clearly, apart from nourishment, shelter is humankind's most essential material need. Housing, however, was selected to be the principal victim of public expenditure cuts during the long period of the Thatcher and Major administrations. Whereas (at 1994–5 prices) public spending on housing amounted to £11.8 billion in 1980–81, by 1995–6 it had plummeted to only £4.7 billion, a decrease of 65 per cent. Housing's share of public expenditure had fallen from 5.1 to a derisory 1.5 per cent over the same period (Treasury, 1995).

As consequences of these cuts, the annual number of housing completions in the social rented sector in the United Kingdom fell from 110,000 in the 1980s to only 37,400 in 1994 (Department of the Environment, 1996); nearly 1.5 million houses were declared unfit in England in 1991 (Department of the Environment, 1993), with equivalent numbers in Scotland, Wales and Northern Ireland; and the number of homeless households in Great Britain accepted by local authorities soared from about 70,000 in 1979 to 179,000 in 1992 (Wilcox, 1996). By international standards, it was evident that far too little was being spent on housing construction. Whereas in Canada in 1980–93 an annual average of 6.1 per cent of the gross domestic product was attributable to gross fixed investment in residential buildings, and 5.8 per cent was invested in France, Germany and Italy over the same period, in the United Kingdom the equivalent proportion was only 3.6 per cent (OECD, 1994, 1995).

Although public policy is clearly instrumental in shaping the quantitative and qualitative attributes of the housing stock, a number of professions are important actors in the functioning of housing markets, reflecting the multi-disciplinary nature of housing education, training and practice. Within the parameters set by government, economists, valuers, sociologists, town planners, builders, building surveyors, environmental health officers, lawyers and housing managers all play an important role against difficult odds in attempting (with varying degrees of success) to ensure that the supply of housing matches the demand or need for accommodation.

A number of professional institutions thus require elements of housing to be included within the syllabuses of accredited degree courses or their equivalent. This book is intended to introduce students to material central to the concerns of the Chartered Institute of Building, the Chartered Institute of Housing, the Chartered Institute of Environmental Health, the Royal Institution of Chartered Surveyors and the Royal Town Planning Institute. The book should also be useful, in part, to prospective law practitioners, to undergraduates and postgraduates on non-vocational courses where housing is subject to, for example, economic, political or sociological consideration, and to practising professionals as a source of reference.

Although there are many books in print on different aspects of housing, reflecting discrete professional and academic interests, there is an absence of a wide-ranging text covering the essential foundations of housing studies. As a collaborative venture, this book is intended, in large part, to fill this vacuum but, since each relevant profession or academic school is concerned with housing from a different perspective, there is no generally acceptable order in which the appropriate subject-matter should be studied. Inevitably any selected sequence will not suit all professionals or academics. There is nevertheless an attempt to cover the necessary material in a logical order, as far as possible. The book begins with a chapter by Paul Balchin which, in order to set the scene for an interdisciplinary examination of housing in the United Kingdom during the last two decades of the twentieth century, provides a brief history of housing policy from the nineteenth century to the onset of Thatcherism in 1979. In Chapter 2, Gregory Bull seeks to explain the micro- and macro-economics of housing – to shed light on the rationale and some of the shortcomings of housing policy. The relationship between housing policy and finance during the Conservative government's period of office, 1979–97, is critically assessed by Paul Balchin, David Isaac and Maureen Rhoden in Chapter 3. Then, in Chapter 4, Maureen Rhoden considers equal opportunities in relation to housing – looking specifically at issues of concern to the elderly, to women, to black and ethnic minority households, and to sufferers of ill-health. John O'Leary explores the interface of housing and town planning in Chapter 5. Jane Weldon, in Chapter 6, focuses on housing development and housing design, and in Chapter 7, Pauline Forrester assesses housing renewal within an environmental health context. In Chapter 8, Mark Pawlowski undertakes a detailed examination of legal studies, property and housing law. In Chapter 9, Shean McConnell discusses management and organisation within the housing arena, and in Chapter 10, Maureen Rhoden analyses policy-making and politics at central and local levels of government and within housing associations. In Chapter 11, Paul Balchin, Maureen Rhoden and John O'Leary conclude by predicting some future developments and causes of concern across the spectrum of housing studies. Although each chapter focuses on a different aspect of housing studies, 'pointers' to where associated subject-matter can be found elsewhere in the book are provided at the end of each chapter, ensuring an element of cohesion. Where appropriate, case studies are presented to relate some of the material examined within the text to more detailed and 'real life' situations. Questions for discussion are also included, together with recommended reading.

We must acknowledge a very great debt we owe to other present and past colleagues and to a wide range of people in the various professions concerned with the built environment who have stimulated and advised us in the preparation of the book. We would like particularly to thank Sayeed Rahman and Gary Holt for technical advice, plans and drawings, and Andy Bradley and Simon Clark for photographic material presented in Chapter 6. In addition, we are very grateful to Pauline Newell, Gwen Oldfield and Cherie Apps who painstakingly typed and retyped most of the manuscript and to Sue Lee and Pete Stevens who produced most of the artwork within the text. Last but not least, we would like to thank our respective families for their continual encouragement and patience.

Paul Balchin
Maureen Rhoden
London, summer 1997

REFERENCES

Department of the Environment (1993) 'English House Condition Survey, 1991', HMSO, London.

—— (1996) 'Housing and Construction Statistics, 1986–95', HMSO, London
OECD (1994) 'National Accounts', Organisation for Economic Co-operation and Development, Paris.
—— (1995) 'National Accounts', Organisation for Economic Co-operation and Development, Paris.
Treasury (1995) 'Public Expenditure', Cm 2821, HMSO, London.
Wilcox, S. (1996) 'Housing Review 1996/97', Joseph Rowntree Foundation, York.

1

AN OVERVIEW OF PRE-THATCHERITE HOUSING POLICY

Paul Balchin

The purpose of this introductory chapter is to provide a review of the evolution of housing policy from the nineteenth century to 1979. It was during this period that present-day housing markets evolved and present-day causes for concern emerged. During the latter years of this period, owner-occupation succeeded private renting as the dominant form of housing tenure, and the number and proportion of local authority dwellings reached its peak, but housing needs were far from satisfied in each of the housing sectors. This chapter specifically examines:

- The Victorian origins of housing policy.
- The development of housing policy, 1914–39, emphasising rent control and decontrol in the private rented sector, the introduction of housing subsidies in the local authority sector, and the gradual expansion of owner-occupation.
- Post-war housing policy, 1945–51, focusing on the extension of rent control and the increase in subsidies to local authority housing.
- The 'consensus years', 1952–79, reviewing rent decontrol and the introduction of rent regulation, house-building in the local authority sector, private sector rehabilitation, housing associations, and the substantial growth of owner-occupation.
- Housing finance reform.
- The changing pattern of tenure, 1913–79, and its regional distribution.

INTRODUCTION

Housing issues are frequently examined within the context of recent economic, social and technological change, and housing problems are often attributed to the policies of the current or previous government. But in any of the older industrial countries, the present state of housing is in large part a legacy of policies reaching back to the period of rapid industrialisation and urbanisation in the nineteenth and early twentieth centuries.

Since its inception, housing policy in the United Kingdom has been conditioned by the dominant political philosophy of the time. The virtual absence of housing legislation in the first half of the nineteenth century was a reflection of a liberal or *laissez-faire* approach to most matters – economic and social – and of an adherence to the free market. By the late nineteenth century, however, it was increasingly recognised that legislation was necessary to enhance environmental health in our towns and cities and that intervention into housing markets was essential if living conditions were to be improved and the productivity of urban workforces raised.

After the First World War, but more particularly after the Second, a mixed economy developed whereby state intervention and the market coexisted. For a few years in the inter-war period, and for much longer interludes after 1945, corporatism replaced a *laissez-faire* adherence to the free market.

The Beveridge Report (1942), heralding the expansion of the welfare state, highlighted the need to protect 'from the cradle to the grave' all individuals and the family from, among other inflictions, the squalor of poor housing. There was also a broad consensus that policies, fairly even-handedly, should further the interests of all households, whether they were private or social sector tenants or owner-occupiers. Although, of course, there were periods in the decades following the Second World War when Conservative governments, to a limited extent, reverted to market criteria in formulating policy, it was not until after 1979 that Conservative administrations adopted a vigorous neo-liberal approach to housing – characterised most notably by reduced public investment in house-building, the marketisation of rents in both the private and social sectors, and the privatisation of social housing (see Chapter 3).

THE VICTORIAN ORIGINS OF HOUSING POLICY

The introduction of housing policy in the nineteenth century was a direct response to the economic and social legacy of the industrial revolution. The population of Great Britain increased from 11.9 million in 1811 to 40.8 million in 1911 and it had become largely urban. Poverty and squalor were manifested in the condition of housing in our towns and cities. The increase in the supply of labour enabled employers to keep wages to the minimum – often to subsistence level – but urban landlords, in their desire to maximise profit on their property, developed housing at a very high density and of appalling quality. Overcrowding and disease were the inevitable results throughout the industrial areas of Britain.

During the first half of the nineteenth century, *laissez-faire* attitudes in government largely prevailed. Within a free market, almost all working class housing was privately rented. Landlords needed to raise about two-thirds of the value of their property on mortgage, and if interest rates increased, landlords passed on the cost as much as possible in higher rents in order to maintain profitability. Gauldie (1974) has shown that rents rose steadily in the period 1780–1918 (even when the general price trend was downwards), and that in the nineteenth century the average working class family paid 16 per cent of their income in rents in contrast to 8–9 per cent paid by middle class families. The majority of private landlords were relatively small capitalists content with a secure return on their capital – housing comparing very favourably with other forms of investment. Until the extension of limited liability in the late nineteenth century, investment in joint stock companies was unattractive to those with modest means. But with the development of the stock exchange and building societies, the expansion of government and municipal stock, and increased investment opportunities overseas, private rented property became much less attractive as an investment.

On a very limited scale there was, however, an improvement in the condition of working class housing. 'Not-for-profit' housing associations originate from 1830 when the Labourer's Friendly Society was formed. The society built very few houses but these were of a higher standard than most low-income dwellings at the time. However, throughout the rest of the century, poor households failed to attract financial backing. Charitable trusts therefore attempted to show that private enterprise could provide decent housing for the working classes. Bodies such as the Guinness Trust, the Peabody Donation Fund, the Joseph Rowntree Trust and the Sutton Dwellings Trust, formed in the nineteenth century, are still active today in supplying general family housing. Higher-paid workers, however, often showed an interest in owner-occupation and set up building societies – 'self-help' organisations established initially to divert the savings of members into house-building.

In the second half of the nineteenth century, a *laissez-faire* approach to environmental and social problems gradually became discredited. Not only was an improvement in housing deemed necessary for health reasons, but it was thought that it would indirectly raise productivity at work and alleviate political agitation at a time when the majority of the population was disenfranchised. Trailing the introduction of public health legislation (the Public Health Act of 1848), housing legislation was added step by step to the statute book.

The Labouring Classes Lodging Houses Act and the Common Lodging Houses Act (both of 1851) were targeted at mobile labour, and in turn permitted local authorities to provide temporary housing and controlled and monitored private common lodging houses. The success of these Acts was, however, thwarted by the unwillingness of ratepayers to provide the necessary revenue. Subsequently, the Artizans and Labourers Dwellings Act of 1868 (the Torrens Act) and the Artizans and Labourers Dwellings Improvement Act of 1875 (the Cross Act) were intended to promote slum clearance, but because ratepayers were reluctant to finance clearance, and as most slum housing was sited on high-value land in the inner urban areas, the Acts were ineffectual. Authorities also had the problem of having little or no accommodation to offer displaced households.

The Public Health Acts of 1875 and 1890 were nevertheless having a favourable effect on the quality of new private and later public sector housing and on residential environments in the emerging inner suburbs. It was unfortunate, however, that increased public intervention from the 1848 Act onwards further reduced the attraction of housing investment. Controls often resulted in either higher rents (to compensate landlords for improvement costs incurred) or a decrease in the supply of accommodation if investment became less profitable.

Plate 1.1 By-law housing developed under the Public Health Acts of 1875 and 1890
Two-storey terraced housing developed at the end of the nineteenth century, London Borough of Merton

Public sector house-building thus became essential if affordable housing for lower-income households was to be provided. In 1869, a local authority in Liverpool was the first to build municipal housing for rent and, following the Housing of the Working Classes Acts of 1885 and 1900, local authorities and particularly the newly constituted London County Council and London's boroughs developed a number of large housing estates often with their own workforces – the direct labour organisations. Under these Acts, local authorities (without any financial assistance from central government) thus became the suppliers of housing for general needs and partly usurped the role of self-help organisations and charities.

THE DEVELOPMENT OF HOUSING POLICY, 1914–39

Rent control and decontrol

Despite the introduction of local authority housing at the end of the nineteenth century, private rented accommodation still constituted 90 per cent of the nation's housing stock at the outbreak of the First World War (with owner-occupation accounting for most of the rest). There was, moreover, still a tendency for private landlords to raise rents to their highest possible level – a practice particularly prevalent during the first year of the war, when housing shortages were exacerbated by a dramatic reduction in the rate of house-building. Following rent strikes across the country (but most notably in Glasgow), the government introduced rent control in 1915 by the Increase in Rent and Mortgage Interest (War Restrictions) Act. Rents were controlled at 1914 levels on property where rateable values were less than £35 in London, £30 in Scotland and £26 elsewhere in the United Kingdom (Table 1.1). In the years that followed, the Act inevitably discouraged investment in rented property. The Increase in Rent and Mortgage Interest (Restrictions) Act of 1920 substantiated these fears. Rent control was continued into peacetime and applied to properties with rateable values of less than £105 in London, £90 in Scotland and £78 elsewhere, but the increase was more a reflection of increased property values and rerating than any significant extension of control. The

Table 1.1 Rent control and decontrol, 1915–38

Year of Rent Act	*Major provisions*		*Rateable value (£)*		
			London	*Scotland*	*Elsewhere*
1915	Rents controlled at 1914 levels	Not exceeding	35	30	26
1920	Rents controls continued	Not exceeding	105	90	78
1923	Decontrol by possession; letting freed from control when tenant left	Not below	45	45	45
1933	(a) Decontrol of houses	Not below	45	35	35
	(b) Decontrol of possession	Not below	35	20	20
	(c) Decontrol on registration of possession	Not below	35	20	20
	(d) No decontrol by possession unless decontrolled 1912 to 1933, and registered				
1938	(a) Decontrol of houses	Not below	35	20	20
	(b) No decontrol by possession or self-contained dwellings	Not exceeding	35	20	20

Onslow Report (1923) confirmed that rent control deterred investment in new housing. It stated that the 1915 and 1920 Acts had made private enterprise reluctant to perform its traditional function of supplying working class housing, but warned that instant decontrol would cause hardship.

The Rent and Mortgage Interest (Restrictions) Act of 1923 generally continued the policy of rent control, but there was immediate decontrol if the landlord gained possession, or if sitting tenants accepted a lease of two years or more, or if a lease was granted fulfilling certain conditions. The Act remained in force for ten years. The Marley Report (1931) investigated the working of the Act, and showed that of the 1.5 million houses built from 1918 to 1931, 600,000 local authority dwellings constituted virtually all the new accommodation for the working classes. Rent control was clearly deterring investment in low-income housing, although the 1923 Act had worked well for middle-income housing where a large measure of decontrol had not caused hardship to tenants, while it encouraged private developers/landlords to increase supply. The Report proposed that rents should be immediately decontrolled where supply exceeded demand (usually in the case of large houses); rents should be decontrolled when landlords obtained vacant possession (in the case of medium-size houses where supply equalled demand); and rents should continue to be controlled where demand exceeded supply (usually in the case of small houses). Following these proposals, the Rent and Mortgage Restrictions (Amendments) Act of 1933 divided houses into three classes. Class A houses (the most expensive properties) were decontrolled immediately; Class B houses (those intermediate in price) were decontrolled on vacant possession; and Class C houses (those with rateable values less than £20 in London and £13 elsewhere) remained controlled, regardless of whether there was a change of tenant. The Act was to remain in force until 24 June 1938 and no longer.

The Ridley Report (1938) examined the working of the 1933 Act and was critical of the effects of the control of Class B houses. The Increase of Rent and Mortgagee Interest (Restrictions) Act of 1938 consequently decontrolled the higher-rent houses in Class B (those with rateable values above £35 in London and £20 elsewhere), but abolished decontrol by vacant possession of the lower-rent self-contained dwellings in that class.

In the period 1923–38, approximately 4.5 million dwellings had been decontrolled, and investment in the development of medium- and high-rent housing had become attractive. There were still, however, 4 million controlled dwellings, and at the lower end of the market properties were mainly pre-1914 in origin, usually terraced, in poor condition and lacking basic amenities.

The introduction of housing subsidies

It was not until after the First World War that local authority housing really 'took off'. In 1919, 610,000 new houses were needed in Britain as house-building had virtually ceased throughout the duration of the war, and at the 1919 general election, Lloyd George promised to supply homes 'fit for heroes' to attract the ex-serviceman's vote (Swenarton, 1981). After his election win, his coalition government introduced a housing programme in which local authorities and public utility societies (akin to housing associations) were to build 500,000 houses within three years. This was incorporated in the Housing and Town Planning Act of 1919 (the Addison Act), a watershed in British social history. Under the Act, local authorities initially had the duty of surveying housing needs in their area (an innovation) and then, having quantified the shortage, generous 'bricks and mortar' subsidies were introduced to help meet the needs of working class families. Local authority losses in house-building were limited to the product of a penny rate with the Exchequer automatically meeting all additional losses – losses which inevitably would be high as rents were to be pegged to the level of prevailing 'working class' rents in the area, adjusted to the means of the tenant. Many rents were therefore equal to controlled rents. The Act also fixed standards for new housing well above the normal conditions of working class houses. Addison 'more than any other man thereby established the principle that housing was a social service, and later governments had to take up his task' (Taylor, 1965: 148).

But the 1919 Act gave little incentive for local authorities to economise, and the capacity of the construction industry was strained, pushing up costs and exacerbating post-war inflation. Subsidies simultaneously increased as houses costing, for example, £400 to build in 1918 were costing over £900 by 1920. Exchequer grants were therefore sharply restricted in 1921 and stopped in 1922. The Addison Act nevertheless resulted in 213,000 houses being built.

A new subsidy system was devised and included in the Conservatives' Housing Act of 1923 (the Chamberlain Act). Chamberlain believed that the rising cost of housing was the result of Addison's open-ended subsidies rather than a cause, and introduced a subsidy which in form was to continue through to the Housing Finance Act of 1972. It consisted of a fixed annual Exchequer payment of £6 per dwelling for twenty years – available to both the public and the private sectors. The government showed a preference for the latter sector as it built houses for sale, and local authorities would only qualify for the subsidy if they built small and sub-standard houses in areas where private enterprise could not meet demand.

Soon after the first Labour government took office, insisting that more and better houses be built, it repealed the Chamberlain Act and replaced it with the Housing Act of 1924 (the Wheatley Act). The subsidy was raised to £9 for forty years, and the Act

Plate 1.2 Local authority housing developed under the Housing Act of 1924
Two-storey terraced housing developed on the St Helier Estate, London Borough of Merton, by the London County Council in the early 1930s

transferred the main responsibility for housing back to local authorities, which did not now have to demonstrate that private enterprise could not meet local needs before they could proceed with building. Rents were to be equal to 'appropriate normal rents', interpreted as being equal to controlled rents in the private sector. The difference between this rent level and market rents was to be offset by a minimum rate fund contribution of at least half of the Exchequer subsidy.

Although 503,000 dwellings were built under the 1923 and 1924 Acts, it was doubtful whether the needs of the poorest working class families had been met. Council housing was regarded by many as prestigious and it was going mainly to the lower middle classes, such as clerks, teachers and shop-workers. The main working class areas – the inner cities – had an insufficient rate base to take advantage of Exchequer grants. It was also in these areas that slum clearance was necessary, but only 11,000 unfit houses were demolished in England and Wales in 1923–39. Although the 1924 Act provided 50 per cent Exchequer grants for slum clearance and rehousing, the complexity of the way in which this subsidy was calculated was an inhibiting factor, and in 1929 it was withdrawn by the Conservative government (Chamberlain again having responsibility for housing).

The return of a Labour government produced the Housing Act of 1930 (the Greenwood Act). Generous subsidies were granted for slum clearance, based not on the number of homes demolished or provided but on the number of persons displaced. Extra subsidies were available if displaced families were rehoused in blocks of flats (of over three storeys) on expensive sites within the inner urban areas. Rents, although still approximately based on controlled rents, were differentiated according to the means of tenants, and a system of rent rebates was introduced.

The National government's Housing Act of 1933 discontinued the Wheatley subsidies and all government provision for new public housing for general needs – even though the 1931 Census had shown that there was a deficit of 1 million dwellings in relation to households. The Conservative-dominated government believed that new council housing should be confined to those households displaced by slum clearance, and that the private sector should satisfy the needs of the rest. This was a confirmation of the Conservative Party's long-held belief that council housing was a restricted welfare service and not a facility to meet a general need for rented accommodation.

Under the Housing Act 1935, the emphasis again shifted to the problems of low-income housing. The Act charged local authorities with the duty of relieving overcrowding (defined as an occupancy rate of more than two persons per room), and further legislation – the Housing Act of 1936 – pooled the subsidy and rent provisions of previous Acts, giving local authorities greater discretion in fixing rent levels and giving rent rebates. From 1935 until the outbreak of the Second World War, local authorities concentrated on slum clearance that reduced overcrowding, 400,000 houses being constructed for these purposes in Great Britain in this period. Even more replacement houses might have been built, but the Housing Act of 1938 reduced the level of subsidy to local authority housing, though special grants were available for high flats. It may have been thought that the problems of low-income urban housing had eased, and that 1.3 million council dwellings in 1939 (about 11 per cent of the total housing stock) was the maximum which should be developed, taking into account that the sector was not intended by the government to meet general housing needs.

The gradual expansion of owner-occupation

In the first three decades of the twentieth century, owner-occupation was by no means considered by most households to be the 'ideal' or 'natural' form of tenure. During the 1920s, except for the council houses produced under the Housing Act of 1923 (the Chamberlain Act), local authority dwellings were 'in every sense the ideal, being better produced

Plate 1.3 Local authority housing developed under the Housing Act of 1938
Four-storey flats in the London Borough of Richmond, developed by the London County Council immediately before the Second World War

at high standard for the better-off members of the working class' (Clarke and Ginsburg, 1975: 5).

Owner-occupation only began to be popularly attractive when local authority housing became generally restricted to the displaced families of slum clearance schemes in the 1930s. The desire for home ownership was more of a response to a lack of choice, than a reaction against renting. The housing policy of the 1930s was 'directly associated with the drive to make the better-off members of the working class into owner occupiers' (Clarke and Ginsburg, 1975: 5).

During the inter-war period, the increase in home-ownership was facilitated by the tenure-conversion of private rented housing into owner-occupation as a consequence of rent control; the provision of subsidies (under the Chamberlain Act) for private construction; falling building costs and interest rates in 1919–35; local authority guarantee of mortgages; and the lengthening of mortgage repayments from 15 to 20–25 years. Increased car-ownership, moreover, and the lack of effective suburban planning, meant that cheap land could be used for extensive speculative house-building, often in the form of ribbon development.

POST-WAR HOUSING POLICY, 1945–51

During the Second World War there was virtually no house-building, and 208,000 dwellings were

Plate 1.4 Owner-occupied suburban housing of the 1930s
Speculative ribbon development of inter-war private housing, London Borough of Merton

completely destroyed, 250,000 made uninhabitable and over 250,000 seriously damaged (equal in total to 5 per cent of the housing stock). As much as 33 per cent of the stock had been damaged and, together with the rest, remained largely unrepaired or unmaintained throughout the six years of the war. In this period, the population had grown by 1 million: the total housing shortage was therefore about 1,460,000, not including unfit and obsolete housing which needed replacing. The construction industry was unable to meet this demand in the immediate post-war years. The workforce had fallen to a third of its size in 1938, and materials (many of which had to be imported) were scarce and costly.

After the war, it was clear that the private rented sector would play little part in satisfying house needs. The Interest Restriction Act of 1939 had abolished decontrol on vacant possession and extended rent control to over 10 million dwellings with rateable values of less than £100 in London, £90 in Scotland and £75 elsewhere (Table 1.2). Until 1957 the rents of these properties were frozen at their 1939 level but the general price level had increased by 97 per cent by 1951. Although this helped to ensure affordability at a time of war and subsequent uncertainty, supply was severely constrained since there was little or no incentive to invest in new housing in this sector, and landlords considered it advantageous to sell off their properties for owner-occupation whenever they had the opportunity.

Public sector house-building was thus to dominate the period 1945–51 and, together with the National Health Service, National Insurance, education reform, a comprehensive system of town planning

Table 1.2 Legislation controlling and regulating rents, 1939-65

Year of Rent Act	*Major provisions*		*Rateable value (£)*		
			London	*Scotland*	*Elsewhere*
1939	Rents controlled	Not exceeding	100	90	75
1957	Rents decontrolled	Not below	40	40	30
	Owner-occupied houses partly let				
	New unfurnished dwellings				
	Remaining tenancies had rents fixed at twice their 1939 rateable value				
1965	Rent regulation	Not exceeding	400	200	200
	Rent control continued	Not exceeding	110	80	80

and Keynesian economic policy, council housing became a pillar of the welfare state. As in the years immediately following the First World War, the emphasis at first was on building for general need to meet an acute housing shortage – one which seemed likely to remain, as marriages in 1945–8 increased by 11 per cent over the period 1936–9, and the number of births increased by 33 per cent over the pre-war rate.

The Housing (Financial Provisions) Act of 1946 – a personal triumph for Labour's Minister of Health (and minister responsible for housing), Aneurin Bevan – provided a generous basic subsidy for local authority housing of £16 10s (£16.50) per dwelling per annum over sixty years, a sum which varied according to the needs of different authorities. Building licences were introduced in the private sector so that house-building would respond to 'need' rather than exclusively to the ability to pay. It was hoped that this would ensure that materials and labour would be available for local authority house-building – in contrast to the situation in 1919. It was necessary, however, for a system of building quotas to be imposed on public sector building.

From the outset of the post-war housing programme, it was stressed that council housing was a 'general-needs' tenure – it was not intended solely for the poor, the underprivileged or the population of traditional working class areas. Bevan's Housing Act of 1949 incorporated this view into legislation. It removed the 'ridiculous inhibition' restricting local authorities to the provision of houses for the 'working class'. Instead they could attempt to meet the varied needs of the whole community.

Bevan was also concerned with improving the quality and increasing the floor space of public sector housing. The minimum size of a three-bedroom house had been fixed at 750 ft^2 (75 m^2) in the 1930s. In 1944, the Ministry of Housing manual prescribed 800–900 ft^2 (80 m^2–90 m^2) and in the same year the Dudley Committee recommended 950 ft^2 (95 m^2). Bevan accepted the latter proposal and encouraged local authorities to adopt even higher standards where this was possible.

Between 1945 and 1951 a total of 1.01 million houses were built in Great Britain – 89 per cent being local authority dwellings. Output accelerated from 55,400 completions in 1946 to 184,230 by 1948. Overall this was a great achievement in view of post-war materials shortages and the need to reconstruct industry, curb inflation and correct balance of payments deficits. More importantly, the quality of new housing was improved and it was re-established that the public sector had a role to play in satisfying general housing need.

Housing rehabilitation, however (or 'patching-up' as it was disparagingly called), was discouraged in the immediate post-war period – the White

Plates 1.5 and 1.6 Low- and medium-rise post-war local authority housing
Five- and eight-storey flats developed by the London County Council under the provisions of the Housing (Financial Provisions) Act of 1946, London Borough of Wandsworth

Paper, *Capital Investment in 1948* (Treasury, 1947), setting out the government's intentions to steer resources to council house-building. The Housing Act of 1949 nevertheless introduced improvement grants for private owners and improvement subsidies for local authorities.

THE UNEASY CONSENSUS, 1952–79

Although housing policies sometimes diverged, for example in respect of the private rented sector – Conservative governments favouring decontrol and Labour administrations implementing regulation – there was for three decades cross-party agreement that the local authority stock should be expanded (albeit by different forms of subsidisation and rent regimes), that public expenditure on housing rehabilitation should broadly match that on new house-building, that housing associations should be encouraged to supplement the local authorities as providers of social housing, and (by means of tax relief, tax exemption and cash subsidies) the growth of owner-occupation should be supported.

Rent decontrol, 1957–65

Landlords obviously wanted the 1939 Act to be repealed, and within the Conservative government there was a desire to return to free market rents. The Housing (Repairs and Rent) Act of 1954 permitted landlords to raise controlled rents if proof of recent repair expenditure could be produced and the dwelling was subsequently in a good state of repair. But the Act was complex and generally not successful. The Rent Act of 1957 consequently set out to decontrol 5 million dwellings in an attempt to increase the supply of private rented accommodation. Its aims were to enable landlords to afford to repair and maintain their properties and to remove the incentive to sell for owner-occupation. The main provisions of the Act were as follows:

1 Dwellings with rateable values greater than £40 in Greater London and Scotland and £30 elsewhere were completely decontrolled on vacant possession or by 'agreement' with the tenant. It was forecast that this would free approximately 750,000 dwellings, but the Act actually decontrolled only 400,000.
2 Owner-occupied houses (about 4.75 million) and any houses falling vacant were immediately decontrolled.
3 New unfurnished dwellings were freed from control.
4 The remaining controlled dwellings had rents fixed at twice their 1956 gross rateable value if the landlord was responsible for repairs. Rents therefore only increased to twice the 1949 rateable value (since rateable values were still at pre-war level), but by 1957 the general price level was 156 per cent higher than in 1939 and house prices were 200 per cent higher.

Despite the 1957 Act decontrolling only 2.5 million rented houses by 1965 (half of the intended number), the Milner Holland Report (1965) highlighted the many undesirable effects upon tenants of decontrol and prompted a redirection of rent policy.

Rent regulation, 1965–70

The Rent Act of 1965 (consolidated by the Rent Act of 1968) was one of the most important pieces of legislation introduced by the 1964–70 Labour government. Rent regulation was to apply to unfurnished dwellings where the rateable value was less than £400 in Greater London and £200 elsewhere (Table 1.2). Those properties which had not been decontrolled by the 1957 Act remained controlled. Regulated rents were 'frozen' at the amount payable in 1965 and for new tenancies rents were to be equal to the amount payable under the previous regulated tenancy. The Act implemented a proposal of the Milner Holland Report that there should be security of tenure for unfurnished tenants.

Machinery was set up to fix and review rents for regulated tenants. Rents were to be assessed and registered by a Rent Officer after an application by a tenant, a landlord or both. The rent officer was to objectively assess a 'fair rent', although there was no fixed formula available to enable him to determine what was 'fair'. Officers were to have regard 'to all circumstances (other than personal circumstances) and in particular to the age, character and locality of the dwelling house and its state of repair'. Scarcity value had to be disregarded, therefore the 'fair rent' was to equal the hypothetical market rent which would result if supply and demand were in equilibrium in the area concerned.

Landlords of controlled tenancies (with rents pegged to 1939 or 1957 levels) were permitted to increase their rent if they rehabilitated their properties up to a twelve-point standard with the aid of improvement grants under the Housing Act of 1969 (Table 1.3). They could either charge an annual rent equal to the new gross rateable value plus 12.5 per cent of their share of the authorised improvement cost or, in consultation with the local authority, fix a rent equivalent to a hypothetical market rent less the inflationary effect of any local shortage of similar accommodation. But the rent could be subsequently assessed by the rent officer (if the tenancy was unfurnished) or the rent tribunal (if it was furnished).

Although the 1965 Act was intended to benefit both landlords and tenants, and to enable the market to function efficiently, within the major cities fair rents were being assessed well below hypothetical market rents. Rent regulation failed to take supply and demand into account. Because of the resulting low returns on investment in unfurnished

Table 1.3 Legislation regulating rents, 1969–74

Legislation	*Major provisions*		*Rateable value (£)*	
			London	*Elsewhere*
Housing Act of 1969	Controlled dwellings rehabilitated up to a twelve-point standard to be decontrolled and regulated	Not exceeding	400	200
Housing Finance Act of 1972	All controlled tenancies to be decontrolled and regulated	Not exceeding	400	200
Rent Act of 1973	Rent regulation extended to higher rateable value properties	Not exceeding	1,500	750
Rent Act of 1974	Rent regulation extended to furnished tenancies	Not exceeding	1,500	750

rented housing, landlords were deterred from continuing to supply accommodation in this sector. The 1965 Act, like the 1939 and 1957 Acts, failed to safeguard this sector for the working classes.

Rent decontrol and regulation, 1970–79

With the return of a Conservative government in 1970, legislation was soon drafted to convert the remaining controlled tenancies into regulated tenancies – notwithstanding the weaknesses of the fair rent system. The Housing Act of 1972 acknowledged that more and more private dwellings had fallen into disrepair, to the serious disadvantage of the tenant. Some had become unfit and had been lost to the housing market altogether.

Under the Act therefore, rents were to rise up to fair rent levels, by £1 per week from 1 January 1973, and up to a further £2 per week in each of the succeeding two years. If a tenant and landlord agreed between themselves to a rent increase, the rent would have to be registered with the local authority, but if they failed to agree, the rent officer would have to assess the current level. Alternatively local authorities could refer proposed registered rents to the rent officer if they seemed unreasonably high. The higher rateable value properties were to be the first to be converted to regulated tenancies, and only those dwellings statutorily unfit and scheduled for clearance were to remain controlled. After three years, the landlord or tenant could apply for the cancellation of the registration, and a new fair rent could either be negotiated or assessed by the rent officer – in either case being subsequently registered.

The second main provision of the Act was that from 1 January 1973 unfurnished private tenants were able to apply to local authorities for rent allowances. A tenant was assessed as having a 'needs allowance' for himself and his wife and for each child. When this was the same as his gross income he would pay 40 per cent of the rent on his dwelling. If his income was more than the needs allowance he would pay 40 per cent of his rent plus 17p for every pound his income exceeded his allowance. If his income was less, he paid 40 per cent of his rent minus 25p for every pound it fell below the allowance.

With Labour being returned to office in 1974, rent policy was soon under review. The resulting Housing Rent and Subsidies Act of 1975 replaced the 1972 Act and introduced new measures concerning fair rents. Rents were to be raised in three stages spread over two years, but landlords could apply for a further increase in the third year. In 1973, the upper limit to Rent Act protection was raised to £1,500 in Greater London and £750 elsewhere in England and Wales. This was incorporated into the 1975 Act. Only about 2,000 privately rented dwellings in Greater London and fewer elsewhere were above these limits. Rent officers and rent assessment committees retained their previous functions.

The incoming Labour government also extended rent regulation to the furnished sector (previously providing accommodation at market rents). The Rent Act of 1974 (consolidated by the Rent Act of 1977) enabled tenants to apply for a fair rent and security of tenure was granted – both provisions applying to properties with rateable values of up to £1,500 in London and £750 elsewhere. The rent tribunal's rent assessment function was taken over by the rent officer. In 1974, these provisions covered nearly 90 per cent of the 764,000 furnished lettings in the United Kingdom. A landlord could only regain possession (through the courts if necessary) if he had been temporarily letting his own home, or eventual retirement home; letting holiday accommodation out of season; letting student accommodation out of term; and temporarily letting accommodation to a number of different categories of occupiers such as agricultural workers. Possession could also be regained if the tenant failed to pay rent, damaged the property or furniture, caused a nuisance to neighbours, or undertook unauthorised sub-letting.

The golden ages of local authority housebuilding, 1951–4 and 1964–70

When the Conservatives returned to office in 1951, Harold Macmillan was appointed Minister of Housing. At the Conservative Party Conference the previous year he had pledged that the party, if elected to power, would have a house-building target of 300,000 houses per annum – a figure cautiously included in the Conservative manifesto at the 1951 election. By 1952 this target had been achieved and the number of houses built continued to increase, albeit at the expense of space standards. In 1951, the number of completions in Great Britain had been 185,000, of which 88 per cent were in the public sector. In 1954, when Harold Macmillan was transferred to the Ministry of Defence, completions numbered 347,000, 74 per cent being council houses. Subsequently, house-building in the local authority sector diminished, as Conservative governments placed increasing emphasis on owner-occupation.

Many other consequences of the government's housing policy in 1951–4 can be identified. Few of the new local authority dwellings were earmarked for low-income inner city dwellers – the 'upper working classes' and middle classes being the main beneficiaries; more slums were being created by poor maintenance and lack of repairs than were being removed by clearance (extensive compulsory purchase, public sector redevelopment and local authority allocation could have helped with these problems but failed to do so); resources were tied up in house-building and insufficient were available for investment in industry or road construction (the United Kingdom road programme lagging behind that of most other industrial countries); and imports of timber and other building materials put a strain on the balance of payments. In total, these may have been an acceptable price to have paid for an increase in the size of the housing stock – the greatest proportion of the increase being council housing. For the first and last time (to date) a house-building target had been achieved.

There was a minor shift of emphasis in 1953–6 towards the rehabilitation of unfit houses, grants being made available for this purpose, and private landlords were permitted to raise (controlled) rents in relation to their contribution to the cost of renewal (see Chapter 7). The 1930s provided a precedent for this: there had been a switch to slum clearance away from local authority house-building for general needs. Between 1955 and 1961 subsidies to local authorities were reduced to decelerate the average rate of increase in the number of public sector houses, the proportion reaching only 35 per cent of new house-building by 1960.

In the early 1960s, the high rate of inflation meant that rents based on 1956 values were no longer appropriate. Over 4.2 million local authority houses were let at rents which failed to cover the cost of repairs, maintenance and administration, or loan-servicing charges. Authorities with large amounts of low-cost pre-war houses were asking lower rents than others which had built most of their houses

Plate 1.7 Medium-rise local authority housing of the late 1950s
Two of twenty-five eleven-storey blocks of flats developed by the London County Council on the Roehampton Estate, London Borough of Wandsworth

since the war. Rural authorities and the new town development corporations which had few or no pre-war houses had to charge the highest rents. After the years in which Harold Macmillan had been Minister of Housing, public sector activity diminished as the Conservative government placed increasing emphasis on owner-occupation.

After the Labour Party's general election win in October 1964, Richard Crossman, Minister of Housing and Local Government, was determined to achieve the party's manifesto pledge of building 500,000 houses a year by 1970. There was an emphasis placed on increasing house-building in the public sector, and in the period 1964–70 nearly half of the completions were in that sector (Table 1.4). But output over the whole period was not at the expense of standards – in contrast to the Conservative period of office 1959–64. After 1964, Parker Morris standards (relating to room sizes, fittings and amenities) became mandatory. The decrease in public sector starts after 1967 can be mainly attributable to cuts in public expenditure to combat the inflationary effects of currency devaluation in 1967.

In an attempt to encourage house-building at a time of inflation, the Housing Subsidies Act of 1967 modified the basic form of subsidy which had been in existence since 1924. The Exchequer subsidy would henceforth meet the difference between loan charges incurred by the local authority on borrowing for the financial year and charges which would have been incurred had interest rates been 4 per cent. At a time of rising interest rates, the local authorities' commitment therefore remained stable. The Exchequer's commitment was, however, far from being open-ended (unlike its commitment under the 1919 Act). The central government was indirectly able to control spending though the 'approved cost element' measured by the 'cost yardstick' and subsidies were conditional on the adoption of Parker Morris standards. The 1967 Act also provided for additional subsidies where blocks of flats of four or more storeys were built.

HOUSING FINANCE REFORM

An initial attempt

In its White Paper, *Fair Deal for Housing* (Department of the Environment, 1971), the Conservative government argued that although existing policies for subsidising new buildings prevented an overall shortage of houses, they hindered a solution to the problems that remained. It claimed that the prevailing system of historic cost rents was fundamentally unfair since local authorities with the

Table 1.4 Houses started, Great Britain, 1964–70

	Public sector		*Private sector*		*Total*
	(000s)	*(%)*	*(000s)*	*(%)*	*(000s)*
1964	179	42	248	58	427
1965	181	46	211	54	392
1966	186	49	193	51	379
1967	214	48	234	52	448
1968	194	49	200	51	394
1969	177	51	167	49	344
1970	154	48	165	52	319
Total	1,285	48	1,418	52	2,703

Source: Ministry of Housing and Local Government, Scottish Development Department and Department of the Environment, *Housing and Construction Statistics*

greatest need to provide new dwellings (for example in an area of slum clearance) would, by necessity, put the biggest rent burden on the tenants; there was too little help for people in need; and many housing authorities received subsidies but did not need them, while authorities with the worst problems received too little.

Based on the White Paper, the Housing Finance

Plate 1.8 High rise local authority housing of the 1960s
Twenty-one-storey local authority flats developed with the use of industrialised systems, London Borough of Wandsworth

Act of 1972 therefore introduced:

1 Fair rents for all local authority tenants who could afford them (fair rents having been applied to unfurnished private letting in 1965).
2 A rent rebate for those who could not afford the full fair rent (tenants of unfurnished private lettings unable to pay the full fair rent were already eligible for rent allowances).
3 A concentration of Exchequer subsidies on authorities with the worst housing (private sector housing in poor condition was eligible for improvement grants).

To an extent, local authority housing was thereby put on the same footing as the private rented sector, with the intention of establishing an element of competition between the two types of tenure.

Despite or because of the 1972 Act, the Labour government (returning to office in 1974) was faced with a declining public housing sector in absolute terms. The number of housing starts had fallen almost to a post-war low, and investment in the sector had dropped by 60 per cent in 1968–73. Subsidies had been too narrowly targeted at the worst housing, higher rents were paid by 2.5 million council tenants in April 1974, with a further 2.5 million liable for an increase on 1 October, although local authorities in England and Wales had made an overall profit of £20 million on housing in 1973.

One of the first measures which the new government introduced was a freeze on rents until the end of 1974 (later extended to March 1975), and Labour's election pledge to repeal the 1972 Act was fulfilled by the Housing Rent and Subsidies Act of 1975. By this, local authorities again had the legal right to decide council rents; while the 'no-profit' rule was restored to council housing. 'Reasonable rents' were re-introduced (a level of rent not too low to impose a burden on tax and rate payers, and not too high to provide a profit), and the system of statutory rent increases was abolished.

At a time of high rates of inflation and interest rate, the 1975 Act ensured that subsidies were increased in order to increase house-building and hold down rents.

In the late 1970s, however, council housing became a victim of public expenditure cuts. The number of public sector house starts in Great Britain fell from 173,800 in 1975 to 80,100 in 1979, and improvements fell from a peak of almost 193,300 dwellings in 1973 to only 52,250 in 1979. Cuts became more vicious with a change of government. In 1979, the incoming Conservative government cut Exchequer grants by £300 million (1979/80), and many councils were consequently faced with cuts of 40 per cent, with drastic effects upon their house-building programmes.

The 1977 Green Paper

Although, in the mid-1970s, it was clear that there were still major problems relating to housing finance, the average subsidy to council tenants was £195, and house purchasers received an average of £185 in the form of tax relief on mortgage interest and option mortgage subsidy; house purchasers received increased assistance as their incomes rose whereas tenants received broadly the same subsidy over a wide range of incomes (Table 1.5). But although the Green Paper, *Housing Policy: A Consultative Document* (Department of the Environment,

Table 1.5 Financial assistance to mortgagors and council tenants, Great Britain, 1974/5

Income (£)	*To mortgagors (£)*	*To council tenants (£)*
Less than 1,000	59	120
1,000–1,499	73	132
1,500–1,999	91	152
2,000–2,499	104	137
2,500–2,999	101	147
3,000–3,499	129	154
3,500–3,999	129	148
4,000–4,999	148	164
5,000–5,999	179	154
6,000 and over	369	

Source: DoE (1977) *Technical Volume*

1977), suggested that general housing assistance should be distributed more fairly and evenly, it dismissed all means of doing this. It ignored the evidence that there was a growing concentration of poorer households in the public sector, with 55 per cent of tenants being eligible for means-tested rent rebates and a growing number receiving supplementary benefits (income support).

While upholding the status quo in the owner-occupied sector, the Green Paper proposed some major changes in the public sector in the areas of resource effectiveness and subsidisation. Each local authority was to be obliged to formulate a Local Housing Strategy showing the need for new council building, improvement grants, lending to housing associations, lending for home-ownership, and the need for private building for sale. From this they would then submit Housing Investment Programmes (HIPs) according to the shortages and needs not catered for in the private sector, and the Department of the Environment would then allocate spending permission (although of course this was not required for private building for sale). Allocations were decided on the basis of total demand and the total amount which the Treasury thought the country could afford. The Green Paper thus made local authority housing a 'residual' rather than a general needs tenure, thereby undermining a basic principle on which public sector housing was largely based. Moreover, government funds to local authorities were no longer dependent upon the number of council houses being built – local authorities could instead finance the private sector (for example through improvement grants) and housing associations.

Private sector rehabilitation, 1950s–70s

Following the White Paper, *Housing – The Next Step* (Ministry of Housing and Local Government, 1953), the value of improvement grants (introduced in 1949) was increased by the Housing Repairs and Rent Act of 1954, the Housing Act of 1957, the House Purchase and Housing Act of 1959, and the Housing Acts of 1961 and 1964. Most of this legislation linked improvement to rising rents since normally grants were limited to 50 per cent of the cost of rehabilitation.

Because of the decline in house-building in the late 1960s and stemming from the White Paper, *Old Houses into New Homes* (Ministry of Housing and Local Government, 1968), the Housing Act 1969 was intended to hasten the pace of rehabilitation. Rehabilitation, or, more specifically, 'improvement', was to be encouraged by increased grants and a relaxation of the conditions attached to them. For private housing, standard grants were to be raised from £155 to £200 to assist in the provision of the standard amenities. Discretionary grants were also increased – from £400 to £1,000 – to enable such work as essential repairs, damp-proofing and rewiring to be done in addition to the installation of the standard amenities. Discretionary grants could also be used for conversions – up to £1,000 per dwelling being available for a conversion into two flats or up to £1,200 per dwelling for conversion into three or more flats. In all cases, the grants would have to be matched pound for pound by the applicant (i.e. grants were at 50 per cent of authorised expenditure).

Further encouragement was given to improvement by the rent provisions of the Housing Act of 1969. If grant-aided improvements were carried out in the case of regulated tenancies, then the rent could rise to a new fair rent level as certified by the rent officer (under the provisions of the Rent Act of 1968), with the increase phased over three equal annual stages.

The 1969 Act placed great emphasis on the declaration of General Improvement Areas (GIAs) – areas which in scale could vary widely from small and compact areas of, say, 300 houses up to larger areas of between 500 and 800 houses. Government grants of one-half the cost of environmental improvement were to be available up to a limit of £50 per house in the area. The Act enabled local authorities to improve amenities and to acquire land for this purpose.

The government contribution towards the

improvement of local authority and housing association properties was (as in the case of private sector housing) £1,000 per dwelling and was likewise calculated at 50 per cent of the cost of improvement and acquisition. The government also provided local authorities with the equivalent amount of assistance for the purchase of houses for improvement as they received as a subsidy for building new houses under the Housing Subsidies Act of 1967. The Housing Act of 1971 adjusted the amount of the improvement subsidy available to private owners, local authorities and housing associations. Discretionary grants paid by local authorities to private owner-occupiers and landlords in Development and Intermediate Areas were increased to 75 per cent of the approved expense of improvement works, but the government's share of the cost of improving local authority housing fell to three-eighths of the cost, the local authority having to meet the remaining five-eighths. Housing associations could receive either cash grants in the same way as private owners or contributions in the same way as local authorities, but outside of Development and Intermediate Areas, the proportion of government contribution was higher than that to local authorities – one-half instead of three-eighths of the total approved cost.

Based on the White Paper, *Better Homes: The Next Priorities* (Department of the Environment, 1973a), the Labour government's Housing Act of 1974 prevented recipients of improvement grants from selling their properties (or leaving them empty) within five years unless the grant was repaid to the local authority at a compound rate of interest. A seven-year restriction within the newly introduced Housing Action Areas (HAAs) was also imposed to ensure that housing in very poor condition would remain available for its existing tenants after renovation. Local authorities were empowered to demand the improvement of individual rented properties – a nine-month period being imposed on landlords for this purpose. If landlords failed to improve their tenancies, the local authority was able to purchase the properties with a compulsory purchase order if necessary and then to hand the municipalised accommodation back to the original tenants. To alleviate the 'disincentive' effect on owners of the new conditions to grant approval, grants and limits of eligible expenses were increased. Improvement grants were raised from £1,000 to £1,600, or, where a building of three or more storeys was being converted, from £1,200 to £1,850. These amounts represented 50 per cent of the increased level of eligible expenses – £3,200 and £3,700 in respect of the above cases. Within the GIAs improvement grants were increased up to £1,920 and £2,220, and in the HAAs they were raised up to £2,400 and £2,755.

Housing associations, 1960s–70s

During the 1960s and 1970s, under both Labour and Conservative administrations, housing associations played a small but important part in supplementing local authorities as providers of social housing. Since the 1960s, housing associations have been funded less from charity than from local and central government, and registered associations have been eligible for loans and grants from the Housing Corporation – established under the Housing Act of 1964 – for the purchase, rehabilitation and conversion of old houses or the development of new dwellings. Until the early 1970s, housing associations were able to charge cost rents to cover the cost of construction and maintenance, but realistic cost rents were becoming too high for low-income tenants, reaching £30–£40 per week for an average association dwelling.

Associations traditionally provided accommodation for special groups, such as single-parent families, the elderly, former mental patients and discharged prisoners, who fail to qualify for local authority housing. Many associations are technically still charities under the Charities Act of 1960 and can therefore supplement funds by voluntary donation. Associations are non-profit-making bodies run by voluntary committees, and sometimes they had an advantage over local authorities in that they could rehabilitate a few houses at a time or build small infill schemes which contrasted with large council estate development.

Under Part IV of the Housing Finance Act of 1972, housing associations became eligible for subsidies from the Exchequer. On new construction, the government gave subsidies on a sliding scale to cover the gap between income from rent and 'reckonable expenditure', which included running costs and capital expenditure, and on conversion subsidies of £5 per week payable for twenty years. Associations could borrow from the Housing Corporation under Section 77 of the Act and obtain mortgages from local authorities up to 100 per cent of the cost of acquiring property including loan costs. In return for the mortgage, associations were usually required to take a number of families from local authority housing lists.

Subsidisation was becoming increasingly necessary as cost rents were rising very rapidly. The Housing Finance Act of 1972 therefore brought all housing associations into the fair rent system, but Section 5 of the Rent Act of 1968 continued to exempt housing associations from the Act's provisions on security of tenure, although tenants of unregistered associations gained security in 1974. Most association tenants therefore not only had to pay higher rents than council tenants (although they were eligible for rent allowances), but lacked the security of most private tenants.

Government support for housing associations was nevertheless emphasised by the White Paper, *Widening the Choice: The Next Steps in Housing* (Department of the Environment, 1973b). A further White Paper, *Better Homes: The Next Priorities* (Department of the Environment, 1973a), proposed that housing associations (helped by the Housing Corporation and National Building Agency) should play a major role in acquiring, improving and managing properties in newly declared Housing Action Areas. It was hoped that housing associations and local authorities would work closely together to provide accommodation to replace the dwindling supply of private rented housing.

In 1974 house-building slumped to its lowest level since the early 1950s, and more people were homeless or on housing waiting lists than ever before. It was recognised that local authorities could not by themselves deal with the problem of housing need. The voluntary housing movement was therefore strengthened. Although introduced by the incoming Labour government, the Housing Act of 1974 had been substantially drafted by its Conservative predecessor, and generally enjoyed bi-partisan support. It was intended to encourage the expansion of housing associations under public supervision. The Housing Corporation's powers of lending and control were greatly extended. Its main function was to promote housing associations and to intervene in their activities where it appeared that they were being mismanaged. The registration of housing associations was introduced, administered by the Housing Corporation – associations having to be non-profit bodies and registered as charities or under the Industrial and Provident Societies Act of 1965. Only registered associations were able to receive loans and grants from public funds – mainly under a new subsidy system, a major element of which was the housing association grant (HAG). The HAG was to be paid on completion of a housing association scheme and allowed the association to pay back 75–80 per cent of the loans it had received from the Housing Corporation and local authority to finance the scheme. The annual loan charges on the remaining debt (together with management and maintenance costs) should then be recovered from rent income, but if there was a shortfall this could be offset by a revenue deficit grant (RDG). The Housing Corporation was also granted powers to provide dwellings for letting by means of construction, acquisition (by compulsory purchase if necessary), conversion and improvement, and it could borrow up to £750 million for this purpose.

Encouraged by the government and the Housing Corporation, housing associations rapidly increased their activity. The type of activity was largely determined by *Circular 170/74*, which emphasised that housing associations should play an important role in relieving housing stress or homelessness by their operations in General Improvement Areas and Housing Action Areas; provide housing for those with special needs such as the single and elderly; design schemes to maintain the stock of rented accommodation in areas where there were severe shortages; acquire properties from private landlords who were failing in their duty towards their tenants or property; and make provision for key workers.

Owner-occupation in the 1950s–70s: a period of substantial growth

After the Second World War, the building licensing system (introduced in 1939) was continued – severely restricting private development. But resources concentrated in the public housing sector became available for private house-building when building licences were abolished in 1952, one year after the Conservative government had replaced Labour. The Conservatives have pledged their support for owner-occupation to the present day. Between 1954 and 1957 the Conservative government guaranteed loans to mortgagors in excess of the percentage of valuation that the societies would normally advance; the party fought the 1955 general election with a pledge that it would create a 'property-owning democracy'; between 1959 and 1962 the government lent £100 million to building societies to fund the purchase of pre-1919 dwellings; in 1963 it abolished tax on imputed income (Schedule A) from owner-occupied property; in 1971 its White Paper, *Fair Deal for Housing*, and in 1972 the Housing Finance Act both continued the system of tax relief on mortgage interest, and in the period 1971–4, and then from 1977 (at first at a local government level) sales of council housing have been encouraged.

In the 1960s Labour governments also attempted to extend home ownership. *The Housing Programme 1965–1969* (Ministry of Housing and Local Government, 1965) outlined Labour's aims for the late 1960s, one of which was 'the stimulation of the planned growth of owner-occupation'. This aim was achieved by encouraging leaseholders to buy their freehold under the Leasehold Reform Act of 1967, the introduction of the Option Mortgage Scheme in 1968 which made low-income house-buyers eligible for subsidies (equivalent to tax relief on mortgage interest payments), and by the continuation of tax relief to other mortgagors. There was little change of policy during 1974–9, although interest relief was withdrawn in 1974 on mortgages in excess of £25,000. In the same year mortgagors were protected from rising interest rates by government loans of £500 million to building societies to offset a shortfall in funds.

By the mid-1970s, serious concern was expressed that housing subsidies were becoming more and more inequitable, prompting the government to produce the Green Paper, *Housing Policy: A Consultative Report* (Department of the Environment, 1977). A major consideration of the report was the extent of income tax relief on mortgage interest. Table 1.6 shows that the total tax relief and option mortgage subsidy to owner-occupiers is broadly comparable to Exchequer grants and rate contributions (the subsidy to council housing). The Green Paper reported that in 1976/7, the average subsidy to a mortgagor was £205 and the average subsidy to a council tenant was £210. But whereas the amount of assistance to mortgagors increased with income, assistance to council tenants remained fairly constant over a wide range of incomes (Table 1.5).

The Green Paper, however, left mortgage interest relief alone since it did not wish to 'upset the household budgets of millions of families' and expressed unease about the effects of its reduction or withdrawal on the house-building industry.

Table 1.6 Mortgage tax relief and option mortgage subsidy, and general subsidies to council housing, Great Britain, 1972–7

	Owner-occupiers (£ million)	*Council housing (£ million)*
1972/3	802	592
1973/4	1,044	722
1974/5	1,213	1,188
1975/6	1,239	1,236
1976/7	1,333	1,320

Source: Treasury (1977) *The Government's Expenditure Plans 1977/78 to 1980/81*, HMSO

CHANGING PATTERNS OF TENURE, 1913–79

Although a large number of economic and social factors determine tenurial preferences, government housing policy (as reviewed in this chapter) has had a major impact on tenure throughout most of the twentieth century. Had market forces alone prevailed, a very different pattern of tenure would undoubtedly have emerged. It is almost certain that the local authority housing sector would not have emerged, and owner-occupation and private renting might have been split fairly evenly – with unquantifiable social consequences for low-income households excluded from the market.

From 1915 until the 1970s, however, rent control played a significant part in the substantial decline in the private rented sector, while almost simultaneously housing subsidies facilitated the expansion of local authority housing; and since the Second World War in particular, bi-partisan fiscal support for home-ownership has ensured the dramatic increase in the size of the owner-occupier sector. Whereas private renting was unable to compete successfully with public sector renting until the 1960s, subsequently both private and public landlords were faced with competition from owner-occupation as the dominant type of tenure.

By 1979, therefore, the private rented sector had declined to only 13 per cent of the total housing stock of Great Britain (compared to 90 per cent in 1913); the local authority sector had expanded from negligible beginnings in 1913 to 21 per cent of the total stock by 1979 – Britain having the largest municipal housing sector in Western Europe; while owner-occupation had increased its share of the stock more than five-fold, from 10 to 54 per cent in 1913–79 (Table 1.7).

The regional distribution of tenure

At a regional level, there were marked variations in tenure distribution (Table 1.8), reflecting spatial differences in the political, economic and social attributes of decision-makers and households, and to some extent the targeting of funds (for example to the inner cities or to deprived regions). Whereas 63 per cent of the stock of housing in the South East (outside Greater London) was owner-occupied in 1979, the proportion of owner-occupied housing in Scotland was as low as 35 per cent in the same year. The proportion of housing rented from local authorities was, conversely, particularly high in Scotland, at 54 per cent, and also high in the North, at 40 per cent. Private rented housing and housing association dwellings accounted for a small proportion of housing nationwide, varying from 20 per cent in Greater London to only 9 per cent in Northern Ireland.

Clearly, market forces and government policy had a very variable effect on the spatial pattern of tenure, but in all regions owner-occupation had become the dominant tenure by 1979.

Table 1.7 Housing tenure, Great Britain, 1913–79

	Owner-occupied %	*Local authority* %	*Private rented and housing associations (%)*
1913	10.0	—	90.0
1950	29.0	18.0	53.0
1961	42.3	25.0	31.9
1971	50.6	30.6	18.8
1979	54.4	32.1	13.4

Source: Department of the Environment; Scottish Office; and Welsh Office, *Housing and Construction Statistics* (various editions)

Table 1.8 The regional distribution of tenure, United Kingdom, 1979

	Owner-occupied %	*Local authority* %	*Private rented and housing associations* %
South East (excl. Greater London)	63	24	13
South West	63	22	15
East Midlands	59	29	12
North West	59	30	11
East Anglia	58	26	15
West Midlands	57	33	10
Yorkshire and Humberside	56	32	12
Greater London	48	32	20
North	47	40	13
England	57	29	13
Wales	59	29	12
Scotland	35	54	11
Northern Ireland	52	39	9

Source: Central Statistical Office (1987) *Regional Trends 22*, HMSO

CONCLUSIONS

From Victorian times to the late 1970s, housing policy evolved gradually, with comparatively few changes in direction. Throughout the whole period, the overall outcome of policy was that the private rented sector diminished while the local authority and owner-occupied sectors expanded. The housing association movement, throughout, provided only a negligible proportion of the total housing stock. There were, however, inevitable differences in emphasis. In the twentieth century, Conservative governments undoubtedly showed a predilection towards policies which aimed to free the private sector from the excesses of rent control and to limit the availability of local authority housing to specific groups of households (for example, families affected by slum clearance), while Labour administrations, to be sure, showed a greater inclination to protect private sector tenants by rent control and to recognise local authority housing as a 'general needs' tenure. However, both parties when in office presided over massive house-building programmes (the Conservatives in the early 1950s and Labour in the late 1960s) and both parties furthered the expansion of owner-occupation.

By the 1970s, housing finance was increasingly a cause for concern. It was apparent that housing subsidies were becoming inequitable and inefficient, and failed to offer households a real choice between renting in the public or private sectors, or between renting and buying. Despite the Housing Finance Act of 1972 and housing policy Green Paper of 1976, the problems of subsidisation were not resolved. Although, of course, market forces were also instrumental, housing policy, by the 1970s, had turned much of the United Kingdom into a 'nation of home-owners', a phenomenon pre-dating the onset of Thatcherism with the Conservative Party's general election victory of 3 May 1979.

This overview of pre-Thatcherite housing policy provides the reader with a historical context for many of the subsequent chapters of this book, particularly Chapter 3, on housing policy and finance, and Chapter 5, on town planning and housing development.

QUESTIONS FOR DISCUSSION

1 Assess the role of 'bricks-and-mortar' subsidies in facilitating the growth in the stock of both public sector and owner-occupied housing, 1919–79.

2 'Until the 1970s, rent control or regulation was the principal reason for the decline in the private rented sector.' Critically discuss this statement.

3 Provide an overview of the factors which have determined the changing pattern of housing tenure, 1945–79.

4 Examine the reasons why there was a shift of emphasis from new house-building to rehabilitation in the late 1960s, and discuss some of the social and economic consequences of this shift.

5 'The reform of housing finance in the 1970s was a failure in terms of both efficiency and equity.' Discuss.

RECOMMENDED READING

Gibson, M. and Langstaff, M. (1982) *Introduction to Urban Renewal*, Hutchinson, London.

Holmans, A. (1987) *Housing Policy in Britain: A History*, Croom Helm, London.

Merrett, S. (1979) *State Housing in Britain*, Routledge & Kegan Paul, London.

Merrett, S. with Gray, F. (1982) *Owner-occupation in Britain*, Routledge & Kegan Paul, London.

REFERENCES

Beveridge Report (1942) *Social Insurance and Allied Services*, Cmd 6404, HMSO, London.

Clarke, S. and Ginsburg, N. (1975) 'The political economy of housing', in *Political Economy and the Housing Question*, Political Economy of Housing Workshop, London.

Department of Environment (1971) White Paper, *Fair Deal for Housing*, Cmnd 4728, HMSO, London.

—— (1973a) White Paper, *Better Homes: The Next Priorities*, Cmnd 5339, HMSO, London.

—— (1973b) White Paper, *Widening the Choice: The Next Steps in Housing*, Cmnd 5280, HMSO, London.

—— (1977) Green Paper, *Housing Policy: A Consultative Report*, Cmnd 6851, HMSO, London.

Gauldie, E. (1974) *Cruel Habitations*, Allen & Unwin, London.

Marley Report (1931) *Report of the Committee on the Rent Restriction Act*, Cmd 3911, HMSO, London.

Milner Holland Report (1965) *Report of the Committee on Housing in Greater London*, Cmnd 2625, HMSO, London.

Ministry of Housing and Local Government (1953), White Paper, *Housing – The Next Step*, Cmd 8996, HMSO, London.

—— (1965), *The Housing Programme, 1965–1969*, Cmd 2836, HMSO, London.

—— (1968), White Paper, *Old Houses into New Homes*, Cmnd 3602, HMSO, London.

Onslow Report (1923) *Report of the Departmental Committee on the Rent Acts*, Cmd 1803, HMSO, London.

Ridley Report (1938) *Report of the Inter-Departmental Committee on the Working of the Rent Restriction Act*, Cmd 5261, HMSO, London.

Swenarton, M. (1981) *Homes Fit for Heroes*, Heinemann, London.

Taylor, A.J.P. (1965), *English History, 1914–1945*, Oxford University Press, London.

Treasury (1947) *Capital Investment in 1948*, Cmd 7268, HMSO, London.

2

THE ECONOMICS OF HOUSING

Gregory Bull

This chapter introduces the basic theories of micro- and macro-economics in order to provide a foundation for analyses of housing issues, and to apply theory to specific housing policies, notably rent control in the private rented sector and rent-setting in the social rented sector. The chapter focuses on:

- Demand and supply and the determination of market equilibrium.
- Price, income and supply elasticities.
- Price controls and the market: an example of the effects of rent control and the effects of short- and long-run supply elasticities.
- The macro-economy.
- A case study – an economic analysis of the problems of rent-setting in the social rented sector.

INTRODUCTION

Economic problems facing consumers and society result from the fact that resources in any economy are scarce, so it is necessary to choose between the alternative uses to which such resources could be put. Solutions to these problems therefore consist of allocating finite resources to satisfying the potentially infinite wants of society.

With fixed incomes and resources, consuming or producing more of one commodity always involves consumption or production being reduced elsewhere. Say we have a consumer with a fixed income; he/she has fixed earnings of £150 a week and spends £50 on rent, heating, lighting, etc. for a flat measuring 40 square metres. Of the remaining £100, £40 is spent on travel and transport, £40 on food and clothing and £20 on cinema tickets. If the individual decides to rent a better flat (say, 50 square metres) costing £60 per week, then a £10 per week saving has to be made elsewhere. Assuming no economies are possible on transport, food and clothes (although some would, in practice, be expected), the individual would choose to reduce spending on cinema visits by £10. If each cinema ticket costs £5 then the cost of a larger flat (specifically 10 square metres larger) can be expressed as two cinema tickets forgone. The term for costs expressed as forgone alternatives is *opportunity cost.*

Equally, in production and with fixed resources and technology, it will always be necessary to reduce output of one good or type of goods in order to free resources for increased production of another good or type of goods.

Collectively the resources available to an economy are known as *factors of production.* They are used to produce goods and services which are themselves regarded as a means to an end; that is, the satisfaction of wants. There are four kinds of resources. First, all natural resources such as land, forests and minerals are known collectively as land. Second, all human resources, both physical and mental, inherited and acquired (for example through training and education), are known as labour. Third, all man-made aids to production, such as tools, machinery and factories, which are used up in the process of making other goods and services, are known as capital. Fourth, those who organise the other factors of production and take risks in the markets are known as entrepreneurs and this factor is known as entrepreneurship.

Figure 2.1 shows what is known as the production possibility frontier; that is, the combinations of goods that can be achieved using the resources (factors of production) currently available to society. With fixed resources the economy is able to produce any combination of goods along the frontier *r*, *s*, *t*, *u*, *v* in Figure 2.1. For example, society might choose to produce and consume h_1 units of housing and g_1 units of other goods (point *s*). Or they might choose a point such as *u* where housing production has approximately doubled, but in order to achieve this, production of other goods has had to be reduced to nearly a quarter of its previous level. Points *r* and *v* represent zero production and consumption of housing and other goods respectively, and although in theory production could take place at either of these points, it is unlikely that society would choose to forgo consumption of either entirely. As we move down the production possibility frontier, factors of production are being transferred from firms and producers in one sector of the economy (e.g. other goods) to another (in this case housing). A move from *v* towards *r* would move resources in the opposite direction. Other points such as *w*, inside the frontier, are achievable but would represent an inefficient use of resources (point *t* for example represents a better outcome, where we have more housing and more of other goods than at *w*). Point *x* is not achievable given the current level of resources available to society, although with an expansion of such resources in the future (e.g. a better-trained workforce, or greater capital investment in plant and machinery), such a position might eventually become attainable.

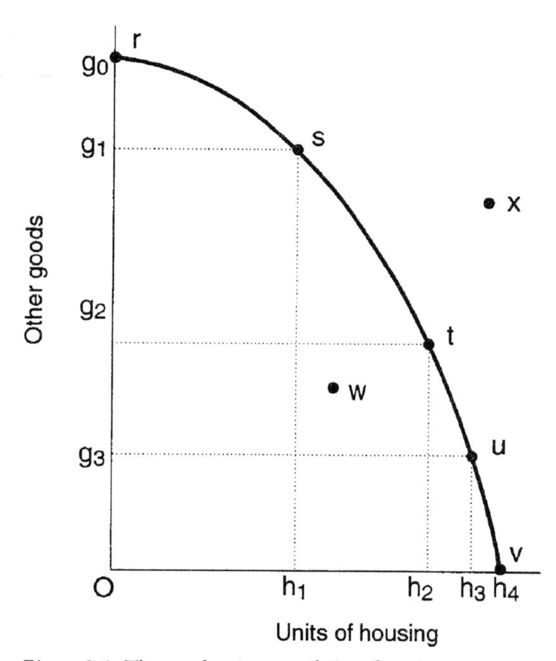

Figure 2.1 The production possibility frontier

Decisions concerning what is to be produced can in principle be undertaken through various channels, although in a market economy this generally occurs through the market mechanism. If consumers demand more of a particular commodity then the market price will rise, encouraging firms and producers to take on more resources and expand output. In the short term, resources in the economy are fixed, so firms expanding production need to attract resources from elsewhere, usually by paying higher wages and prices for factors of production. In sectors where demand is contracting the opposite will occur and output will decline.

In reality most western economies are 'mixed', with an important public sector. Governments raise revenues through taxation of incomes, profits and expenditure, and can provide goods and services either directly (e.g. local authority housing) or by providing grants to leverage spending by the private or quasi-public sectors (e.g. housing associations). In the case of housing, most spending occurs through the market mechanism and the owner-occupied sector, although even here governments often provide some incentives (e.g. tax relief on mortgage interest payments). Government intervention generally aims to encourage production and consumption of certain types of goods and services or to discourage, or otherwise regulate, production or consumption of others. Such intervention is quite common in a modern 'mixed' economy and usually arises for one of the following three reasons.

First, some goods can be termed 'merit' goods. Housing, education and certain areas of health provision are generally considered to fall into this category. Although these goods and services could in principle be provided through the market mechanism, individual consumption generates wider social benefits. In the absence of government intervention, there may be a tendency for individuals to underconsume – either because such wider benefits are not perceived by individuals or because their incomes are simply too low. For example, the construction of poor and unsanitary slum housing in the nineteenth and early twentieth centuries is often ascribed to the fact that many working families were too poor to be able to afford anything better. From 1919 the British government therefore provided subsidies to enable local authorities to build affordable housing for such groups (see Chapters 1 and 4). More recently owner-occupation has been encouraged. Since house purchase is expensive (averaging between three and four times the borrower's annual income) and therefore requires substantial repayment of capital and interest over a long period of time, many governments have provided tax incentives to encourage purchase for owner-occupation.

Second, the production or consumption of certain goods may produce negative effects on the economic well-being of consumers or producers elsewhere in society, and these spillover or external effects are known as negative externalities. Governments usually try to limit these effects either by regulation (e.g. to remove lead in paint used for domestic purposes) or by means of some form of taxation (e.g. the recently introduced tax on waste disposal in landfill sites).

Third, there is a range of goods and services known as 'public goods' where charging to cover full costs of provision would be either impractical or inefficient. For example, charging for entry to the National Parks or for the actual use of streetlighting would either be virtually impossible or prohibitively expensive, and if supply depended on revenue obtained, the market would only rarely, if ever, supply such goods or services.

To summarise, while resource allocation is generally left to the market mechanism, most 'mixed' economies retain varying proportions of government intervention, to either encourage or discourage some types of production/consumption or, more directly, in the public provision of certain goods or services. The next section looks in more detail at the underlying factors determining supply and demand for particular commodities and at the operation of the market mechanism.

DEMAND AND SUPPLY AND THE DETERMINATION OF MARKET EQUILIBRIUM

Within a freely operating market, the price of any good will depend on the interaction of demand and supply. All commodities that are bought and sold, whether on a fruit stall or the stock market, are priced to reflect current conditions of supply and demand. Movements in the market price can occur if these underlying conditions change.

The demand curve for a commodity

Demand can be defined as the quantity of a commodity consumers are willing to purchase at a particular price. The demand for most commodities will rise as their price falls and fall as their price rises and there is therefore an inverse relationship between price and quantity demanded. This can be understood more clearly when we realise that the market demand for a commodity is itself made up entirely of the individual demands of consumers or producers. Individual consumers will have a choice of alternative products in any single category (for example, different makes of cars), and they will also be constrained by fixed levels of income. If the market price of a commodity rises, consumers will have to either reduce their consumption of the good or find a suitable alternative or they will have to reduce their consumption of another good. A similar pattern will emerge for commodities which are demanded mainly by producers. For example, if there is a rise in the price of bricks, producers (e.g. construction firms) will try to substitute other materials such as timber, cement or steel (the prices of which we assume have not risen) for bricks in the production process (in this case construction).

This situation is illustrated in Table 2.1, which shows a hypothetical demand schedule for double glazing, giving the number of units sold at a range of prices. By plotting these points on a graph we can also show the level of demand at any intermediate point (see Figure 2.2).

The demand curve as shown only measures the relationship between price and quantity demanded. It assumes that the underlying conditions of demand remain unchanged. We can see, for example, that as the price falls towards £50 per unit, the level of sales can be expected to expand. With a rise in price, households' demand will contract and fewer units will be sold. With fixed incomes, households will substitute other goods for double glazing as the price of the latter rises and vice versa if the price falls.

Table 2.1 Hypothetical demand schedule for double glazing

Price per standard window unit	*Number of units demanded per week*
300	5,000
250	6,000
200	7,500
150	10,000
100	15,000
50	25,000

For any commodity, the effect of changes in the underlying conditions of demand will be to alter the demand schedule. There are three variables (apart from price) which can influence the demand for a commodity:

1 The purchasing power of the household, which is

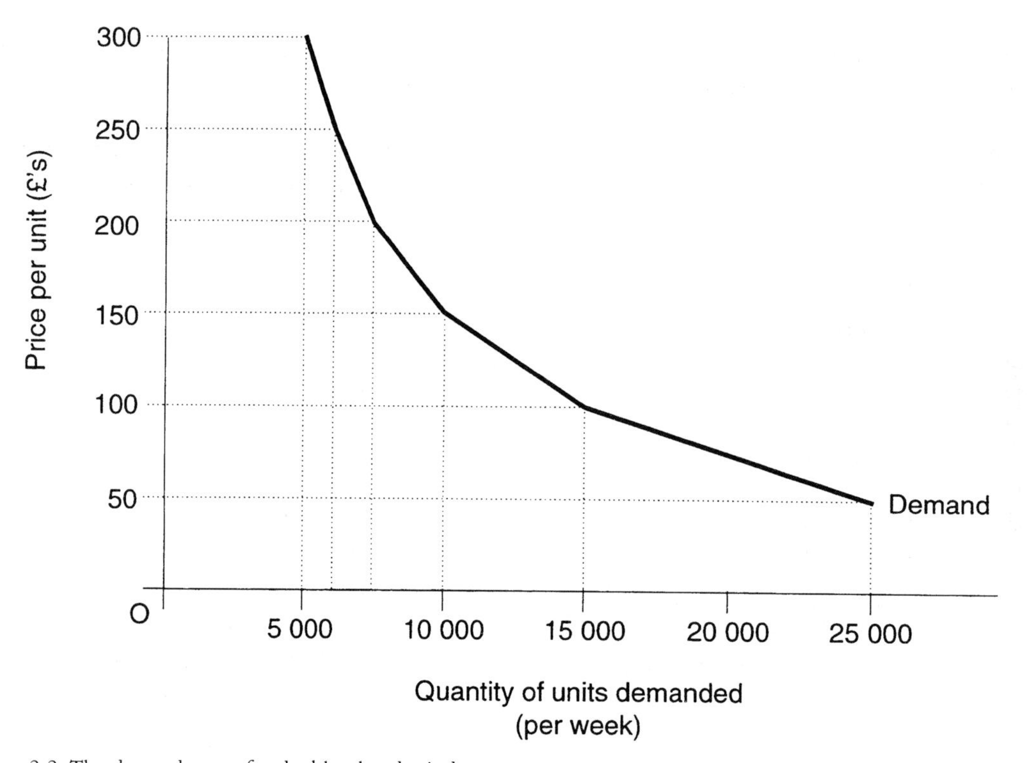

Figure 2.2 The demand curve for double-glazed windows

mainly dependent on the level of income. With a rise in income, households can potentially purchase greater quantities of all of the goods they currently consume. In practice, however, demand for some commodities will rise faster than others as incomes increase (see 'income elasticity of demand', p. 37).

2 A change in tastes may influence consumption patterns. For example, a trend towards 'green' awareness may encourage households to consume more energy-efficient items – demand for roof insulation, double glazing, low-energy lighting, etc. may subsequently rise.
3 The prices of other goods. Some goods may have close substitutes, for example tea and coffee, so that if there is a sudden rise in the price of either, consumers are able to switch and consume more of the (now relatively cheaper) alternative. Many goods (complements) are consumed jointly with other goods (for example tea and milk, petrol and cars, bricks and cement) so that a rise in the price of one could result in a downturn in demand for the other. A rise in electricity prices, for example, will reduce the demand for all items which consume electricity. However, the purchase of some products may actually save energy (e.g. double glazing, roof insulation, etc.) and demand for these will therefore increase if energy prices rise.

The effect of changes in any of the above variables will be to shift the demand curve for a good. Figure 2.3 illustrates the potential effects. Figure 2.3a shows the effect of a rise in demand where the demand curve shifts to the right. Consequently, at any price (e.g. £150 per unit) a greater quantity is demanded. This could occur due to changes in any of the variables mentioned above, namely, a rise in the purchasing power of households, a shift in tastes and consumer trends in favour of the good, a rise in the price of substitutes (e.g. single glazing) or a fall in the price of complements. Figure 2.3b illustrates a fall in demand where there is a shift inwards, and to the left of the demand curve, and this can occur for exactly the opposite reasons.

In order to produce greater quantities of any commodity, more resources are required in the production process. Existing firms and producers will need to expand production and new firms may enter the market.

In order to attract more resources away from other uses, firms will generally have to pay higher prices for factors of production. They may also have to

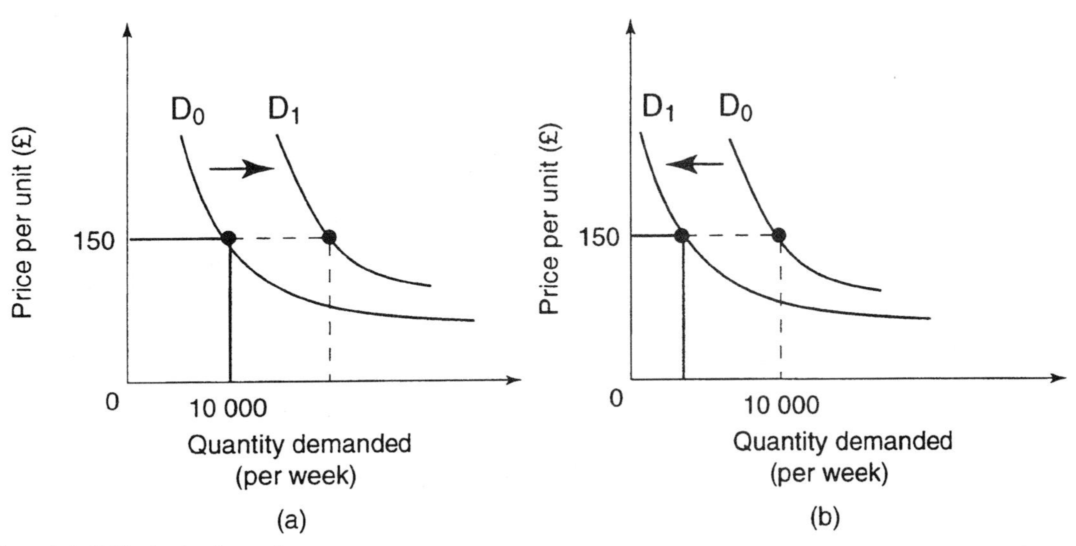

Figure 2.3 Shifts in the demand curve

operate at levels of production which are not the most efficient and where extra production costs are incurred. Nevertheless, in general, the higher the market price, the greater the profits that can be earned and the greater will be the incentive to produce the good and offer it for sale.

Figure 2.4 shows the market supply curve for double-glazed units. This is itself determined by the production decisions of all the firms making up the industry. At a price of just under £50 per unit, no production at all will be undertaken, as at this point firms would not be able to cover their costs. On the other hand, at a price of £250 per unit, 25,000 units would be supplied.

As with demand, the supply curve may shift if there is a change in the underlying conditions of supply facing producers. This may occur for three main reasons:

1 If there is a change in the state of technology which lowers the cost of production. For example, it was hoped that the introduction of industrialised building techniques in the 1960s would greatly reduce the cost of local authority housebuilding.
2 If there is an upwards or downwards shift in the price of factors of production. For example, a rise in the price of oil would raise the cost of producing plastics used in windows and many building materials. It would also increase transport costs, thus affecting most industries in the economy in some way.
3 If there is a change in the objectives or goals of producing firms. Normally, we assume firms have the same goals and aim to maximise profits. However, there may be other objectives which can override profit maximisation; for instance, firms may accept lower prices to achieve higher market share, or they may accept higher costs to ensure security of supply of certain factors of

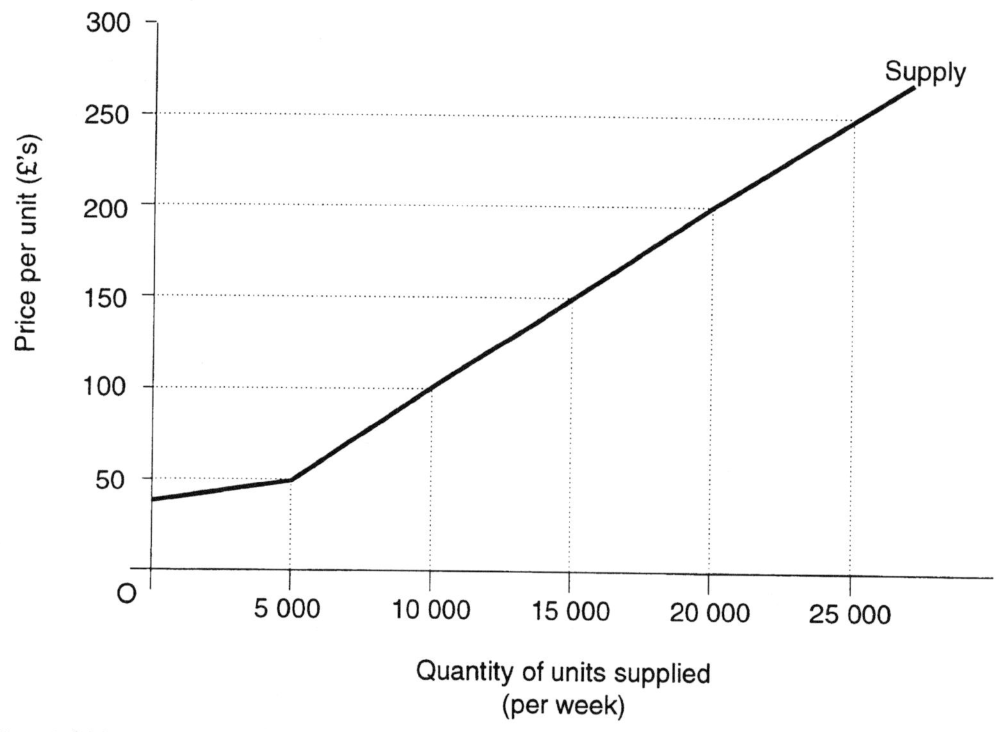

Figure 2.4 The supply curve for double-glazed windows

production (e.g. raw materials). Other firms may accept lower prices or higher costs in order to achieve social or environmental objectives.

An improvement in technology or a fall in factor costs would shift the supply curve to the right as in Figure 2.5a (i.e. more would be supplied at any price), whereas a rise in factor costs (a loss of technology is unlikely) would reduce the quantity supplied at any price (Figure 2.5b).

To answer the question 'How much will be produced and at what price?', we need to determine the market equilibrium for the industry. This is the point at which firms will have no incentive to either increase or decrease production. We can find this point by combining the demand and supply schedules for the industry as shown in Figure 2.6.

Point *e* is the point of market equilibrium where, at price of £120 per unit, the quantity of units supplied and demanded is the same, at 12,000 units. At any higher price, say £200, producers would wish to supply 20,000 units but consumers would only purchase 7,500 units.

The quantity produced and supplied to the market will therefore decline until point *e* is reached and excess market supply is eliminated.

At a price of £100 per unit, demand at 15,000 units exceeds the amount producers are willing to supply (10,000 units). Stocks held by firms are quickly depleted and as the market price rises, firms will increase production until point *e* is reached, where demand and supply are once more in equilibrium.

Changes in the underlying conditions of supply or demand will produce shifts in existing supply or demand curves and lead to the establishment of a new market equilibrium position. Such changes are often referred to as the 'laws of supply and demand'. A number of outcomes are possible and these are summarised as follows. Figure 2.7a shows the effect of changes in demand. A rise in demand for a good will lead to more being produced, but at a higher price (e to e_1). A fall in demand will lead to less being produced, but at a lower price (e to e_2). Figure 2.7b shows that with an increase in supply, equilibrium price will fall and quantity supplied and demanded will expand (e to e_1). With a fall in supply, equilibrium price will rise and the quantity supplied and demanded will fall (e to e_2).

While the above analysis is essentially short term and assumes that factors of production available to the economy are fixed in quantity, in the long run it may be possible to expand the capacity of the economy, for example through higher investment and

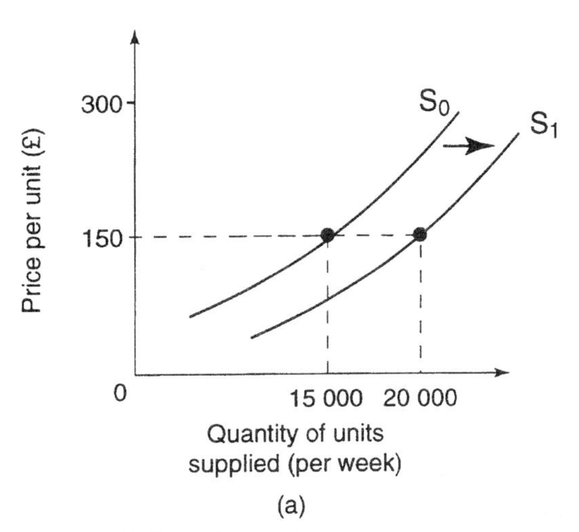

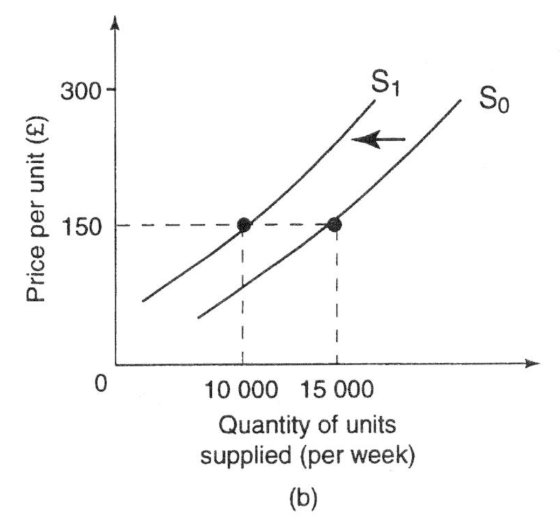

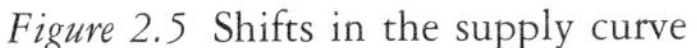
Figure 2.5 Shifts in the supply curve

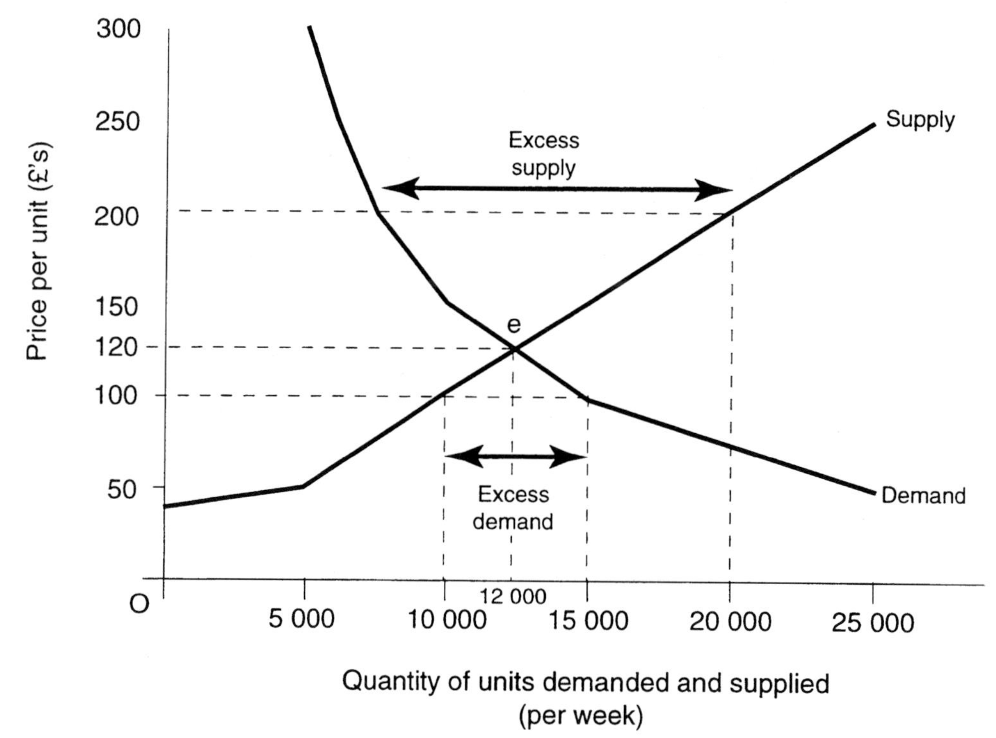

Figure 2.6 Market equilibrium and disequilibrium

improvements in technology. In this case, the supply curve may continue to move right over time so that conceivably (as in Figure 2.8) in the long run a rise in demand could actually result in lower prices for consumers. Henry Ford's introduction of production line manufacturing greatly reduced production costs of motor cars in the early twentieth century, and more recent examples of mass production would include pocket calculators and computers.

Figure 2.9 illustrates how a shift in demand in one housing sector can influence market conditions in another. Demand in the owner-occupied sector shifts down from D_0 to D_1, possibly due to an actual or expected rise in the cost of borrowing (mortgage rates, or other costs of borrowing, e.g. mortgage protection policies), or to a change in lending policy by banks and building societies (e.g. requiring a higher initial deposit or greater security of income), or due to pessimistic expectations of future growth of house prices.

Although not all of this reduction in demand would be passed on to other housing sectors – some people would simply stay where they are, or remain with friends and relatives – a significant shift would nevertheless occur to the other (rented) sectors. As these sectors represent a much smaller proportion of the housing stock (the private rented sector is less than one-sixth the size of the owner-occupied sector), the effect on equilibrium rents (to r_1 in Figure 2.9b) could be considerable. Clearly, a change in the opposite direction caused, for example, by a shift downwards in the demand for rented accommodation would only have a relatively small effect on the owner-occupied sector and house prices.

The effect of supply shifts can be analysed in a

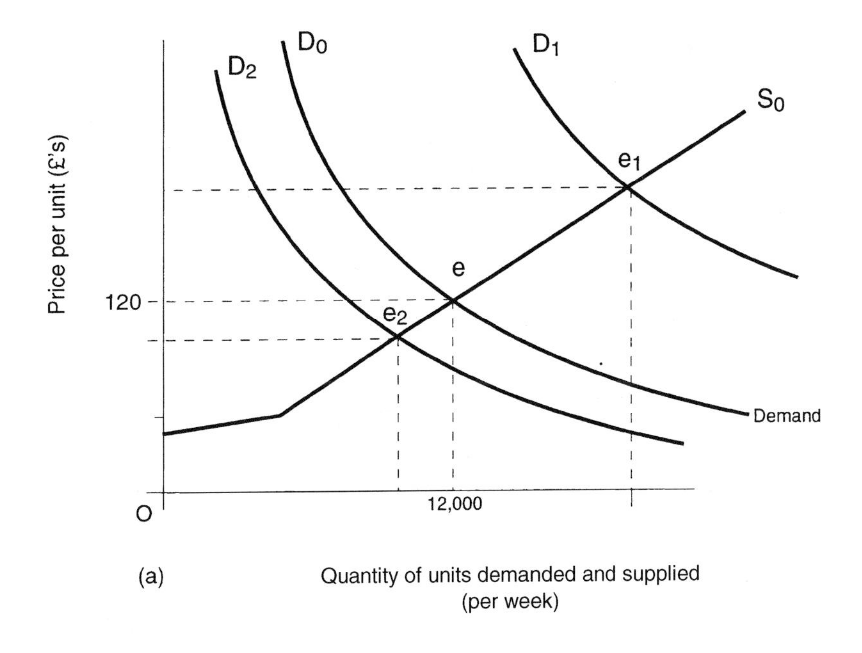

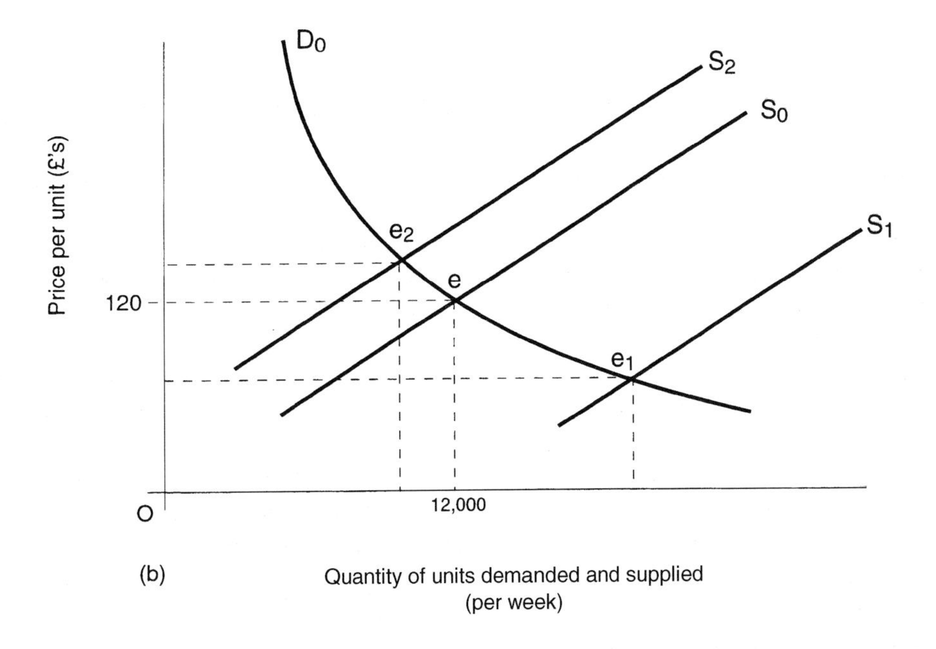

Figure 2.7 Shifts in demand and supply

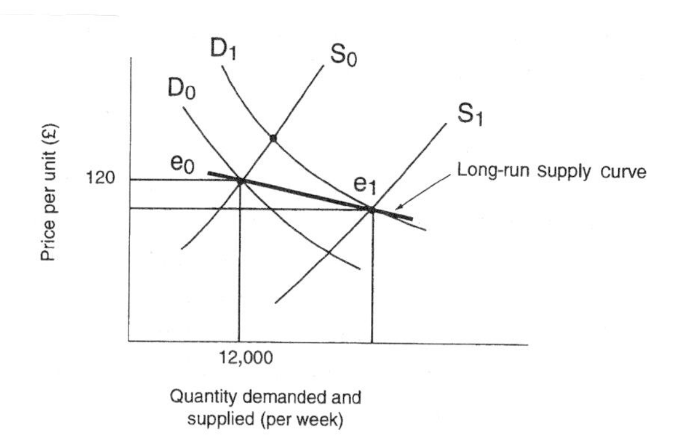

Figure 2.8 The long-run supply curve for an industry

similar way. For example, a withdrawal of landlords from the private rented sector (a shift left and upwards of the supply curve to S_1) would cause a rise in equilibrium rents. Some former tenants would now see owner-occupation as a better alternative and some (relatively small) shift to the right of demand for owner-occupied housing would occur, putting some downward pressure on equilibrium price. However, since these former privately rented houses would not actually disappear but would most likely be sold for owner-occupation, there would also be a shift right of the supply of owner-occupied

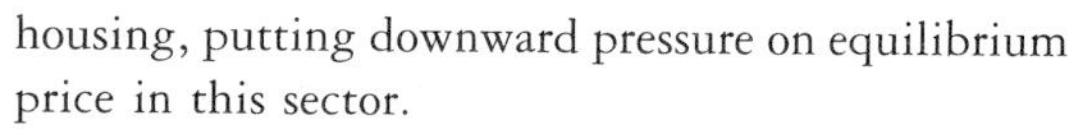
housing, putting downward pressure on equilibrium price in this sector.

PRICE, INCOME AND SUPPLY ELASTICITIES

From our previous analysis, we know that price influences both the quantity consumers demand and the quantity producers are willing to supply. The quantitative effect on demand or supply may range from quite small to very large. The economic term to describe such responsiveness is *elasticity*. A high degree of elasticity implies a high level of responsiveness and vice versa.

Price elasticity of demand

This represents a measure of the responsiveness of demand to changes in the price of a commodity, as we move along a demand curve. It is defined as follows, where *Ed* stands for elasticity of demand.

$$Ed = \frac{\text{percentage change in quantity demanded}}{\text{percentage change in price}}$$

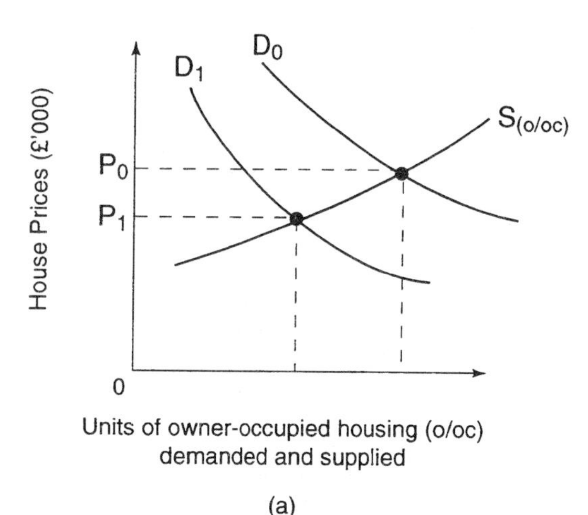

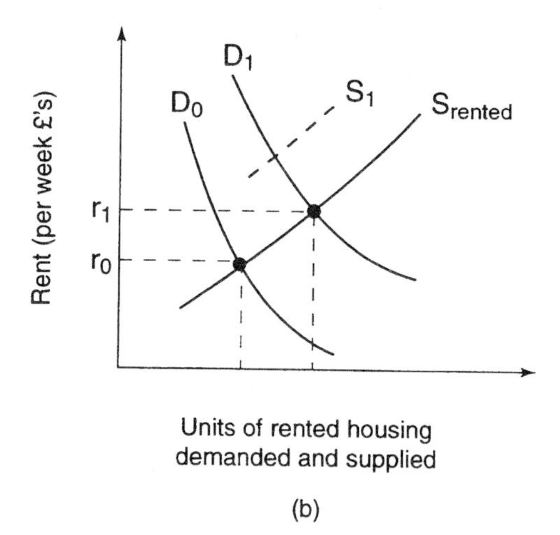

Figure 2.9 Demand and supply in the housing market

For example, if a 10 per cent rise in price results in a 20 per cent fall in quantity demanded, the value of elasticity can be calculated as:

$$Ed = \frac{(-)20}{10} = (-)2.0$$

The possible values of *Ed* vary from zero to (−) infinity. Examples of elastic and inelastic demand curves are given in Figure 2.10. Figure 2.10a shows an inelastic demand curve. A fall in price from P_0 to P_1 causes a proportionately smaller rise in quantity demanded (q_0 to q_1). A rise in price (P_1 to P_0) would have produced a proportionately smaller fall in quantity demanded (q_1 to q_0).

However, in Figure 2.10b, demand is elastic and a small reduction in price (P_0 to P_1) produces a relatively large increase in quantity demanded (q_0 to q_1). Conversely, a rise in price (P_0 to P_1) would have resulted in a substantial decline in quantity demanded (q_0 to q_1).

It is important to note that price changes will have opposite effects on total expenditure (and therefore producer revenue) in these two cases. Where demand is inelastic, a price rise will cause expenditure/revenue (i.e. price × quantity demanded) to rise, but a fall in price will cause expenditure to fall. With elastic demand, a price rise will produce a relatively large fall in demand and expenditure will fall. A fall in price will produce the opposite effect and expenditure will rise.

In conclusion, price changes will have opposite implications for expenditure and revenue, depending on whether demand is elastic or inelastic, and this information is of clear interest to producers. Of course, it is possible that demand elasticity could be equal to (−) unity, and hence neither elastic nor inelastic. Consequently, total expenditure would remain unchanged whatever the change in price. An example is shown in Figure 2.11a.

Finally, in reality, elasticity for any commodity may vary depending where we start from on the demand curve. Figure 2.11b, for instance, shows that for a straight-line demand curve demand will be inelastic at high price levels and elastic at low price levels. This can be seen from the fact that the same fall in price (P_0 to P_1) causes a relatively large rise in quantity demanded at the higher point on the demand curve, but a relatively small rise at the lower end. Elasticity takes a value of (−) unity only for very small price changes around the centre point of the demand curve.

Estimates of demand elasticity have been made for most commodities and Figure 2.12 gives exam-

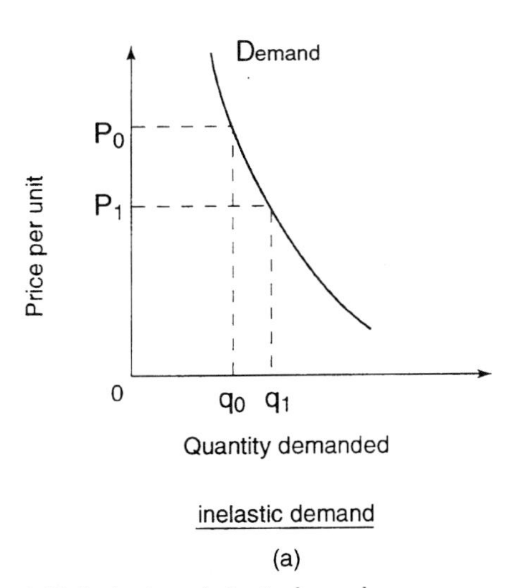

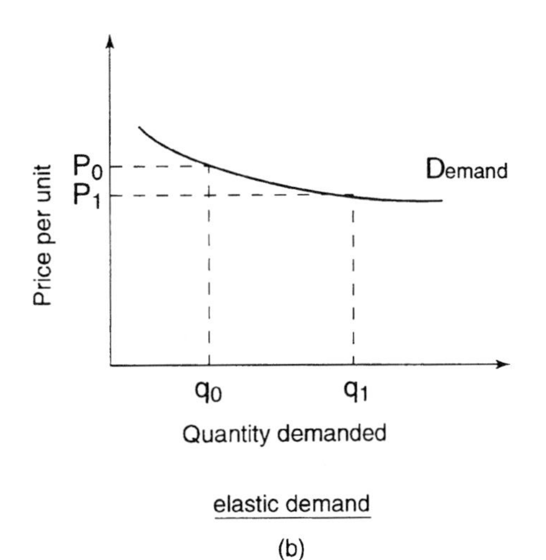

Figure 2.10 Inelastic and elastic demand curves

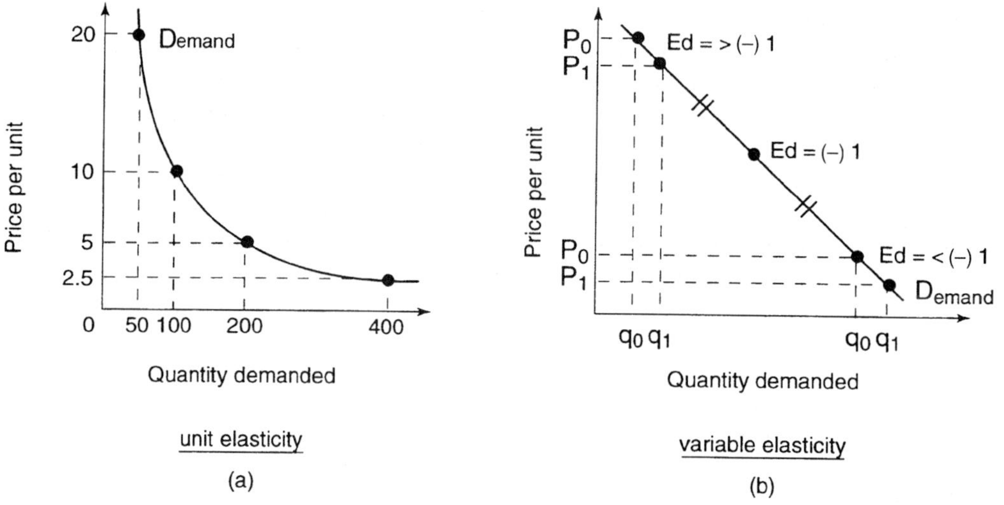

Figure 2.11 Elasticity along a demand curve

ples of the values we might expect to find for a selection of goods.

The main factor influencing price elasticity is the availability of substitutes; food products for example are a necessity and there are no overall substitutes for food. This is especially the case for bread and potatoes, for example. Nevertheless, the demand for various types of meat (e.g. beef, pork, lamb, etc.) is clearly elastic as each represents a potential substitute for the other, and because more basic foodstuffs (e.g. vegetables) can provide substitutes for meat as a whole if there is a rise in price of all types of meat.

The demand for housing is clearly price inelastic, as there are no substitutes for housing and it is often considered to be something of a necessity. With a value of around (–) 0.4, a 50 per cent increase in price can be expected to lead to only a 20 per cent reduction in demand. Overall, housing expenditure by households would therefore rise. Some observers might expect elasticity to be even lower. After all, it is clearly difficult, costly and time-consuming to

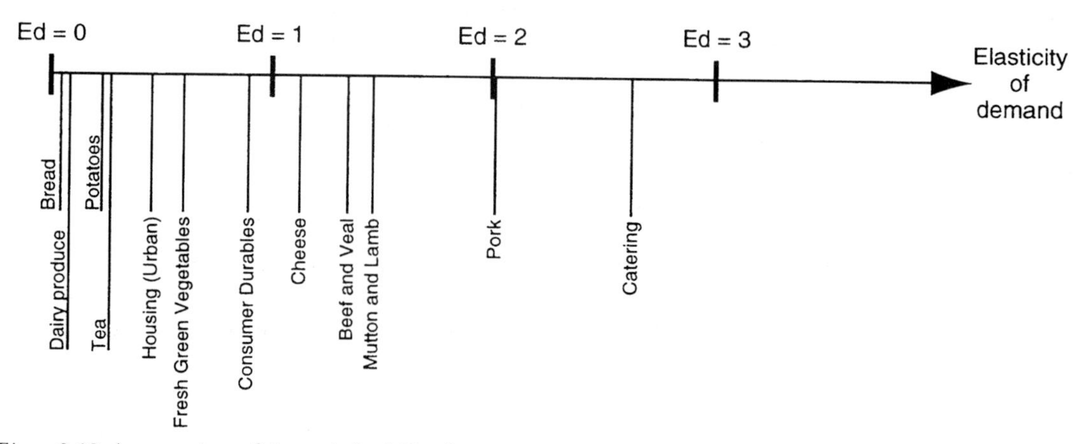

Figure 2.12 A comparison of demand elasticities for various goods

adjust housing expenditure in this way (i.e. by moving), and one might expect households to simply stay put and reduce expenditure elsewhere. Housing, however, represents a large share of the average household budget and a significant rise in rent or mortgage payments cannot therefore easily be met by adjusting other areas of expenditure (unlike tea, bread or potatoes, for example, which take up only a small percentage of the average household budget).

Other types of elasticity

While price elasticity measures reactions along a demand curve, changes (other than price) in the underlying conditions of demand can produce shifts in the demand curve for a commodity. For example, the *cross-elasticity of demand* measures the responsiveness of the quantity demanded of one commodity to a change in price of another. For goods which are substitutes, such as bricks and timber (for house construction), a rise in the price of one (say bricks) will increase demand for the other (timber). A fall in price will have the opposite effect. For goods which are complements (e.g. bread and butter, CDs and CD players), the opposite will occur and a rise (fall) in the price of one will cause a fall (rise) in the demand for the other.

Income is another underlying condition of demand that can rise (or fall) and cause shifts in the demand curve for a commodity. Assuming prices of all goods remain unchanged, a rise in consumers' incomes will enable households to increase their expenditure on all types of commodities that they are currently in the habit of purchasing (as well as some they are not). In practice, the demand for some goods will rise much faster than the demand for others, and demand for yet other goods could even fall if they are considered to be somehow inferior to more expensive alternatives (e.g. electric fires and central heating). A measure of this change is the *income elasticity of demand* (Ey) which can be defined as:

$$Ey = \frac{\text{percentage change in quantity demanded}}{\text{percentage change in income}}$$

The value of Ey can be negative (for inferior goods) or positive. For positive values below unity, demand is said to be inelastic (it rises with income, but at a slower rate). For values above unity, consumption rises proportionately faster than income and demand is said to be income elastic. Goods with high income elasticities are often considered to be luxury items (e.g. wines and spirits and recreational goods), but many everyday goods such as consumer durables also have high income elasticities. Retailers are keen to exploit such sectors, as over time retail sales will clearly increase faster in product lines with high income elasticities than where income elasticity is close to or below unity.

In the case of housing, recent estimates for the United Kingdom have arrived at a value of about 0.5 for income elasticity (Ermisch, Findlay and Gibb, 1996). This would suggest that, say, a 10 per cent rise in (real) income (over several years) would result in only a 5 per cent increase in (real) housing expenditure. Clearly such estimates are not particularly good news for the house-building industry. Furthermore, if we look at households ranked by weekly household income, we would expect to see a declining share of expenditure on housing as income rises if, as the above study suggests, housing expenditure is in fact income inelastic. This indeed turns out to be the case; figures for 1991 from *Household Expenditure Survey* (Office of Population Censuses and Surveys, 1992) show that, whereas households in the top two deciles of income spend only around 17 per cent of their income on housing (net expenditure), the proportion spent by households in the second and third deciles is nearly 25 per cent.

Price elasticity of supply

The elasticity of supply (Es) measures the degree of responsiveness of quantity supplied to changes in the market price of the commodity. As such it can be defined as

$$Es = \frac{\text{percentage change in quantity supplied}}{\text{percentage change in price}}$$

The value of *Es* ranges from zero to infinity (shown by S_0 and S_1 respectively in Figure 2.13). Zero elasticity is quite rare (it implies there is no change in quantity supplied to q_0 whatever the price), as is infinite elasticity (it implies that an infinite quantity can be supplied at P_0). In reality most supply schedules fall somewhere in between. Figure 2.13b shows that any straight-line supply curve passing through the origin (0) will have an elasticity equal to unity. To confirm this we can see that doubling the price from P_0 to P_1 results in a doubling of quantity supplied in both supply curves, S_2 and S_3.

For some commodities, such as housing (which takes months or even years to plan and construct), there may be a considerable contrast between inelastic supply in the short term and elastic supply in the longer term. This may indeed provide one of the main explanations for the inherent instability of the owner-occupied housing market as illustrated in Figure 2.14. A shift in demand for owner-occupation (D_0 to D_1) can occur quite rapidly for a number of reasons: a fall in interest rates, population growth, expectations of future price rises, or a rise in government subsidy being the most obvious. In the short term the supply cannot be expanded quickly and the market equilibrium will move towards e_1 and a market price of P_1. Higher house prices will encourage house-builders to expand output and eventually a new long-run equilibrium will be achieved at e_2. Clearly it would be advantageous all round for the market to move directly from e_0 to e_2 (to avoid prices rising and falling again from P_1 to P_2). This can, however, only be achieved if either the shift outwards of the demand curve from D_0 to D_1 is slowed down (e.g. cuts in mortgage interest tax relief to offset lower interest rates), or the process of planning and construction is speeded up so that the short-term supply curve becomes more elastic. The latter is clearly difficult to achieve, given obvious constraints on the construction process. Governments

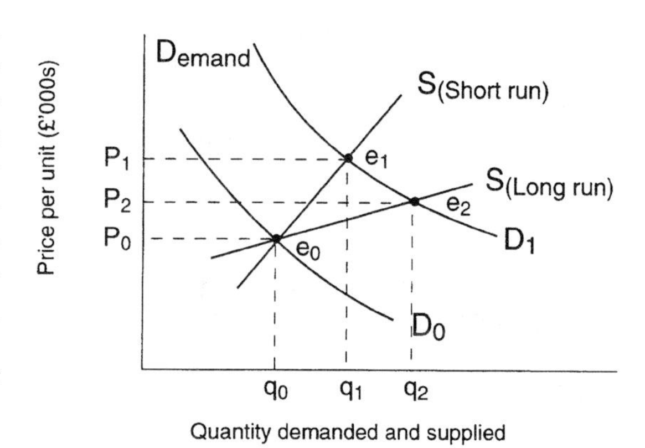

Figure 2.14 Market equilibrium in the housing market in the short and long run

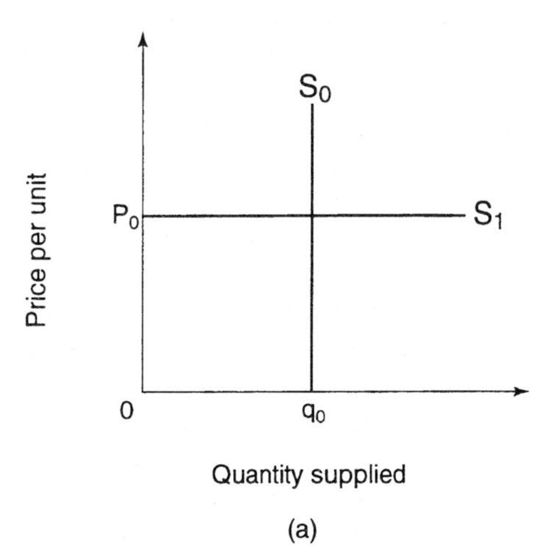

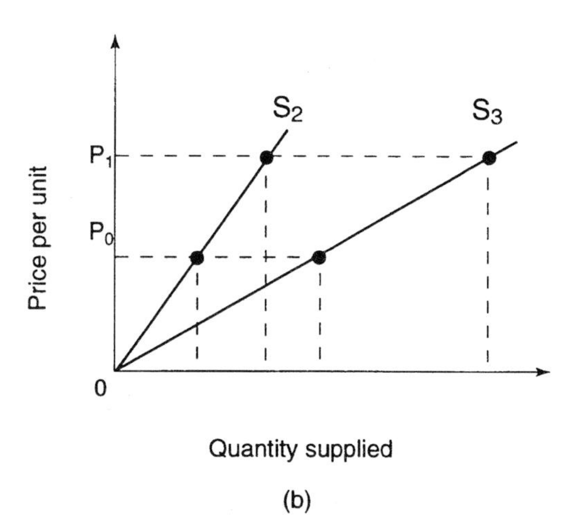

Figure 2.13 Elasticity of supply

now appear to be more aware of the cyclical nature of the housing market. After the boom and bust situation of the late 1980s, a greater willingness to introduce policies to dampen the demand side of the housing market can be observed.

PRICE CONTROLS AND THE MARKET: AN EXAMPLE OF THE EFFECTS OF RENT CONTROL AND OF SHORT- AND LONG-RUN SUPPLY ELASTICITIES

In many European countries rent control was an important characteristic of the housing market throughout much of this century, often up until the 1980s. In the private rented sector rent controls have generally been weakened or restricted in scope (e.g. to certain high-cost areas, or to restrain the annual rate of increase in rents) but in the social housing sector such controls are more common, as governments which provide subsidies to this sector usually expect such housing to be more affordable and to be available at below-market rates. Figure 2.15 illustrates the effects of maximum price-setting (i.e. rent control) in the private rented sector. In the United Kingdom landlords are able to charge market rents, so the initial equilibrium rent would be r_m at point *a*, where f_0 units would be provided. Rent controls represent a price ceiling. If they were to be set at a level of r_m or higher they would have no effect on the market, which would continue to clear at point *a*. Assuming the maximum rent level is set at r_c, it can be seen that in the short term the quantity of rented accommodation provided will fall to f_1 as some landlords who are in a position to do so (e.g. because tenants' leases have just expired) withdraw their properties from the market. In the longer term, more landlords will be able to withdraw from the market, and this will occur especially if the return available on private letting falls below returns which can be achieved from other alternative investments. At this point supply will fall to f_3 as vacant houses and flats are sold for owner-occupation.

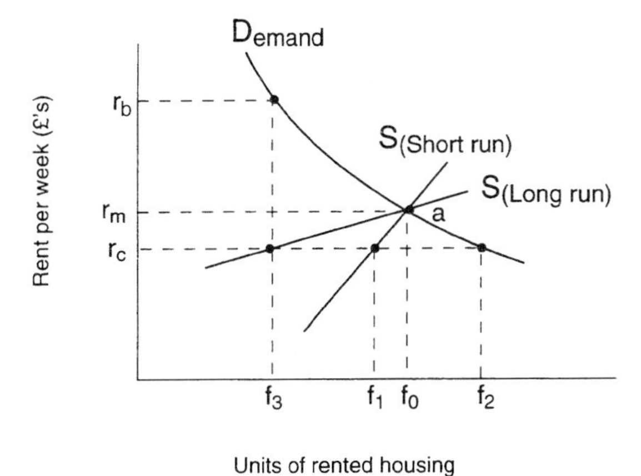

Figure 2.15 The economic effects of rent controls

Lower rents will, however, increase demand in this sector (to f_2), so that in the longer term a market deficit appears, equal to $f_2 - f_3$ (or $f_2 - f_1$ in the short term). This outcome can produce a black market for flats where either landlords are prepared to risk letting outside the Rent Acts or tenants are able to sub-let (again illegally) part or all of the property. In this way rents as high as *rb* can be achieved!

If rent controls are then lifted, the supply of rented accommodation should increase (from f_3 to f_0) as the market moves back to equilibrium. However, a number of points should be noted.

First, much of any increase could result from existing owner-occupied properties (especially those which are hard to sell) being bought up by landlords, rather than through new build. As the majority of landlords own only one or a few properties (and include former owner-occupiers unable to sell), they are more likely to buy existing properties 'off the peg' than to undertake major investment in new developments. Consequently properties will tend to transfer from one type of tenure to another with little net increase in the housing stock.

Second, both demand and supply for privately rented accommodation can shift greatly over time, and such changes may be more important in explaining market trends than the actual slope and initial position of the supply and demand curves. For instance, changing preferences in favour of owner-occupation could shift the demand curve to the left. An improvement in rates of return of other types of investment (e.g. shares) would also shift the supply curve to the left. Equilibrium supply would then decline, although the market rent could move either way depending on the relative importance of each shift. It seems quite likely that such trends may at least partly explain the decline in the share of the private rented sector in most European countries over the post-war period.

Third, the demand curve for private renting measures not only willingness but also ability to pay. At rent levels of the early 1990s, 44 per cent of tenants in this sector spent over 25 per cent of their income on housing, yet nearly two-thirds of all private tenants could not have afforded to purchase a house in the cheapest part of the country. Around 1 million private tenants claimed housing benefit to enable them to pay market rents (out of a total of 2 million).

THE MACRO-ECONOMY

Macro-economics is the branch of economics that is concerned with the economy as a whole. It deals with the study of broad economic aggregates such as total output of the economy, levels of employment (and unemployment) and the rate of change of prices (inflation). These aggregates are themselves interrelated so that, for example, a fall in production and output of the economy will tend to reduce the number of workers required by firms and unemployment will rise. Levels of these aggregates will depend on the actions of many different groups such as consumers, governments and firms. Finally, macro-economics attempts to analyse long-run trends in these aggregates as well as the causes and significance of short-run fluctuations (known as business cycles).

The circular flow of income

In any economy the consumers are also producers, in the sense that they provide firms with their labour in exchange for earnings which they can then spend or save. The government sector in any economy is also important, spending on capital investment (e.g. roads, social housing, etc.), providing services (e.g. education) and providing monetary transfers to individuals, principally in the form of state pensions, support for loss of earnings through unemployment, support for low-income families and help towards payment of housing costs (e.g. housing benefit). In order to finance this expenditure, governments need to raise money through taxation of individuals and firms.

At any point in time (say over a year) total expenditure on goods and services must equal the value of production (output) of goods and services by firms in the economy, and this in turn must equal the payments made by firms to acquire factors of production, which then become factor incomes. Aggregate expenditure (AE) must therefore be equal to the total output of the economy (O) and this again will be equal to the level of national income (Y), i.e.

$$AE \equiv O \equiv Y$$

This can be seen more clearly from the flow diagram shown in Figure 2.16, which represents a simple economy with no government sector or external trade. Households can either spend all their income (Y) on domestically produced goods and services or they can save some portion of it (S). Firms produce goods and services (O) which are purchased by households (AE). Furthermore, it can be seen that an additional form of expenditure on the output of firms occurs through investment expenditure (I) from other firms.

Whether economic activity is measured from the level of income (Y), production (O) or expenditure (E) the same result is obtained, since the same flow is being looked at from different points.

Savings constitute a withdrawal from this circular flow (since they represent income not spent on currently produced goods and services) but investment represents an injection into the flow (i.e. additional expenditure not dependent on households or household incomes). Under stable conditions and with a

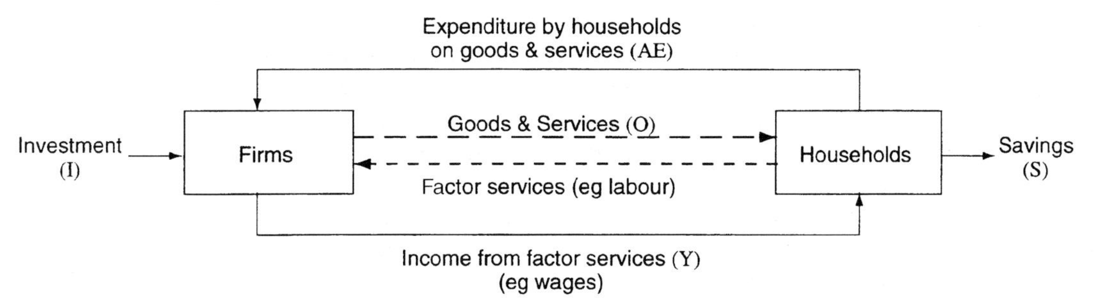

Figure 2.16 A simple model of the circular flow of income

constant level of national income, withdrawals must equal injections (otherwise the circular flow would be expanding or contracting). Hence savings must equal investment.

So how could a change in the level of income and production come about? To see how this could occur let us assume that households decide to spend a higher proportion of their income than is currently the case. Savings will decline but expenditure (*E*) will increase. Firms will increase output (*O*) to meet the new level of demand and to do this they must have more factors of production, so factor incomes (*Y*) increase.

By how much will national income increase? To answer this question we need to know how much of any increase in income is spent on domestically produced goods and services, and how much is saved. Let us assume that households typically save 20 per cent of earnings but spend 80 per cent. Therefore, of every additional £100 earned, £80 will be spent. This in turn will lead to a further increase in production and further growth of income and so on, as shown in Figure 2.17.

In this figure an initial increase of £100 in spending produces a final effect on aggregate expenditure found by summing the figures on the top row. Alternatively we can look at the proportion of income which is passed on as spending (known as the marginal propensity to consume – MPC) or the proportion which is saved (marginal propensity to save – MPS). If the latter figure is taken (i.e. 20 per cent or 0.2) we can find the sum of the series using the formula 1/MPS, i.e. 1/0.2, giving a value of 5. The figure derived from this process is known as the *multiplier* (*k*) as it can be used to find the overall impact on aggregate income and expenditure of any

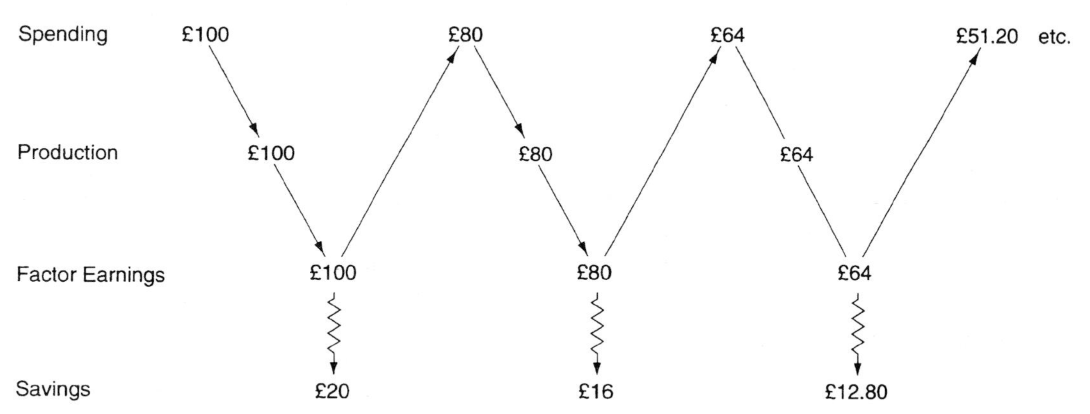

Figure 2.17 The multiplier effect of an increase in expenditure

initial change in spending. In the above example, we have:

$$£100 \times k\ (5) = £500$$

However, if the MPS had been as high as 40 per cent or 0.4, the value of the multiplier would have been 1/0.4, i.e. 2.5, and an initial increase in spending of £100 would have raised aggregate expenditure by only £250 (i.e. £100 × 2.5). Clearly, the higher the proportion of income that is withdrawn from the circular flow in each 'round', the lower the level of the multiplier. In either case, a new equilibrium in the circular flow of income will be achieved once savings and investment (or more generally, injections into and withdrawals from the circular flow) are once again equal.

In reality, the macro-economy is much more complex than the above analysis would suggest. Nevertheless, the basic principles apply. Figure 2.18

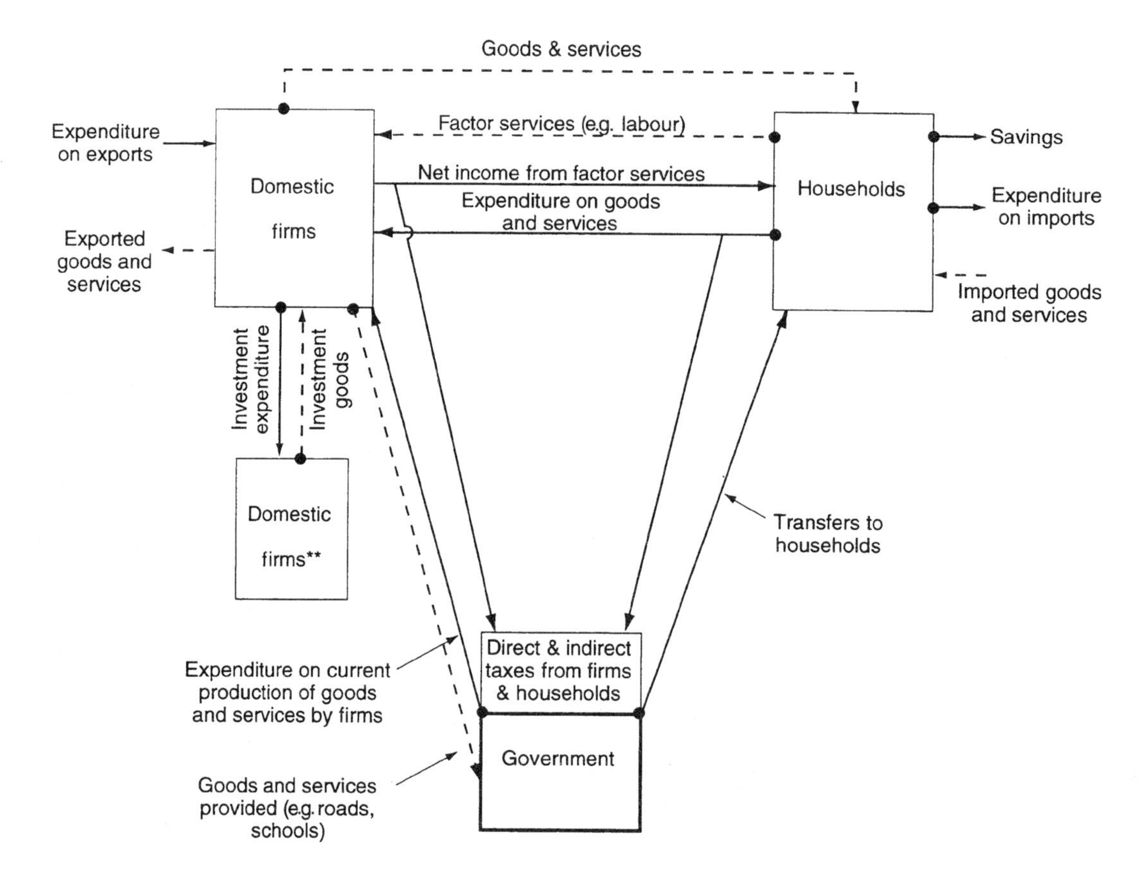

Figure 2.18 The circular flow of income

provides a more detailed picture of the circular flow of income in a real-world economy, with a public sector (government) and international trade.

Households make withdrawals from the circular flow of income through the purchase of imports as well as savings, and further withdrawals take place in the form of government taxes on expenditure. The government sector receives these taxes as revenues (as well as receiving revenue from direct taxes on individuals and firms) and may choose to withhold such revenue (in which case it remains a withdrawal from the circular flow) or spend it, either to purchase currently produced goods and services (e.g. new public housing, roads, etc.) or to raise the level of income of certain low-income groups (e.g. unemployed) who might not otherwise be able to afford a decent standard of living or adequate housing.

Domestic firms may find they can avail themselves of savings generated in the economy through the intermediary of banking institutions. They can thus borrow money from banks to finance new capital investment in plant and machinery, which as shown is supplied by other domestic firms. Such investment expenditure counts as an injection into the circular flow of income.

Aggregate expenditure (AE) in this figure now also includes investment expenditure (I), government expenditure (G) and (net) exports ($X-M$) and can be defined as

$$AE = C + I + G + (X-M)$$

An increase in any of these will still produce a multiplier effect on aggregate expenditure and output, but the value of the multiplier will now be much smaller than in our previous model (where savings were the only form of withdrawal), since in addition to savings, some proportion of earnings will be withdrawn as tax and a further proportion will be withdrawn as expenditure on imported goods and services.

The national accounts

Figure 2.19 shows how the national income accounts measure (annual) flows of production, income and expenditure. Moving from left to right, the measure of total output of the economy (gross domestic product or GDP) is broken down by value added for each sector. These values are themselves calculated from the factor earnings attributable to each sector (i.e. at factor cost). The second column shows the share of total production that goes as income to the factors of production, and again this is referred to as GDP at factor cost. Nearly two-thirds of GDP is accounted for by income from employment in the form of wages and salaries.

However, if you earned £100 in a week and spent all of it on currently produced goods and services, producers would not necessarily receive the full £100 as approximately 17.5 per cent (the current value added tax rate) would be collected as tax. Thus, the market price of a commodity is greater than the cost of factors going to produce it. Adding this indirect tax revenue (net of subsidy) therefore gives us GDP at market prices, and this is the figure we obtain in the third column, starting from an expenditure-based approach to the national accounts. In this case, GDP has been calculated by summing all the expenditures made to purchase the final output of the economy. These come from consumption expenditure (on all goods and services sold to their final users), government expenditure, gross investment (on capital goods used to produce other goods/services) and net exports (i.e. gross exports minus imports). Adding in income from overseas assets by UK residents (deducting income received by non-UK residents from assets owned in this country) gives us a measure of gross national product (GNP). Finally, making an allowance for the value of capital that has been used up in the course of production during the year (i.e. depreciation) would give net national product (NNP). This is the figure which used to be referred to as national income in the national accounts. However, the term national income now refers more generally to any of the standard measures just described.

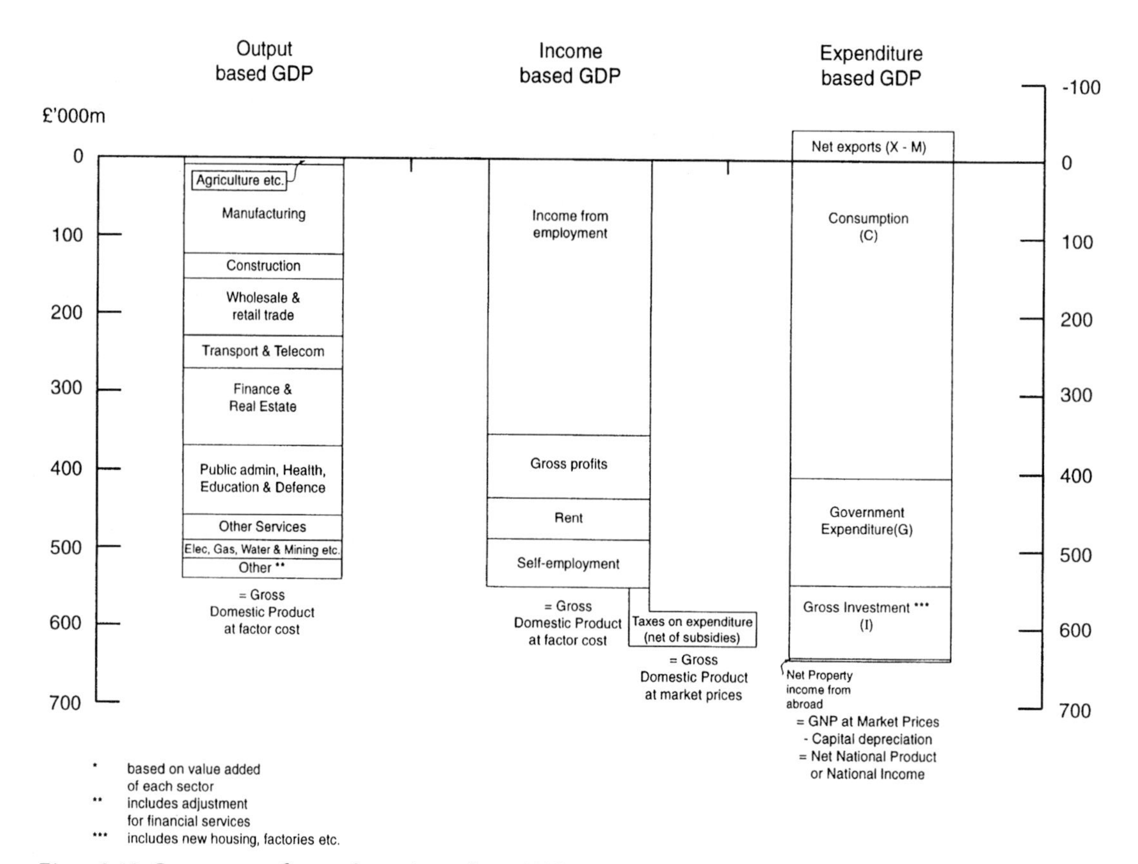

Figure 2.19 Components of gross domestic product, 1993

Aggregate demand and government policy

By manipulating the components of aggregate expenditure and demand, the government can influence the level of economic activity in the country. This could, in practice, be achieved in a number of ways.

First, governments can increase the level of government expenditure (*G*) not covered by additional tax revenue. This gap between government spending and revenue is known as the public sector borrowing requirement (PSBR). Alternatively, the government could leave government expenditure unchanged but reduce tax revenue instead (by lowering VAT, income tax, etc.), although this would again raise the PSBR.

Second, the government could take steps to raise or lower interest rates (the cost of borrowing). Since a high proportion of consumption and investment is financed via borrowing (e.g. mortgages), lower interest rates should stimulate these components of aggregate expenditure.

Third, the government could try to raise exports or lower imports. Most governments now have little manoeuvre for implementing trade policy to penalise imports (there is effectively free trade within the European Union, for example), but by allowing the exchange rate to float down (i.e. depreciate) relative to other currencies, the domestic price of exports can be reduced and that of imports increased to provide some competitive advantage for domestic producers. However, were Britain to join a single European currency, this option would no longer be available (in spite of any other benefits the British economy might experience).

Figure 2.20 shows how aggregate expenditure is determined. Consumption will generally rise at higher levels of income, so that consumption expenditure (*C*) is upward sloping. Investment, government expenditure and net exports are assumed to be autonomous and not directly related to the level of income. Summing these components, we find that the aggregate expenditure line (*AE*) is itself upward sloping. From the earlier discussion, we know that in equilibrium, income (*Y*) must equal expenditure *AE* – that is, there is no pressure for the circular flow of income and production to increase or decrease. Points that meet this requirement are found along the line $AE = Y$ in Figure 2.20. Changes in any element of *AE* will result in a new point of equilibrium for national income and output and employment in the economy.

However, this does not mean that the level of economic activity can be varied at will by central governments for, as we have seem, the resources available to the economy are essentially fixed in the short run. Where most resources are fully employed, an increase in aggregate expenditure may exceed the ability of the national economy to meet a new higher level of demand, and prices will tend to rise. Eventually, further increases in expenditure (and therefore demand for goods and services) may result in no increase in output being achievable and further spending would be entirely inflationary. The relationship between real national income and the price level in the economy is given by the aggregate supply curve (*AS*). Starting with Figure 2.21a, an increase in government spending (*G*) is seen to move aggregate expenditure from AE_0 to AE_1. With no change in the price level, equilibrium

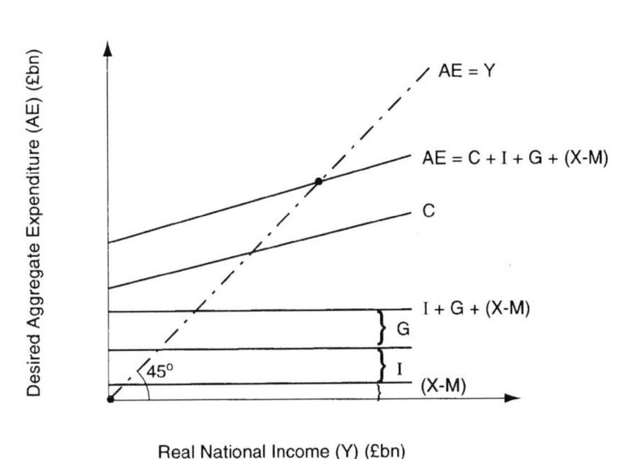

Figure 2.20 Determination of equilibrium national income

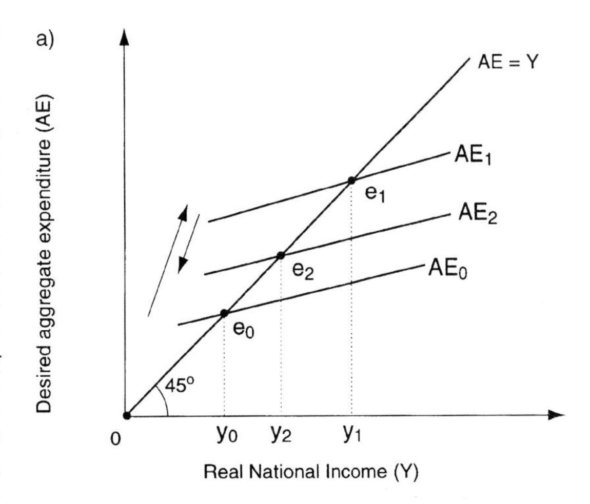

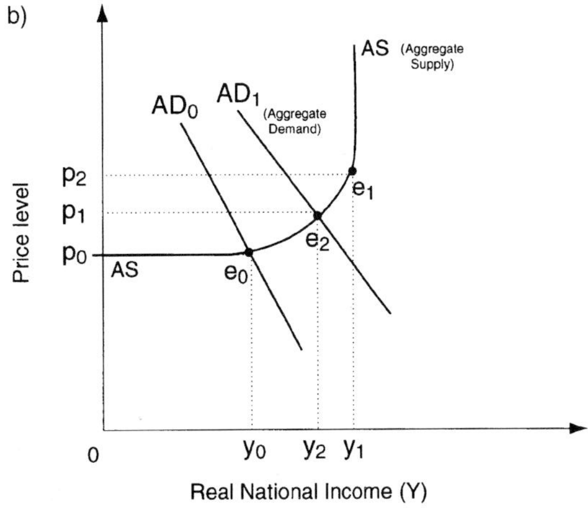

Figure 2.21 Changes in aggregate demand, national income and prices

national income would move to Y_1. However, where prices rise (to P_1 in Figure 2.21b), the real level (i.e. price adjusted) increase in national income (to Y_2) is of course much lower, corresponding to AE_2 in Figure 2.21a and a level of aggregate demand (AD) of AD_1 in Figure 2.21b.

Beyond Y_1, all resources are fully employed and no further increase in real output is possible. After this point is reached, growth of aggregate expenditure will only serve to raise prices rather than output. Because the short-run aggregate supply curve is generally assumed to be rather flat where the economy is running with considerable spare capacity (below Y_0), governments may be able to increase aggregate expenditure to achieve a rise in national income and output with little effect on the price level. Above this point, however, further growth of aggregate expenditure will increasingly raise prices, and the trade-off between output and inflation will gradually worsen.

Governments may of course seek to expand or contract aggregate expenditure either directly through changes in government expenditure (or taxes) or indirectly through changes in interest rates. Furthermore, aggregate expenditure may change for other reasons related to patterns of consumer spending or investment by firms and we shall consider these in turn. A decision by households to spend a higher proportion of earnings (and to save less) will lead to higher levels of consumer spending and aggregate expenditure will rise. If the economy is initially in a position such as AE_0 and AD_0 in our previous figure, the effect of a small change may be to increase output rather than inflation. However, if the economy was already operating close to Y_1, the government may take steps to reduce consumer demand by raising interest rates. During the 1980s housing boom, the personal savings ratio fell from just over 12 per cent to under 6 per cent (between 1984 and 1988). Households felt wealthier because the value of their houses had increased and they were more inclined to spend, and even increased their borrowing to do so. After the peak of the housing boom in 1989, the opposite occurred and personal savings began to rise, reaching over 12 per cent again in 1992. Unfortunately, higher savings led to lower spending and the recession in the economy continued until the end of 1992 – perhaps longer than might otherwise have been the case. More generally, however, consumer spending is always likely to change throughout cycles in the economy. Coming out of a recession, spending and borrowing are likely to rise as confidence returns and interest rates are often low. In a recession, savings will rise and spending will fall and consumers will be more likely to pay off debts than to take on new borrowing.

Investment expenditure is also subject to cyclical changes which are generally more pronounced than for consumer spending. The process which causes such fluctuations is called the *accelerator mechanism*. This process is illustrated in a simple form in Table 2.2. If it takes one machine to produce ten units of some consumer durable (e.g. cars) and on average each machine lasts ten years, then there will be demand for one new machine each year (year 1). In a poor year (year 2) if demand for the consumer durable falls by 10 per cent, one machine is not replaced and the demand for new machines falls to zero. In the following year, demand returns to normal yet the demand for new machines rises to two (one for new capacity and one for replacement). Clearly, in proportionate terms, the change in demand over the three years has been much greater in the capital goods industry (machinery) than in the consumer goods industry. Since, overall, investment expenditure represents around 15 per cent of GDP (at market prices) it is clear that substantial fluctuations in this component of demand can have considerable multiplier effects on the level of economic activity in the economy as a whole. In a downturn, firms may tend to cut back on new investment, but coming out of a recession invest-

Table 2.2 The accelerator mechanism

	Output	*Machines*	*New machines*
Year 1	100	10	1
Year 2	90	9	0
Year 3	100	10	2

ment levels may rise substantially. Forms of investment affected may include factories, shops and offices as well as new machinery and capital equipment, and new housing can also be included in this category. The effect, particularly for firms in construction and civil engineering, is to produce an upturn in workload from both public and private sectors at the same time.

CASE STUDY 2.1

AN ECONOMIC ANALYSIS OF THE PROBLEMS OF RENT-SETTING IN THE SOCIAL RENTED SECTOR

Providers of social housing, who aim to make renting affordable for low-income groups, often face a dilemma. The lower the level of rents they are able to achieve, the greater the affordability. However, the cost of providing new additions to their stock (the marginal cost, i.e. for each additional unit) is always likely to be high in relation to costs in their older stock of dwellings. This can occur mainly because either loans or debt on older stock have been largely paid off, or there has been a rise in the real cost of construction. Other factors such as a rise in land costs or interest rates may also be important. Conversely, it must be recognised that serious repair problems in the older stock may tend to raise costs here over time.

Figure 2.22 illustrates the problem; the older stock of housing is assumed to represent an average cost of $0y^1$. Additions to the stock would see marginal costs rising, and to cover costs on the last units built, the provider would have to charge a rent of $0a^1$ in the diagram. However, since the new, more expensive housing represents only a proportion of

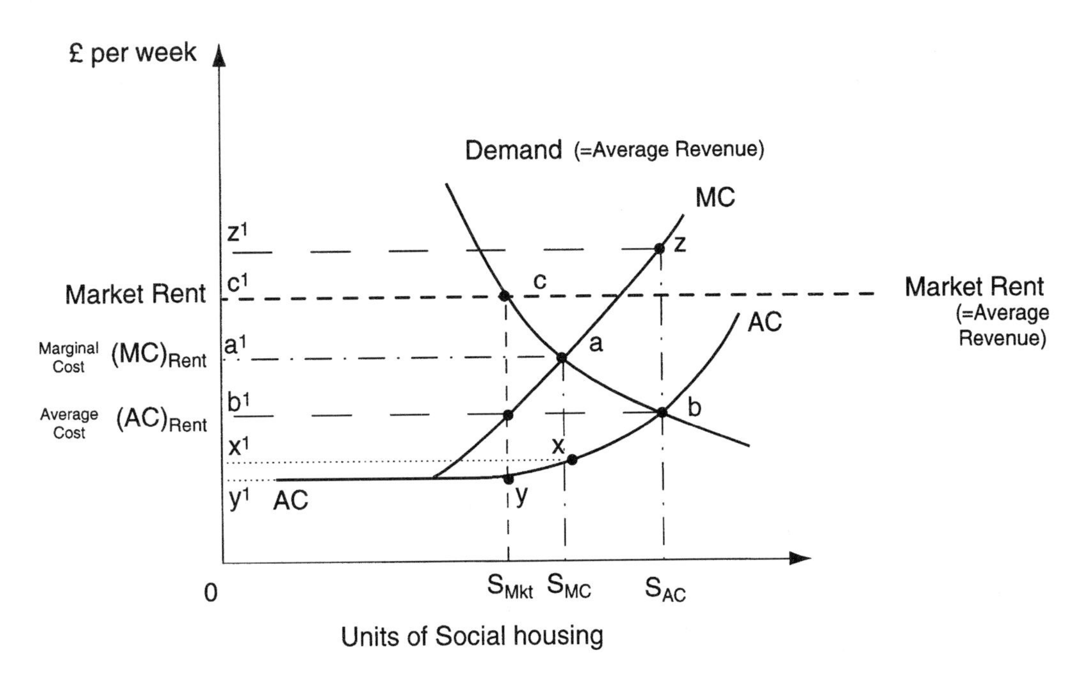

Figure 2.22 Costs and rent setting in the social rented sector

the stock held, averaging total costs throughout the stock would produce a cost equivalent to $0x^1$ per week (i.e. average or unit cost). The various alternatives can now be considered in more detail:

a = a position of marginal cost (MC) pricing – costs are covered for more recent units with a rent of $0a^1$. Since all units are charged this rent, a profit appears over the whole stock as average cost pricing would necessitate a rent of only $0x^1$ (i.e. covering average cost, with a stock of S_{mc}). At point a total costs are therefore $ox^1 \times S_{mc}$, and total revenue $0a^1 \times S_{mc}$. The difference between these two (i.e. $0a^1 - 0x^1 \times S_{mc}$) represents total profit of the organisation although, as can be seen, more recent additions to stock are only just breaking even with rent levels just covering (marginal) cost.

Overall 'break-even' pricing over the whole stock = b. Average revenue equals average costs ($AR = AC$) and therefore total revenue equals total costs ($TR = TC$). The provider is therefore non-profit-making. While similar rents are charged for similar properties (i.e. $0b^1$) the cost of more recent properties is as high as $0z^1$ compared to $0y^1$ for older properties. Rents on older properties (where costs are lower) are therefore used to cross-subsidise rents on newer properties (which would otherwise come to as much as $0z^1$ per week), so that rent levels are the same for similar properties across the whole stock. Since in practice the average cost (AC) levels may be different for different providers of social housing (such as housing associations or local authorities), the degree to which they can effectively cross-subsidise new construction will also vary. For social housing providers unable to do this (either because they are only recently established or perhaps because they have unusually high repair costs on their older stock of housing), only the marginal cost curve is relevant and rents would have to be set very close to $0a^1$ in Figure 2.22.

In some European countries (e.g. Denmark, Germany), however, the ability to cross-subsidise is limited as subsidies are attached to individual dwellings and rents are based on actual construction costs (taking account of the subsidy). As a result older dwellings with lower historic costs are often cheaper to rent than similar, more recently constructed properties. Because social housing providers in these cases are restricted in the extent to which they can raise rents on the older property, rents on newer properties cannot be cross-subsidised (although actual rents are still kept below free market rents because of subsidy).

Point c represents a position where social housing providers might charge open-market rents for new and existing property. Total supply would be much lower than either points a or b. We can also note that the demand for low-cost social housing might fall quite rapidly after this point, particularly if any form of housing benefit, underpinning demand, tapers off rapidly. Total profits would, however, be greater in this situation, at $c^1 - y^1$ per dwelling per week.

The above analysis illustrates why the financial situation of social housing providers can vary enormously depending on circumstances relating to age and condition of stock, extent of new build, location and costs (while construction costs may exhibit relatively small regional differences, land costs may vary considerably, being highest for example in London and the South East).

Some countries (e.g. France, Germany and the USA) also provide subsidies to private developers to enable them to provide low-cost rented accommodation to low/middle-income groups. Rents are then controlled at below-market levels as long as subsidies continue or for a specified time. As such housing would eventually revert to market rental levels (or possibly be sold for owner-occupation), the effect of the subsidy is eventually lost.

CONCLUSIONS

The basic theories of micro- and macro-economics help shed light on the relationship between scarcity and choice. Given that resources are limited, a choice has to be made in order to satisfy consumer, producer or governmental demand. From income, consumers need to choose between one form of consumer good or service and another; producers need to choose between various combinations of land, labour, capital and entrepreneurship (factors of production); and governments need to choose between one form of public expenditure and another. Within the context of housing, consumer choice is reflected in the overall demand for accommodation or in the demand for a dwelling in any one of the different types of housing tenure. Producer choice involves the purchase of appropriate factors of production and the type and number of dwellings to supply. Governmental choice is determined by policy aims, for example whether to expand social housing or owner-occupation, or whether to provide 'bricks and mortar' subsidies or personal allowances.

Prior to a consideration of these choices in Chapter 3 on housing policy and finance, this chapter (as a prerequisite) has analysed demand and supply and the determination of market equilibrium; price, income and supply elasticities; price control and the market in the context of rent control; the essential attributes of the macro-economy; and (as a case study) rent-setting in the social rented sector. Less explicitly, this chapter also provides the economic context for those parts of the book where there are problems of supply and demand disequilibrium, for example in Chapter 4 on equal opportunities and housing, Chapter 5 on town planning and development, and Chapter 7 on environmental health and housing.

QUESTIONS FOR DISCUSSION

1. **In a 'mixed economy' why do governments sometimes intervene in the provision of certain goods and services?**
2. **Explain the difference between movements along a demand curve for owner-occupied housing, and shifts in the same demand curve.**
3. **'Since low-income households spend a higher proportion of their income on housing than high-income households, government housing policies should be aimed mainly at families and individuals in the former group.' Discuss.**
4. **How and to what extent can governments influence output and employment in the national economy?**
5. **Is there a case for free-market rents in social rented housing? What would be the implications of such a policy?**

RECOMMENDED READING

Balchin, P.N. (1996) *An Introduction to Housing Policy*, Routledge.

Balchin, P.N., Bull, G.H. and Kieve, J. (1995) *Urban Land Economics and Public Policy*, 5th edn, Macmillan.

Cooke, A.J. (1996) *Economics and Construction*, Macmillan.

Lipsey, R.G. and Crystal, K.A. (1995) *Positive Economics*, 8th edn, Oxford University Press.

Manser, J.E. (1994) *Economics*, E. & F.N. Spon.

Ruddock, L. (1992) *Economics for Construction and Property*, Edward Arnold.

REFERENCES

Balchin, P.N., Bull, G.H. and Kieve, J. (1995) *Urban Land Economics and Public Policy*, 5th edn, Macmillan.

Ermisch, J.F., Findlay, J. and Gibb, K. (1996) 'Price elasticity of housing demand in Britain: issues of sample selection', *Journal of Housing Economics*, 5.

Office of Population Censuses and Surveys (1992) *Household Expenditure Survey*, HMSO.

3

HOUSING POLICY AND FINANCE

Paul Balchin, David Isaac and Maureen Rhoden

During the period of the Thatcher and Major governments, 1979–97, Conservative housing policy reduced the amount of public expenditure on housing, brought about a marked reduction in house-building in the social sector, was instrumental in raising rents ahead of inflation in both the private and social rented sectors, privatised much of the local authority housing stock and the financial responsibility for housing rehabilitation, replaced local authorities by housing associations as the major providers of social housing, and increasingly left owner-occupiers free to face the vagaries of the market. By the late 1990s, however, owner-occupation was even more entrenched as the dominant tenure (in contrast to its lesser importance in many other European countries), while problems of affordability had become increasingly apparent in each of the housing sectors. This chapter seeks to explain the relationship between policies and housing markets in recent years and specifically examines:

- Housing supply and housing need.
- House-building and the house-building cycle.
- The marketisation of the private rented sector.
- Local authority housing: investment, rents and subsidies.
- The changing role of local authorities: from providers to enablers.
- Privatisation: the selling off of council estates to their tenants, estate privatisation, the privatisation of rehabilitation.
- Housing associations: the new providers.
- Owner-occupation: intervention or liberalisation? A policy of contradiction.
- Affordability and subsidisation.
- Housing tenure.

In conclusion, the chapter contains case studies on the finance of private rented housing and on earnings, rents and house prices in each of the housing sectors.

INTRODUCTION

The housing market in the United Kingdom is composed of three distinct yet interrelated tenures: owner-occupation, private rented accommodation and social rented housing which includes the local authority stock, housing association dwellings and other non-profit housing organisations. These sub-markets are related through a pattern of flows complicated by contractual obligations, ownership, property rights and government intervention: from privately rented housing to owner-occupation or to the social sector; and from the social sector to owner-occupation.

The housing market is dominated by the existing stock of dwellings which represent a high proportion of the total supply of housing. Relative to the size of stock, the net annual addition of newly built housing is small – normally less than 1 per cent per annum. Supply is therefore relatively fixed even in the long term, and prices of the standing stock and its allocation among users are determined primarily by changes in demand conditions. The durability and immobility of housing imposes a brake on the pace of adjustment of supply to changes in demand. In layout and building pattern the market situation is fundamentally one of disequilibrium; since so much of aggregate stock is not of recent construction, its location pattern reflects past distribution of population and economic activity.

House purchase represents a very large capital outlay for the owner-occupier or landlord which can rarely be financed out of income, thus borrowing is necessary and the availability of long-term credit is of critical importance in making demand effective. In the United Kingdom throughout most of the twentieth century, none of the housing sectors were left free to face market forces. Both private and local authority rented housing were subject to extensive intervention, with marked effects upon supply and demand, while tax relief and exemption distorted the appeal of owner-occupation (see Chapter 1). Overall, the price system was not allowed to carry out its function of allocating scarce housing resources between alternative users.

With the return of a Conservative government in 1979 under the premiership of Margaret Thatcher, the uneasy consensus of the previous thirty years (outlined in Chapter 1) broke down. Thatcherism – involving as far as possible, a lurch back to the free market – dominated housing policy for a generation. A neo-liberal welfare regime was created whereby housing investment in the local authority sector plummeted; rents were permitted to rise to market levels within both the private and the social rented sectors; local authority and housing association dwellings were sold off; rehabilitation was forced to rely more and more on private means; housing association investment became increasingly dependent upon bank and building society funding; and subsidisation within the owner-occupied sector was gradually reduced, with home-ownership being dependent more and more on private funding. In general, subsidies were either withdrawn or available only to the poor on a means-tested basis, state funding providing little more than a safety net for those most in need – a welfare service for the seriously disadvantaged.

It is necessary to consider the overall disequilibrium in the United Kingdom housing market and the performance of the house-building industry in recent years to provide a perspective for the examination later in this chapter of the marketisation of each of the housing sectors.

HOUSING SUPPLY AND HOUSING NEED

Housing need in the United Kingdom is largely satisfied but only in very crude terms. For over thirty years after the Second World War, the owner-occupied and local authority housing sectors grew both absolutely and relatively, while the supply of private rented accommodation decreased. By the early 1970s (for the first time since 1938) there was a crude surplus of dwellings over households – the surplus rising to 1,026,000 by 1980 but falling to 822,000 by 1991 (Table 3.1).

The crude surplus in 1991 did not, however, indicate the true relationship between supply and need. Of the 23.6 million dwellings in 1991, there were well over a million unfit dwellings or homes lacking basic amenities, dwellings undergoing

Table 3.1 The number of dwellings and households, United Kingdom, 1980–91

	(000s)			
	1980	*1985*	*1991*	*% change 1980–91*
Dwellings	21,426	22,350	23,622	+10.3
Households	20,400	21,400	22,800	+11.8
Surplus	1,026	950	822	−19.91

Source: Central Statistical Office (various) *Social Survey*, HMSO; Department of the Environment (various) *Housing and Construction Statistics*, HMSO

conversion or improvement, and second homes, while there were about half a million concealed households. Taking these concealments into account, there was a substantial shortage of housing in the United Kingdom of about 3 million dwellings in the early 1990s.

With regard to England alone, Holmans (1995) suggested that in order to meet housing demand and needs over the period 1991–2011, about 240,000 new homes a year will be required with approximately 40 per cent being in the social sector – an estimate compatible with the projected 4.4 million growth in the number of households, 1996–2016 (Department of the Environment, 1996).

The scale of the net housing shortage, and the magnitude of the volume of house-building that will need to be undertaken over the period 1991–2016, are to a significant extent legacies of public policy adopted by successive administrations after 1979. Inadequate government expenditure and an ideological belief in the ability of the free market to satisfy housing need are largely responsible for current deficiencies in supply.

From 1979/80 to 1995/6, government expenditure on housing in real terms plummeted from £11.8 billion to £4.7 billion – a reduction of 60.5 per cent – housing as a proportion of total government expenditure diminishing from 5.1 per cent to 1.6 per cent. In contrast, over the same period, government expenditure on law and order increased by 84 per cent; on social security by 78 per cent; on health and personal social services by 60 per cent, and on education by 30 per cent (Table 3.2) – indicating the very low priority given by the government to housing needs.

HOUSE-BUILDING

Housing need will clearly remain far from satisfied while house-building is at a low level. Whereas in the peak year of 1967, a total of 447,100 dwellings were started in Great Britain – of which 213,900 were in the local authority sector – the total number of starts plummeted to 156,800 in the slump of 1992 with only 36,500 starts being in the social sector (Table 3.3).

Even in the mid-1970s it was recognised that the rate of house-building was failing to satisfy needs. The Green Paper, *Housing Policy: A Consultative Document* (Department of the Environment, 1977), recommended that in order to keep pace with the 'baby boom' of the 1960s, replace unfit housing and

Table 3.2 Government expenditure by function, 1980/81 to 1995/6

	1980/81 £bn	*% total*	*1995/6 £bn*	*% total*	*Real growth 1980/81 to 1994/5 (%)*
Selected services:					
Law and order	8.4	3.7	15.6	5.3	84.4
Social security	51.1	22.3	90.8	30.8	77.7
Health and personal social services	29.8	13.0	47.7	16.2	60.2
Education	27.0	11.8	35.0	11.9	29.6
Housing	11.8	5.1	4.7	1.6	–60.5
Total government expenditure	229.2	100.0	284.5	100.0	28.5

Source: HM Treasury (1995) *Public Expenditure*, Cm 2821.
Note: Expenditure at 1994/5 prices. Other items of government expenditure are not included.

facilitate household mobility, there was a need for 310,000 housing starts per annum until the end of the century.

The housing shortage of the 1990s prompted the *Inquiry into British Housing: Second Report* (Joseph Rowntree Foundation, 1991) to emphasise the need to build 228,000–290,000 houses per annum up to the year 2001, and to stress that, within this total, 100,000 should be built for rent.

The house-building cycle

The cyclical nature of house-building has been evident over the last century but has been particularly pronounced since the 1970s (Table 3.3). Booms have been the lagged outcome of low bank rates whereas, conversely, slumps have been the lagged result of high base rates – increasing and decreasing supply and demand respectively. However, the overall trend in house-building has been downward

Table 3.3 Houses started, Great Britain, 1965–95

	Base rate %	*Starts (000s)*		
		Social sectors (local authority and housing associations)	*Private sector*	*Total*
1971	5.00	136.6	207.3	343.9
1972	9.00	123.0	227.4	350.4 boom
1973	9.33	112.8	214.9	327.7
1974	12.33	146.7	105.3	252.1
1975	10.47	173.8	149.1	322.9 slump
1976	11.71	170.8	154.7	325.4 boom
1977	8.04	132.1	134.8	266.9
1978	9.04	107.4	157.3	264.7
1979	13.68	81.2	144.0	225.1
1980	16.32	56.4	98.9	155.2
1981	13.27	37.2	116.7	153.9 slump
1982	11.93	53.0	140.5	193.4
1983	9.83	48.0	169.8	217.7
1984	9.68	40.2	153.7	193.9
1985	12.25	34.1	163.1	197.2
1986	10.90	32.9	180.1	213.6
1987	9.74	32.8	196.8	229.6
1988	10.09	30.9	221.4	252.2 boom
1989	13.85	31.1	169.9	201.1
1990	14.77	27.2	135.2	162.4
1991	11.70	26.4	135.0	161.4
1992	9.42	36.5	120.3	156.8 slump
1993	7.00	44.1	141.2	185.3
1994	5.50	41.7	160.0	201.7
1995	6.25	34.6	136.2	170.6

Source: Department of the Environment, Scottish Office and Welsh Office, *Housing and Construction Statistics*, HMSO

since 1967 (when there was a total of 447,600 starts), reaching a nadir of 153,900 starts in 1981 (less than in any year since the late 1940s), with output in the social sectors plummeting to a mere 26,400 in 1991 – lower than in any peacetime year since the First World War.

The construction of local authority housing was clearly a victim of public expenditure cuts. With investment (measured by gross fixed capital formation) in new or improved council housing decreasing in cash terms from £2.5 billion in 1980 to £2.2 billion in 1994, there was a reduction in real terms of over 50 per cent.

In the private sector, the house-building industry is mainly speculative. Houses are built mainly in expectation of being sold during or shortly after construction. The private house-building industry – particularly because of the preponderance of small firms – is very sensitive to fluctuations in cost. It is often argued that at times of high interest rates and tight monetary policy, the number of housing starts falls to a relatively low level (for example in 1974, 1980–81 and 1990–92), and that at times of low interest rates and relaxed monetary policy, the number of housing starts rises to a higher level (for example in 1972, 1978 and 1986–8). Housing investment in this sector is thus geared to speculation and, largely due to the housing boom of the late 1980s, increased from £6.1 billion to £18.3 billion in cash terms in 1980–94. Despite this increase, total gross fixed capital formation in dwellings as a percentage of gross domestic product diminished from 3.7 to 3.1 per cent over the same period (Table 3.4).

Table 3.4 Gross fixed capital formation in dwellings, 1980–95

	1980 (£m)	*1995 (£m)*
Public sector	2,559	2,207
Private sector	6,115	19,291
Total	8,674	21,498
Gross domestic product (£bn)	231.8	699.6
Gross fixed capital formation as % of gross domestic product	3.7	3.1

Source: Wilcox (1996)

The resulting reduction in house-building in the local authority sector from 41,500 to 1,700 starts in 1980–94 undoubtedly led to the marked increase in homelessness from 70,038 households in 1979 to 178,867 in 1991 in Great Britain as a whole. Although the provisions of the Housing Act of 1985 and Housing (Scotland) Act of 1987 obliged local authorities to accept households in 'priority need' – specifically pregnant women, families with children, the elderly, the mentally ill, the handicapped and 'disaster victims' – the Act made no provision for other homeless people to be housed, for example the 80,000 single homeless in England and Wales in 1992 (Edwards, 1992). Only through the introduction of the 'Rough Sleepers Initiative' in 1990 with an allocation of £96 million over three years, followed by an additional £86 million in 1992 for an equivalent period, was it feasible for additional hostel places and 'move-on' accommodation to be provided for the non-priority homeless. It was tragically ironic that at the same time as the number of homeless reached record levels in the early 1990s, half a million building workers had been made redundant in the worst slump in the construction industry in peacetime since before 1914.

House-building, land prices and land availability

Because land is heterogeneous, there are enormous differences in individual plot prices. The granting or refusal of planning permission also influences prices. Official data show that from 1971 to 1993 the average price of plots increased from £1,030 to £14,855 (Table 3.5), a rise of 1,345 per cent (at current prices). The average price of new houses, however, increased by only 1,256 per cent over the same period – the price of land as a proportion of house prices having risen from 18.7 per cent to 19.8

Table 3.5 Private sector housing land prices (at constant average density and new house prices), England, 1971–93

	Weighted average price per plot (£)	*Average price of new houses (£)*	*Plot price as a percentage of house price (%)*
1971	1,030	5,510	18.7
1973	2,676	9,630	27.8
1975	1,839	12,124	15.0
1977	1,943	14,343	13.6
1979	3,395	21,455	15.8
1981	4,470	27,910	16.0
1983	5,568	30,943	18.0
1985	8,150	36,295	22.5
1986	10,150	42,319	23.9
1987	14,068	49,435	28.5
1988	19,254	61,551	31.2
1989	18,955	72,256	26.2
1990	19,726	75,403	26.1
1991	15,936	75,119	21.2
1992	12,993	73,093	17.9
1993	14,885	74,857	19.8

Source: Department of the Environment, *Housing and Construction Statistics*

per cent, 1971–93. During the boom years of 1986–90, the average price of land nearly doubled (from about £10,000 to almost £20,000) and house prices rose by 80 per cent, but with the onset of the slump in 1990–92, plot prices fell by 34 per cent and house prices by less than 4 per cent. The market for house-building land is thus more volatile than the housing market, with land prices rising and falling to a greater extent than house prices.

Land prices, of course, vary regionally. Table 3.6

Table 3.6 Regional housing land prices, 1986–93

	Average price per plot			*Plot prices as a percentage of house prices*		
	1986 £	*1990* £	*1993* £	*1986* %	*1990* %	*1993* %
Greater London	24,322	32,128	21,456	44	40	28
South East (excl. Greater London)	18,496	41,600	20,775	38	52	24
West Midlands	7,164	17,027	18,871	25	32	26
South West	8,908	15,395	14,070	23	23	19
North West	5,195	16,150	12,257	19	32	17
East Anglia	7,687	27,020	12,087	21	44	18
East Midlands	5,295	18,859	11,224	18	31	16
North Yorkshire and Humberside	5,107	12,051	10,407	20	25	15
Wales	3,832	14,278	6,783	14	31	10

Source: Department of the Environment, *Housing and Construction Statistics*

shows that during the boom year of 1990, the price of plots as a proportion of average house prices varied from 52 per cent in the South East to 23 per cent in the North and South West. But having reached these inflated levels from the relatively stable year of 1986, the price of plots fell dramatically during the slump of the early 1990s.

But do house prices determine plot prices or does the price of land dictate house prices? The generally accepted view is that the price of residential land is largely derived from the demand for housing, which is substantially determined by mortgage availability and subsidy (see Bassett and Short, 1980; Ball, 1983; Hall *et al.*, 1973; Harloe *et al.*, 1974; and Merrett and Gray, 1982). Based on Ricardian rent theory (Ricardo, 1971), this *residual* view of land prices suggests that, with a given site, location and legal status (for example with outline planning permission for development) and subject to the respective bargaining strength of the landowner and the builder, the maximum price the latter will pay for the land is equal to the expected sales revenue from the completed houses less the combined cost of construction (exclusive of land price) and acceptable profit on the development.

In regions where the demand for housing is greatest, prices are pulled up to a significantly higher level than in those regions where demand is comparatively low. Since there are only minor regional variations in the (non-land) cost of construction and profit on development, it can only be assumed that the regional differences in house prices are reflected in differences in plot prices and land values. Figure 3.1 shows that, for example, in the South East demand (*DD*) is markedly higher than demand (D_1D_1) in the North – the difference in house prices determining the difference in plot prices and land values.

Research undertaken by Evans (1987), however, adds weight to the argument that land prices significantly influence house prices rather than vice versa. In a report commissioned by the House Builders' Federation, Evans suggested that a restricted supply of land resulted in both land prices and house prices being higher than they would otherwise be; that, by restricting the supply of building land, planners were forcing up house prices and *either* squeezing out first-time buyers *or* adversely affecting the size of housing units; and that if agricultural land was converted into residential use it would significantly bring down the cost of new housing. Ambrose (1986) similarly claimed that the planning system adds to Britain's housing problems by being excessively concerned with environmental protection and by not releasing the correct amount of land in the 'correct' areas as assessed by the house-building industry.

In all probability, both explanations are correct. While a decrease in supply can undoubtedly increase land prices, the rate of price increase could be even greater if demand also increased. Figure 3.2 shows that if, for example, supply is reduced from *SS* to S_1S_1 the price of land would rise from p to p_1; but if demand increased from *DD* to D_1D_1 the price would rise more substantially to p_2.

In theoretical terms, it can be shown that there is clearly a relationship between the supply of land, land prices and the level of house-building. In Figure 3.3 the supply of land is completely inelastic (for example, as a result of tight planning policy). With an increase in the demand for housing (from *DH* to DH_1), more land is demanded for development (*DL* shifting to DL_1) but, since the supply of land (*SL*) is completely inelastic, land supply is not increased and land prices rise sharply, preventing the level of house-building from increasing to satisfy the increase in demand. If the supply of development land were to be reduced from *SL* to SL_1 (by government policy or by landowners exercising monopoly power), then land prices would rise even more dramatically and the supply of housing would be reduced from SH to SH_1. (Clearly, the reverse could occur if the supply of land were to be increased.)

The connection between land costs and the rate of house-building in the social sectors is also manifest. Over the period 1975–9 social sector housing starts in Britain decreased from 173,800 to 34,600 per annum, a reduction of 80 per cent, but, due to the competition for sites, social sector land costs increased in line with the land costs incurred by

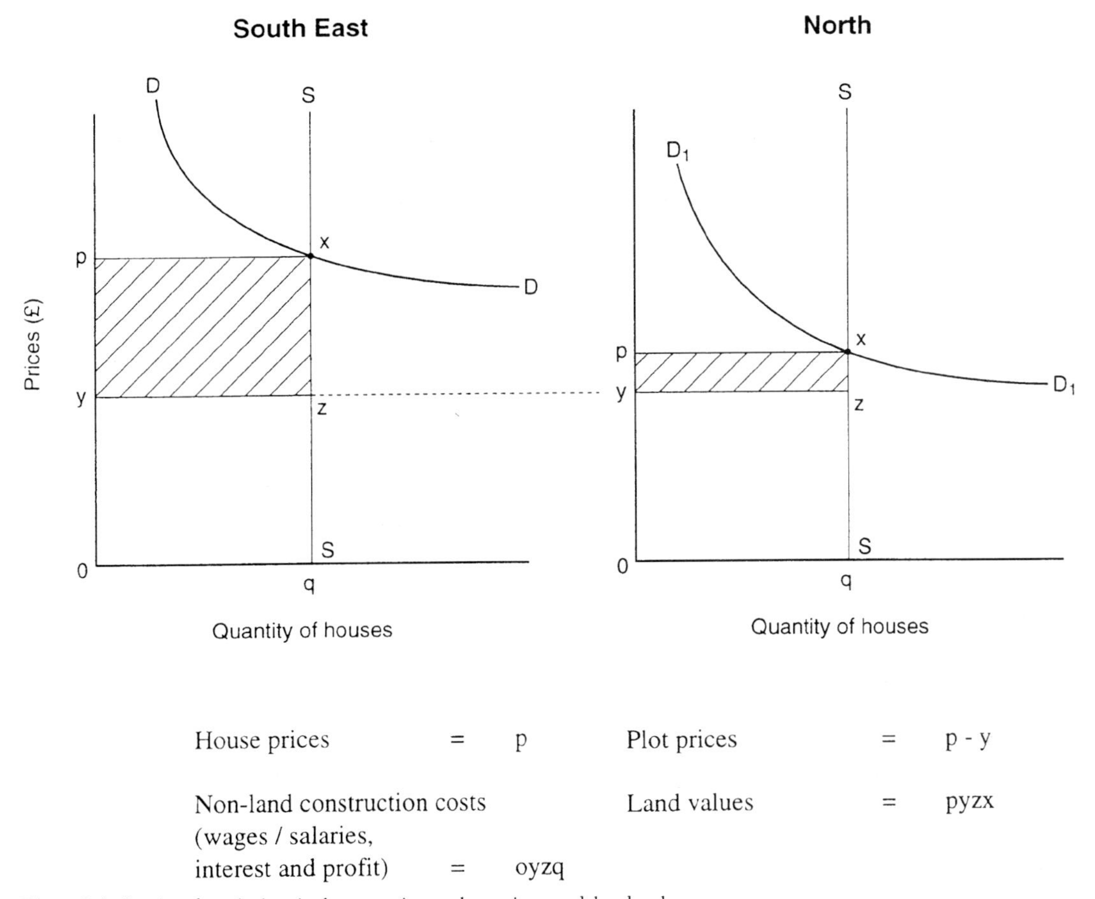

Figure 3.1 Regional variation in house prices, plot prices and land values

private developers. Notwithstanding the fact that many local authorities had reserves of land for development, it must be borne in mind that substantial cuts in public expenditure exacerbated the problem of rising land costs – reducing still further the ability of most local authorities to acquire land for house-building.

Clearly, the shortage of suitable sites for both private and public development, particularly in urban areas, is the principal cause of high land prices – shortages stemming from a combination of market practices and planning policy. The main determinants of land shortage are as follows:

1 Land is a marketable commodity and thus owners have a speculative interest in withholding a proportion of it from the market in order to increase its price – a particularly disadvantageous practice at the lower end of the housing market where land costs represent a disproportionately large slice of house prices (Wolmar, 1985). A large amount of derelict urban land has been deliberately withheld from the market for many decades – an important factor in land values in London, for example soaring in value to over £1 million per acre (£2.47 million per ha) in the late 1980s. Apart from rendering low-cost housing development in urban areas unaffordable, the effect is to divert housing development to the countryside, but this can be severely constrained by planning policy.

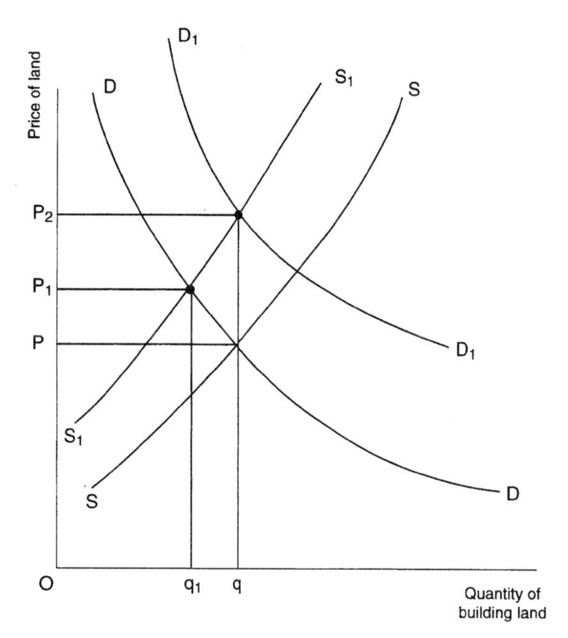

Figure 3.2 The effect of changes in supply and demand upon land prices

2 In response to both the Town and Country Planning Act and the Agricultural Act of 1947 (and subsequent related legislation), governments have applied policies to contain urban growth – most notably through the introduction of green belts – in which over 4.5 million acres (1.8 million ha) is protected from housing and other development and, since 1968, by the implementation and review of County Structure Plans and application of Local Plans.

3 On at least three occasions since the 1950s there have been major property booms, during which house-builders have been unable to compete with office and shop developers for sites – especially when the granting of planning permission for the development of commercial uses might increase the value of a site by as much as sixfold.

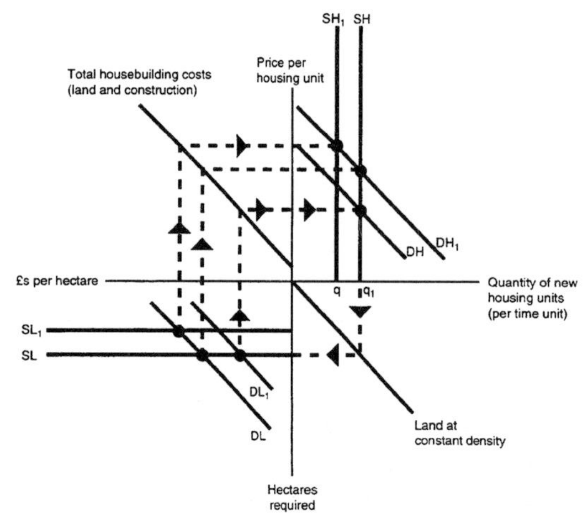

Figure 3.3 The relationship between house-building and land costs under conditions of inelastic supply

The shortage of land is a particular cause of concern in respect of low-cost housing in the inner cities and elsewhere. The government in the early 1990s thus attempted to ensure that the planning system made provision for satisfying social needs. Where there was clearly a shortage of affordable housing and in the case of new housing development on a substantial scale, under *Circular 7/91* planning authorities were obliged to show in their local plans land scheduled for the provision of social housing and were expected to negotiate with developers to ensure that an element of affordable housing would be included in their schemes (Department of the Environment, 1991). In addition, through the use of planning agreements, under Section 106 of the Town and Country Planning Act 1990, local authorities could secure the development of social housing as part of an otherwise private development scheme, and in rural areas an 'exceptions' policy was established allowing local authorities to grant planning permission for local housing needs on land which would not normally be allocated for housing. But, as Barlow and Chambers (1992) showed, during the housing slump of the early 1990s, only about 2,000 social units per annum were developed as a result of 'quota scheme' agreements (that is, where there was a mixture of market and social housing on the same site).

However important the need to supply affordable social housing in the 1990s, total needs in the early twenty-first century are likely to be even greater. (The government had estimated that an extra 4.4

million households, 80 per cent of which would consist of only one person, would need to be accommodated in England over the period 1996–2016). Parts of England (notably Buckinghamshire and Cambridgeshire) could see a 50 per cent increase in house-building over twenty years – but the government anticipated that about half of all new house-building would be within recycled urban areas ('brownfield sites') – although high costs of reclamation might reduce this figure to 30–40 per cent (Town and Country Planning Association, 1996). Clearly, if the supply of land to meet the substantial increase in demand is not fully forthcoming, land prices will soar yet again, choking off the required level of house-building and inflating house prices.

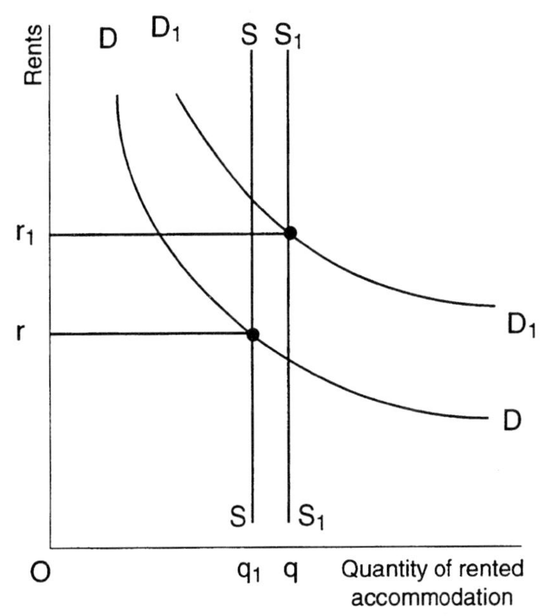

Figure 3.4 The effect of benefit-induced increases in demand upon rents in the social housing sectors

THE MARKETISATION OF THE PRIVATE RENTED SECTOR

Rent control or regulation throughout much of the twentieth century has been justified by Coalition, National and Labour governments on the grounds that free market rents would have a seriously detrimental effect on the standard of living of low-income households – indeed many households might find rents unaffordable and consequently be excluded from the market. As Figure 3.4 shows, an increase in the demand for housing from *DD* to D_1D_1 (as a result of an increase in population and/or increase in the number of households) would pull up rents from r to r_1 without necessarily producing an equivalent increase in supply. Supply might increase, for example from *SS* to only S_1S_1 – the construction industry being generally slow to respond to an increase in demand – while development could be severely restricted by town planning constraints and a shortage of urban building land. Investment in the sector might also remain stagnant, since alternative investment opportunities are often more attractive in terms of rates of return, management and liquidity.

Critics of rent control, however, point out the many economic disadvantages of intervening in the market.

1 Regardless of the condition of the accommodation, whenever a dwelling has fallen vacant, the landlord usually finds it pays to sell it for owner-occupation rather than to relet. However, although it is often argued that rent control (or regulation) has resulted in the withdrawing from the letting market of most privately owned houses suitable for owner-occupation or other uses, Harloe *et al.* (1974) have shown that in both Western Europe and the United States the private rented sector has been in decline whether or not it has been subject to control. In most advanced capitalist countries, owner-occupation is the most heavily 'subsidised' tenure and can outbid all other tenures for a relatively fixed supply of housing land. The long-term effect of rent control/regulation is shown in Figure 3.5. It is assumed that rents are fixed at r_1, thereby creating an initial shortage of accommodation equivalent to $q_1 - q_2$. If landlords sell their properties

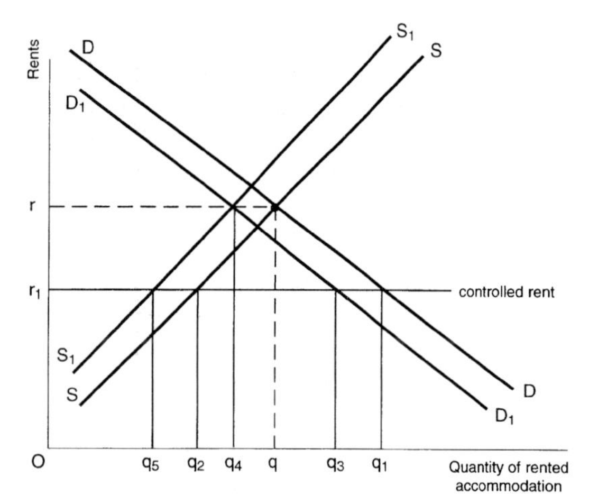

Figure 3.5 The effect of rent control upon the market for private rented dwellings

for owner-occupation, the long-term supply of rented housing will decrease from *SS* to S_1S_1, but if, simultaneously, tenants become owner-occupiers, the demand for rented accommodation will similarly decrease (from *DD* to D_1D_1), the eventual shortage becoming $q_3 - q_4$. However, if demand does not decrease (there might be a growing number of small households in search of housing), the shortage could be as much as $q_1 - q_5$.

2 During inflation, there was a transfer of income from landlords to tenants, as the former money (rent) was fixed and fell in real terms, and the latter's money (wage or salary) normally rose and usually also increased in real terms. Because of rising costs, the landlord's ability and incentive to repair and maintain his property was reduced. Housing consequently deteriorated in quality and whole areas of rented accommodation, overwhelmingly concentrated in the inner cities, degenerated into slums. But as a house was considered to last for ever, landlords could not claim a depreciation allowance to set against taxation, and even payments into a sinking fund to replace the dwelling were not tax deductible.

The process of deterioration is explained in micro-economic terms by Frankena (1975) and Moorhouse (1972). Rented housing, they argue, consists of a combination of services. It is not just accommodation, but includes such items as repairs and maintenance, decoration, and possibly cleaning, lighting and heating – all supplied at a price which in total constitutes rent. When rents are controlled below their market level, landlords' profits will be reduced or eliminated if they continue to provide services in full. They will consequently reduce the supply of services in an attempt to maintain profitability. Figure 3.6 shows that if a controlled rent (*cr*) is set below market rent (*r*) then the price per unit of services falls from *p* to p_1. The landlord might therefore respond by reducing the provision of services and raising their price. But since this landlord's total revenue must not exceed the controlled rent, the price–quantity combination stays the same – a rectangular hyperbola being traced by the *cr* curve. If the provision of services is reduced to q_1 the price per unit of services will be the same under rent control as under free market conditions, and any further reduction in services would be unlikely since demand and supply would be in equilibrium. However, it might not be possible

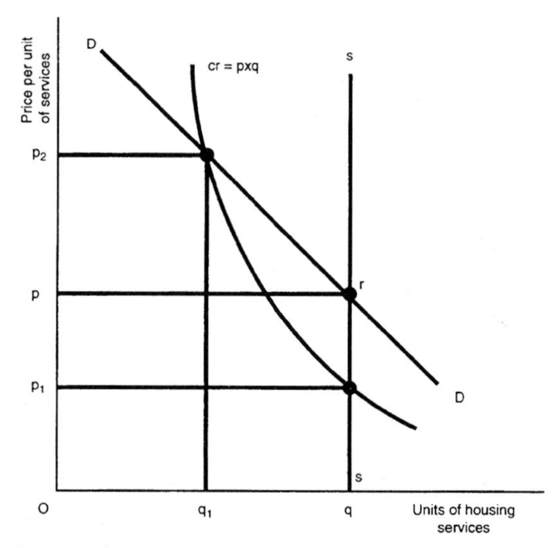

Figure 3.6 Rent control and the decrease of housing services

for the landlord to cut back on services quite to this equilibrium level since standards might fall below those permitted by environmental health law.

3 Because of the above factors, investment in the building of private housing to rent has been negligible since 1939, but it is difficult to quantify to what extent rent restriction was responsible in view of the high returns on investment elsewhere; for example, office and shopping development competitively attracting long-term capital.
4 Rent control led to an under-use of the housing stock. Many small households clung to large dwellings and many high-income tenants benefited from very low rents. Conversely, some large families, often with low incomes, had to settle for small furnished dwellings at high uncontrolled rents. There was, thus, little relationship between household size, income, housing space, amenities and standards. Sometimes a single room in an unfurnished tenancy was let furnished at a rent higher than the controlled rent for the whole.
5 Rent control produced a number of nefarious results such as key money, premiums, licences and 'furniture and fittings' payments – all ways of increasing the landlord's revenue and control over tenants without infringing the letter of the law.
6 Rent control may also have impeded the mobility of labour. Even if the householder was unemployed, secure and low-rent accommodation was often preferable to employment opportunities but problematic housing elsewhere.

Supporters of the free market argue that if rents were permitted initially to rise in response to an increase in demand, this would produce an increase in supply and stabilise the level of rent at approximately the original level. In Figure 3.7, for example, rents might rise from r to r_1 in response to an increase in demand DD to D_1D_1 and this would increase the price of housing from p to p_1. Either more new rented houses would then be built, or more existing houses would be converted into multi-occupied properties – the supply of housing

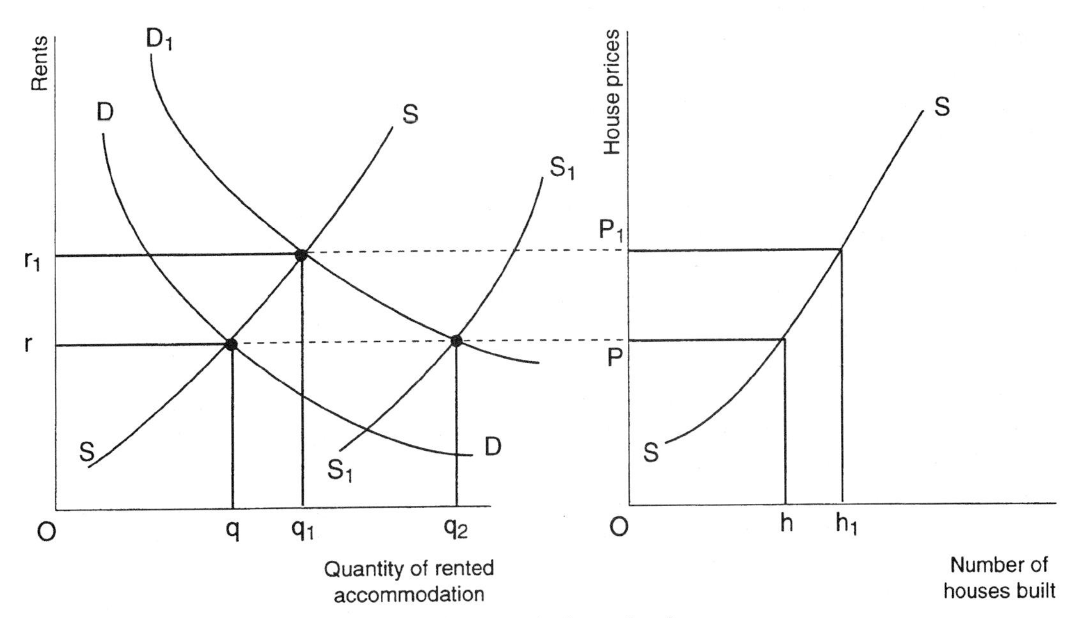

Figure 3.7 Market adjustment following an initial increase in demand and rent

expanding from h to h_1 and the supply of rented accommodation increasing from S to S_1.

After its election victory in 1979, the Conservative government, in an attempt to unleash market forces, soon began to deregulate the private rented sector. Under the Housing Act of 1980, shorthold tenancies were therefore introduced, allegedly because the Rent Acts of 1974 and 1977 had got in the way of landlords and tenants agreeing to a lease for a short fixed period. Shortholds were applicable only to new lettings, and at the end of the fixed-term agreement of one and a half years, landlords had the right to gain repossession. Shorthold was in effect a form of decontrol. If a landlord decided to evict tenants who did not depart at the end of the agreed term, he had the option to issue an eviction notice which gave the tenants three months' notice of his intention of going to court. The court would then have granted an order of possession (normally to have been effected within fourteen days). During its first year of operation, the 1980 Act required that landlords charged fair rents, but after 1981 – if landlords and tenants agreed – market rents on new shorthold tenancies were negotiable outside of Greater London – regulation remaining in the capital, since some 21 per cent of London's housing was privately rented (up to 40 per cent in some boroughs), compared to 13 per cent nationally in 1981.

The Conservatives thought that supply would also be maintained (or increased) if controlled tenancies became fair rent lettings. The 1980 Act therefore, at a stroke, decontrolled 300,000 dwellings and subjected their tenants to fair rents. To ensure that the income of landlords kept more in line with the rate of inflation, fair rents were to be registered every two years instead of every three.

The 1980 Act also introduced 'assured' tenancies, whereby approved landlords are permitted to let their new dwellings outside of the Rent Acts. Building societies, banks and other finance houses and construction firms could be licensed by the government to build homes for rent.

The Housing and Planning Act of 1986 subsequently extended assured tenure to refurbished dwellings. Whereas only 600 assured units had been built from January 1980 to March 1987, in the following six months a further 2,400 (mainly refurbished) assured tenancies were created largely as a result of the 1986 Act.

In its third term in office, the Thatcher government planned to bring back into use 500,000 properties left empty, allegedly because of rent regulation, but it was not prepared to risk complete decontrol.

Based largely on the White Paper, *Housing: The Government's Proposals* (Department of the Environment, 1987), the Housing Act of 1988, and its Scottish equivalent, the Housing (Scotland) Act 1988, aimed to revive the private rented sector by reducing the minimum period of shorthold (renamed assured shorthold) to only six months and extending assured tenancies to the remainder of all new lettings. Assured shorthold lettings were to be at market rents, which were to take account of the limited period of contractual security of the tenant – and the tenant could apply during the initial period of the tenancy to a rent assessment committee for the rent to be determined. Assured tenancies, on the other hand, although being relatively secure, were to be at rents freely negotiated between landlord and tenant and therefore were at market levels. Existing regulated tenants would (ostensibly) continue to be protected by the Rent Acts.

A consequence of the 1980 and 1988 Acts was that private sector rents escalated substantially. Table 3.7 shows that there was at least a threefold

Table 3.7 Private sector rents, 1980–95

Average unfurnished rents per week (£)		*1980*	*1989*	*1995*
England	Fair rents	10.85	24.38	45.27
	Market rents	—	37.42	64.23
Wales	Fair rents	10.10[1]	21.98	34.63
	Market rents	—	29.75	56.92
Scotland	Fair rents	8.06[1]	21.35	29.18[2]

Source: Wilcox (1996)
Note: 1 1981; 2 1983

increase in average fair rents in the unfurnished sector in England, Wales and Scotland over the period 1980–95 and approximately a 50 per cent increase in unfurnished market rents in both England and Wales from 1989 to 1995.

In addition to rent deregulation, the government used fiscal means in an attempt to increase the supply of private rented accommodation. The Budget of March 1988 extended the Business Expansion Scheme (BES) to individuals who invested up to £40,000 per annum in approved unquoted property companies building or acquiring housing for rent under assured tenancy arrangements. (Eligible properties had to have a maximum value of £125,000 in London and £85,000 elsewhere.) Under BES provisions, individuals qualified for tax relief on their investment and companies (in 1988) were able to raise up to £5 million each under the scheme. The BES was clearly intended to kick-start a revival in private renting, but only 16,000 new houses were supplied in four years. From 1989 to 1993, however, there was a small increase in the size of the private rented sector in the United Kingdom (from 2,078,000 to 2,310,000 dwellings), due mainly to the diversion of a limited supply of housing from the owner-occupied market to shorthold renting during the house price recession. Under the Finance Act of 1996, however, and in response to the White Paper, *Our Future Homes* (Department of the Environment, 1995b), housing investment trusts could be set up by institutional investors as housing associations to fund and manage private rented housing. Investors were eligible for capital gains tax exemption and a reduced rate of corporation tax (24 instead of 33 per cent).

Notwithstanding competition from other sectors of the economy, given the right conditions private rented housing could well be an attractive form of investment and provide an adequate rate of return for the investors (see Case study 3.1, pp. 96–102).

The private rented sector currently accounts for 10 per cent of the total housing stock, and houses some 2 million households. Much of the current stock, however, is in relatively poor condition, often lacking basic amenities. Gibb and Munro (1991) state that there are four distinct groups which reside within this sector. There are the elderly and relatively poor who have always lived in the tenure and probably enjoy protection under the Rent Acts. In 1996 unemployment among tenants in the sector was nearly twice the national average and in Great Britain 1,141,000 private tenants were currently in receipt of an average weekly housing benefit of about £50 per week, about one-third of all tenants receiving housing benefit. There are also the business and executive tenants, short-term owner-occupiers, students, single-person households and other transient households, which account for approximately half of the households within the sector. The final group which this sector caters for are the households unable to enter other tenures, such as recent immigrants.

The type of landlords within this sector include the small non-residential individual landlord, which accounts for 60 per cent of all landlords with a portfolio of less than ten properties. Resident landlords and property companies account for the remaining 40 per cent of landlords within the sector. Around a quarter of all lettings consisted of converted flats and average rents were approximately £60 per week in 1996. Although it has been found that 40 per cent of all landlords do not regard rents and rates of return to be adequate to maintain their properties, by the same token 60 per cent of landlords must regard their investment income as at least satisfactory.

Rent control, the demand for owner occupation and the comparatively good quality of accommodation generally available within the social rented sector has accounted for the decline in the private rented sector. The Housing Act 1988, however, has helped to add 300,000 more units to the sector due to the deregulation of rents, the limiting of security of tenure and housing benefit which is available to cover market rents. The increase in the number of owner-occupiers who are letting properties which they are unable to sell further adds to the slow revival of the sector.

LOCAL AUTHORITY HOUSING: INVESTMENT, RENTS AND SUBSIDIES

Council housing is built and paid for through the use of borrowed money. The money is usually borrowed and repaid over a period of sixty years, which is the expected life of a house. Repayment of the loan is calculated to ensure that by the end of the loan period the total debt including interest has been paid. The payments are made by using the sinking fund method whereby the early years of the loan involve paying a large amount of the interest and a small amount of the principal. The amount to be paid each year is increased by 5 per cent each year and involves gradual reduction in the proportion of interest paid with the principal repayment increasing gradually each year.

Local authorities incur capital expenditure in order to provide council housing. The capital expenditure involves the acquisition of permanent assets such as land, roads and the actual properties.

Until 1976 there were no controls by central government over the number of properties which local authorities could build. Although the loans to cover building were subject to loan sanction which was sought from central government, as long as the level of building costs did not exceed central government guidelines then the loan sanction would be agreed.

However, in the wake of the 1977 Green Paper (Department of the Environment, 1977), loan sanctions were scrapped in 1978 due to the economic crisis which existed in the country at that time, when it was felt that tighter control in all areas of public expenditure was required. Loan sanctions were replaced by a local housing strategy which required each council to provide information about housing need in its area. A housing investment programme (HIP) would then have to be produced which would identify what the local authority intended to spend in order to achieve the demand identified in the local housing strategy. Both documents were submitted to the Department of the Environment, which would decide how much the council could borrow.

The Housing Act of 1980 broadly retained the procedure, but block allocations of borrowing permission were replaced by block allocations of permitted expenditure. Expenditure could be supplemented by a proportion of the receipts from the sale of council houses – initially 50 per cent of receipts, then 40 per cent in 1984 and 20 per cent in 1985.

Under the Local Government and Housing Act of 1989, however, there was a reversion to the control of borrowing. HIP allocations, henceforth, were controlled by means of a basic credit approval (BCA) for the following year – the Secretary of State taking into account usable capital receipts. In addition, supplementary credit approval (SCA) could be granted to permit borrowing for estate action (rehabilitation) schemes and initiatives to help the homeless. In addition to credit approvals, HIP allocations included specific capital grants, for example for repairing housing defects, renovation of private dwellings, area improvements and slum clearance. To supplement borrowing, local authorities could continue to use the receipts from the sale of housing and land for capital expenditure. But whereas, before 1989, 20 per cent of sales receipts could be used annually for capital purposes (the remaining 80 per cent 'cascading' to the following year), under the 1989 Act, although 25 per cent could be used for capital purposes, 75 per cent had to be used to repay debt – pre-empting its investment capability.

Largely by means of these controls, central government was able to reduce housing investment in the local authority sector – in England from £5.2 billion in 1979/80 to only £1.3 billion in 1994/5 (in real terms), a decrease of 76 per cent (Table 3.8). New build activity was particularly singled out for reduction – with house-building in the sector (in Great Britain as a whole) diminishing from 107,200 starts in 1978 to 1,100 in 1995. There was a comparable reduction in housing investment in Scotland, although in Wales investment was broadly maintained by a dramatic shift of emphasis from new build to renovation.

Revenue expenditure is incurred once the property has been built and revenue spending is used to

Table 3.8 Local authority housing investment, England, 1979/80 to 1994/5

£ million (1994/5 prices)	*1979/80*	*1994/5*	*% change 1979/80 to 1994/5*
Gross investment			
New build and acquisitions	2,917	69	−98
HRA stock renovation	1,806	1,610	−11
Housing associations	473	360	−24
Private renovation	568	547	−4
Home-ownership	1,562	45	−97
Urban programme	35	39	+11
Total	7,360	2,670	−64
Capital receipts	1,996	1,380	−31
Net investment	5,364	1,290	−76

Source: Department of the Environment (1995) *Public Expenditure Plans*, HMSO
Note: 1 Renovation funded from the Housing Revenue Account (from rents, subsidy, etc.)
2 Urban Programme housing investment under the Inner Urban Areas Act 1978

pay for items such as management and repairs costs. The revenue expenditure is paid for by items such as rents and interest on capital receipts. All revenue transactions are shown in the housing revenue account (HRA).

In the past, rent levels were set by local authorities, which led to variations between council rents in similar locations. During the 1980s central government felt that there was not sufficient difference between the regions and the 1990s have seen the government attempt to ensure that rent levels reflect house prices within the regions.

Rent pooling is used by some local authorities in order to bring some similarity into the rents of similar properties. The local authorities which use this system look at the total outgoings and income for the total stock within the council and then set a rent level for individual properties to ensure that rental income covers outstanding costs.

The housing revenue account also receives two types of subsidies, one aimed at reducing the amount of rent that individual tenants have to pay, and rent rebate subsidy, which repays a local authority for acting on behalf of the government and paying means-tested rent rebates such as housing benefit to poorer tenants.

With regard to rents and subsidies, the Housing Act 1980 introduced a new subsidy consisting of a 'base amount' (equal to the total subsidy paid in the previous year) *plus* a 'housing cost differential' (representing the increase in the total reckonable housing costs over those for the previous year) *less* the 'local contribution differential' (the amount a government expected the local authority to pay towards housing through increased rents or rate fund contributions). In principle, the local contribution differential gave the local authority the choice between increasing rents or increasing the rate fund contribution, but since it was the government's intention to reduce Exchequer subsidies the Department of the Environment had powers to specify the target rate of annual rent increase. This resulted in rent increases considerably in excess of the amount that could be fully met from rate contributions.

Under the 1980 Act, surpluses could thus be made on the housing revenue account, in contrast to deficits, which previously had to be offset by rate fund contributions. Surpluses enabled local authorities to keep rates down, and by 1981/2 fifty local authorities outside London were transferring money from rent surpluses to the rate fund to the benefit of ratepayers. By 1987/8 the number of councils in surplus had increased to 126, and in 1987/8 these authorities (largely in suburban or rural areas) were able to transfer surpluses of £61 million to their general rate fund.

Over the period 1980/81 to 1987/8, change in rent policy enabled the government to cut subsidies from £1,719 million to £498 million – a decrease of 71 per cent; reduce the number of local authorities receiving subsidies from 367 to only 88; raise the level of rents (in England) from an average of £11.42 to £17.70 per week – an increase of 55 per cent (with even larger increases in Scotland and Wales); and, because of these increases, rely less and less on rate fund transfers eliminating any shortfall –

transfers diminishing from £411 million to £329 million from 1980/81 to 1987/8 and diminishing altogether by 1995/6 (Table 3.9).

With regard to the housing revenue system under the Local Government and Housing Act 1989, the housing revenue account (HRA) receives only one subsidy (the HRAS), which is made up of two elements, the rent rebate element and the housing element. The rent rebate element is provided by central government to cover payments such as housing benefit to claimants and is always positive, while the housing element, which is also paid by central government, can be negative and can therefore help to push up rent levels. The HRA is ringfenced so that income received from council tax cannot be used to subsidise rents, as happened in the past. Each year under the Act each local authority now receives a guideline rent which is calculated by central government, and it now receives an annual management and maintenance allowance.

The guideline rents are now set on the basis that the government obtains the total value of properties based on right-to-buy values and each local authority receives the guideline rent which would be a proportion of the total national value of housing stock. Central government decides each year how much the national rent increase should be and this expected level of rent increase is divided between the total number of local authorities based on their share of the total value of their housing stock. In order to prevent the anomaly whereby some local authorities may have to decrease their rents due to the difference in house prices across the country it was decided to ensure that the formula allowed for all local authority rents to increase but not to excessive amounts in political terms.

Table 3.9 Housing subsidies and rents, local authority sector, 1980/81 to 1995/6

	1980/81	*1987/8*	*1995/6*
Exchequer subsidy (£m)	1,719	498	328[1]
Average rents (£ per week)			
England	11.42	17.20	38.31
Scotland	7.67	14.59	27.78
Wales	11.43	17.91	35.50
Rate fund transfers (£m)	411	329	33[1]

Source: Department of the Environment; Scottish Office; and Welsh Office, *Housing and Construction Statistics*
Note: 1 Rent surplus

Despite the guideline rents set by central government, rents have continued to increase at higher rates, due to the need to cover some capital expenditure and to make up the difference between the allowance available for management and maintenance costs, the actual expenditure incurred. Many local authorities have had to cut back on the amount of repairs programmes carried out in order to restrict the increase in rents. In England 2,196,626 local authority houses are in need of renovation. Furthermore, the use of property values to determine rent increase levels has meant that some local authorities with lower house prices have had to introduce increases above the level set. The average local authority rent for England and Wales in 1996 was £40.03, an increase of 5 per cent since 1995.

The Local Government and Housing Act 1989 has resulted in the control of rent levels, management and maintenance costs all being controlled by central government. Furthermore, with a number of local authorities not receiving any housing element of the HRA subsidy, many are finding that the total HRA subsidy is in fact negative, so that council tenants paying rent are bearing some of the costs of housing benefit for other tenants.

THE CHANGING ROLE OF LOCAL AUTHORITIES: FROM PROVIDERS TO ENABLERS

From examining housing policy, it is clear that over the years the function of council housing has been interpreted in two contrasting ways. On the one hand, there was the traditional Labour belief that the public sector should supply housing for 'general needs' – to satisfy the demand from households (irrespective of their income) to rent rather than buy, either through choice or necessity; on the other hand there was the Conservative or neo-liberal view that council housing should fulfil only a 'residual' or 'welfare' role, assisting only those households unable to afford or find any other sort of accommodation (see Chapter 1). This second view was institutionalised by the Housing Act 1988, under which local authorities became 'enablers' rather than the providers of social housing – in 1994 their 1,200 housing starts comparing very unfavourably with the 39,900 starts of the housing associations. Local authorities 'remained responsible for ensuring, so far as resources permit, [that] the needs for new housing are met, by the private sector alone where possible, with public sector subsidy if necessary' (Department of the Environment, 1989). Since the Housing Corporation, however, had become the main conduit through which government funds flowed into the provision of new social housing, it seemed a little odd that the local authority had become 'the key strategic agent in bringing about the provision of social housing' (Bramley, 1993).

PRIVATISATION

The privatisation of housing by Conservative governments in recent years has taken broadly three forms: first, the selling off of council houses to their tenants; second, and in part to facilitate rehabilitation, the disposal of parts or the entirety of council estates to housing associations, trusts and private companies either for renting or resale; and third, the privatisation of the funding of private sector housing rehabilitation.

The selling off of council houses to their tenants

The Housing Act of 1980 and the Tenant's Rights etc. (Scotland) Act 1980 gave council tenants the statutory right to buy the freehold of their house or a 125-year lease on their flat. The Act also allowed tenants to take a two-year option to buy their homes at a fixed price on payment of a £100 deposit. To counter the possibility that a council might delay or impede a sale, the Secretary of State has the right to intervene and complete the sale under Section 23 of the Act. Discounts of 33 per cent were offered to tenants of three years' standing, rising to 50 per cent to those of twenty or more years' standing. Council mortgages of 100 per cent were to be made available, but if the property was resold within five years, the capital gain would be shared between the owner and the local authority.

Under the 'right to buy' (RTB) provisions of the 1980 Act, the number of local authority dwellings initially sold off to their former tenants increased from only 568 in 1980 to 196,430 in 1982, and sales began to exceed local authority housing completions. For the first time the local authority sector began to contract in absolute terms.

By 1986, however, the pace of council house sales was flagging (there were only 82,251 sales in that year compared to 196,430 in 1982). The Department of the Environment was particularly concerned about the small number of flats which had been sold off (little more than 4 per cent, 1980–85). The Housing and Planning Act of 1986 therefore introduced discounts of 44 per cent on the purchase price of flats (for tenants of two years' standing). The repayment rule was reduced from five to three years as an incentive to buy and to assist mobility, but a limit of £25,000 was placed on discounts – a less than helpful constraint in London where even two-bedroom flats could be valued at more than £100,000. In the period 1980–87, however, nearly 800,000 local authority tenants purchased their homes under RTB legislation.

Although local authorities were still the largest landlords in the United Kingdom, owning 4.9 million dwellings or 20.2 per cent of the total housing stock in 1994, the public sector's contribution to the total number of housing starts 1990–94 amounted to a minuscule 2 per cent of the total (compared to 48 per cent in 1964–70), whereas the private sector's share had grown to as much as 80 per cent (compared to only 52 per cent in 1964–70).

Plate 3.1 Former council houses renovated under owner-occupation
Tenants of council houses in suburban locations, rather than inner-city flat-dwellers, were particularly interested in exercising their right to buy, and as owner-occupiers often renovated their properties, sometimes with the aid of grants

In 1988, to boost sales further, maximum discounts were increased to 60 per cent for council house tenants (starting at 32 per cent after two years), and discounts rose to 70 per cent for flat-dwellers after 15 years (starting at 44 per cent on qualification).

A total of 1,569,321 council dwellings were sold in 1980–95, equivalent to 25 per cent of the total local authority stock of 1980. Sales, moreover, greatly exceeded the number of new local authority dwellings built – significantly depleting the size of the council stock (Table 3.10).

By the mid- and late 1980s, former tenants, having previously exercised their RTB, were beginning to put their houses up for sale – the consequential increase in the supply of owner-occupied houses in the market helping to depress house prices during the slump years of 1990–93 (see Figure 3.7).

By the early 1990s, however, it was clear that local authority tenants who had exercised their RTB had been the recipients of the largest housing subsidy of all, averaging £12,094 nationally or as much as £21,675 in London (Maclennan *et al.*, 1991). The RTB process had dwarfed all other privatisation schemes. By mid-1992, the Treasury had received as much as £23 billion from the sale of council dwellings (and a further £3.7 billion was held by local authorities as housing capital receipts). In contrast, the privatisation of Britain's largest public corporations in the period 1980–86 raised only £7.3 billion – a 'property-owning democracy'

Table 3.10 Local authority dwellings sold under right to buy legislation, local authority completions, and the local authority stock, Great Britain, 1980–94

	RTB sales completed	*Local authority completions*	*Local authority housing stock (000s)*
1980	568	86,200	6,400
1981	79,430	54,867	6,387
1982	196,430	33,244	6,196
1983	138,511	32,833	6,060
1984	100,149	31,699	5,959
1985	92,230	26,115	5,864
1986	82,251	21,587	5,779
1987	103,309	18,823	5,661
1988	160,569	19,030	5,483
1989	181,370	16,465	5,270
1990	126,214	15,780	5,105
1991	73,365	9,457	5,031
1992	63,986	4,147	4,811
1993	60,274	2,045	4,710
1994	64,315	1,801	4,605
1995	46,350	1,446	4,504
Total 1980–95	1,569,321	375,539	1,896

Source: Department of the Environment, *Housing and Construction Statistics*

being promoted more effectively than wider share-ownership.

The Conservative government hoped that shared ownership would complement RTB. Introduced in July 1979, shared ownership was intended (very optimistically) to appeal (together with the sale of council houses) to up to six million tenants. At first it was confined to local authority (and non-charitable housing association) new build schemes, but under the Housing and Building Control Act of 1984, shared ownership was offered to tenants and was intended particularly to appeal to low-income households who could not afford to buy outright even with discounts. Tenants could buy at least 50 per cent of the value of their homes (at a discount) and the rest in 'tranches' of 12.5 per cent (but they would have the full responsibility for their own repairs and maintenance). By 1983, the government was aware that council house sales might have peaked (about 500,000 being sold since June 1979), and the 1984 Act was therefore a means of further privatising the public housing stock.

Despite generous discounts, many social sector tenants still found it difficult or impossible to buy or embark upon shared ownership, particularly in relatively high-price or low-income regions. A rent to mortgage (RTM) scheme was thus introduced – initially in Scotland in 1989 where owner-occupation was 20 per cent lower than in England, then in Basildon, Milton Keynes and rural Wales in 1991. Under the Leasehold Reform, Land and Urban Development Act 1993, the government introduced a nation-wide RTM scheme, whereby tenants were able to take a part share in their home by converting their rent payments to mortgage repayments and subsequently either stepping up payments or cashing in their equity if they moved home. Tenants were eligible for a discount on the market price of their homes ranging – in the case of houses – from 30 per cent (after two years' residence) to 60 per cent (after thirty years) and, in the case of flats, from 30 to 70 per cent (after twenty years).

In a further attempt to facilitate home-ownership and at the same time help reduce homelessness, the Department of the Environment permitted local authorities to award grants to council tenants wishing to buy in the open market. Although grants were worth only half the value of the discounts tenants would have received had they exercised their RTB, they were sufficiently large to have enabled tenants to buy in the provinces.

Estate privatisation

Estate privatisation, in its many forms, was motivated by the perceived fiscal need to transfer the responsibility of repairs from government to housing associations and the private sector, and by an awareness that there was a limit to the number of council dwellings that could be sold off under RTB and RTM policy.

Under the Housing and Planning Act of 1986 and Housing Act of 1988, forty local authorities

Plate 3.2 Privatised and renovated local authority maisonettes
Twelve-storey blocks of renovated maisonettes in the Queensbury Estate, London Borough of Wandsworth, privatised under the provisions of the Housing Act of 1986

therefore transferred the whole of their stock (amounting to 178,546 dwellings) to housing associations by 1995 under large-scale voluntary transfer (LSVT) arrangements, and about 240 other local authorities were considering transfer. Tenants of housing transferred to the housing associations were subsequently granted the RTB under the Housing Act of 1996.

LSVTs were clearly a mixed blessing for central government. On the one hand, capital receipts acted as a check on the public sector borrowing requirement, but on the other hand the Department of Social Security (rather than the local authorities) became responsible for the payment of escalating housing benefits.

Based on the White Paper: *Housing, The Government's Proposals* (Department of the Environment, 1987), the Housing Act of 1988 was also directed at the problems of some of the largest council estates which appeared to many to be unmanageable. Under the Act, the government planned to establish a number of Housing Action Trusts (HATs) to repair or rehabilitate housing estates, to improve management, and in general to improve social and living conditions. In July 1988, it was announced that the government would explore the possibilities of setting up HATs (responsible for 24,525 dwellings) in Lambeth, Southwark, Tower Hamlets, Leeds, Sandwell and Sunderland, and introduce in these areas a three-year £125 million renovation programme, scheduled to begin in 1989. But because of considerable opposition from both tenants and local authorities, the Department of the Environment decided to abandon its plans to set up HATs in its initially selected areas (Karn, 1993). It was ironic, however, that HATs were subsequently established in Liverpool, Waltham Forest, Hull North and Castle Vale (Birmingham) largely as a result of local authority instigation or agreement and backed by supporting ballots in the estates concerned.

Under the 1988 Act, council tenants are able to exercise 'tenant's choice' – the second way in which 1988 legislation aimed at transferring housing out of the local authority sector, adding to the rights and choices already given to tenants by the 'Tenant's Charter' of the Housing Act of 1980. Tenants of council houses individually were able to exercise their right to transfer to another landlord but tenants of flats must decide collectively.

In Scotland the Housing (Scotland) Act of 1988, like its English and Welsh counterpart, aimed to encourage alternatives to public sector provision. It introduced a tenant's choice scheme (vis-à-vis council housing) whereby tenants, individually or collectively, were given the right to transfer to another landlord (following a straight majority vote in favour of a transfer). The scheme was open to all council tenants (with minor exceptions) and to any (non-public) landlord approved by Scottish Homes (a new unified central agency) and to Scottish Homes itself. North of the border, however, there was no provision for the establishment of HATs – a role that was subsequently performed by Scottish Homes.

The privatisation of rehabilitation

In an attempt to increase the rate of rehabilitation, the Conservatives, in the Housing Act of 1980, enabled both private and public sector tenants (as well as landlords and owner-occupiers) to apply for improvement grants; abolished the five-year repayment rule in relation to the only or main residence of an owner-occupier; extended the availability of repair grants (with regard to substantial repairs to 1919 dwellings); made it easier for low-income households (owners and tenants) to obtain grants towards the cost of basic amenities – grant aid being paid for less comprehensive improvement than previously allowed; made the grant system more flexible by giving the Secretary of State powers to fix (for different cases) eligible expense limits, percentage grant rates and rateable values, etc.; and enabled the Secretary of State to underwrite part of any losses incurred by local authorities acquiring and improving dwellings for resale.

In December 1980, new grant levels were announced. Improvement grants were raised to £6,375 (in respect of those which were unfit, lacking amenities or in need of substantial repairs) or £2,750 in the case of other dwellings (and up to £8,625 and £3,725 respectively in Greater London) – the two levels of grant amounting to 75 and 50 per cent of eligible expenditure. Intermediate grants were raised to £950 for amenities and £1,250 for 'minor' repairs (£2,500 and £3,500 respectively in Greater London). Like intermediate grants they amounted to 50 per cent of eligible expenditure.

It was therefore considered in government that much greater selectivity in the award of grants was essential. The Green Paper, *Housing Improvements: A New Approach* (Department of the Environment, 1985), thus proposed, first, that improvement and repair grants should be means tested, mandatory and related to renovation up to a new standard of fitness (on the assumption that a third of recipients in 1985 could have afforded to pay for the work themselves, while other less well-off households were ineligible since they lived in properties with rateable values above the qualifying limit – £400 in London and £225 elsewhere in 1985). Second, as an alternative, interest-free discretionary loans for work beyond the new fitness standard should be available, with the local authority taking a financial stake in the improved property – for example, a loan of £5,000 on a house valued at £30,000 would allow the local authority to claim 10 per cent of the eventual sale price of the property. Discretionary loans would also be means tested, but for both forms of assistance the eligibility ceiling on the applicant's income and capital would be low – although in respect of income not as low as the supplementary benefit cut-off level (£3,000 in 1985). Third, grants and loans should be available for the renovation of any pre-war housing (and not just for pre-1919 stock in the case of repair grants).

The *English House Condition Survey 1986* (Department of the Environment, 1988) showed that conditions were far worse within the private rented sector than in owner-occupation or social housing, and that low-income households and the elderly suffered the worst housing. Based on this and on the White Paper, *Housing: The Government's Proposals* (Department of the Environment, 1987), the Local Government and Housing Act 1989 therefore targeted grants towards the worst housing, and to households in greatest need.

The 1989 Act thus replaced the grant system of earlier legislation with a new and largely mandatory regime of grants and reformed the system of area improvement. Renovation grants, housing in multiple occupation (HMO) grants, common parts grants, minor works grants, were introduced and all were means tested. Owner-occupiers and tenants would be unlikely to qualify for grant assistance unless their incomes and savings were no higher than the level which would render them eligible for income support or housing benefit; and landlords would not qualify unless (without a grant) their outlay on improvement and repairs failed to exceed their rental income (but provided that with a grant their return would be sufficient only to service a loan at 3 per cent over base rate for ten years). The Housing Grants Construction and Regeneration Act of 1996, however, tightened up the conditions under which grants were awarded, and made most awards discretionary rather than mandatory.

With regard to social housing, the Department of the Environment (1993a), the Northern Ireland Housing Executive (1993), Scottish Homes (1993) and the Welsh Office (1988) put the total cost of repairs to housing association properties at £2.2 billion.

Unlike the rate of rehabilitation in the private sector, however, renovations in the local authority sector remained at a high level in the late 1980s to early 1990s and public expenditure on local authority (and to a lesser extent housing association) renovation continued to increase. Clearly, there was every incentive for local authorities (and some housing associations) to sell off as much of their housing stock as possible – either to tenants or to other landlords.

HOUSING ASSOCIATIONS AND LOCAL HOUSING COMPANIES: THE NEW PROVIDERS?

The Housing Act of 1980 granted housing association tenants the right to buy (RTB) their homes from their landlords, and mortgages were to be made available to them by the Housing Corporation. (Housing associations registered as charities under the Charities Act of 1960 were also given the right to sell, although tenants were not given the right to buy, an exclusion affecting half of all association households.)

Under the Housing and Building Control Act of 1984, however, tenants of charitable housing associations were offered cash handouts to enable them to buy in the open market. Handouts of up to 50 per cent of the value of a dwelling were granted to tenants of two or more years' standing in respect of an acquisition costing no more than £40,000 in Greater London, £35,000 in the Home Counties or £30,000 elsewhere. To qualify, tenants would have had to have attempted unsuccessfully to negotiate the purchase of their own housing association home.

The RTB was also extended by the Housing Act of 1996 to tenants of registered social landlords, subject to the property being an assumed tenancy and having been provided wholly or in part with public money. The Housing Corporation was empowered to allocate purchase grants to landlords in respect of discounts on disposals – the rate of discount being approved by the Secretary of State for the Environment.

In the 1990s, rather than complementing local authority housing (and notwithstanding the effects of RTB), housing associations became the principal providers of new social housing in Great Britain, as prescribed by the White Paper, *Housing: The Government's Proposals* (Department of the Environment, 1987). Housing association net capital expenditure doubled from £1,156 million to £2,308 million between 1990/91 and 1992/3, while the number of housing starts in the sector increased from 12,684 in 1987 to 33,400 in 1992. Local authority starts, meanwhile, plummeted from 18,849 to only 3,713 over the same period.

Over the three years (1992/3 to 1994/5, the government (in 1992) aimed to spend £2 billion per annum on housing association investment, and to produce a total of 153,000 homes – each association setting out its investment plans in an approved development programme agreed annually by the Secretary of State. But with cuts in the size of the housing association grant (HAG) for each completed dwelling, more had to be borrowed from the private sector. Although public funding continued (involving HAGs and loans from the Housing Corporation,[1] a quango, and local authorities), mixed funding schemes were increasingly undertaken – private finance enabling public funds to be stretched over a much greater volume of housing than hitherto.

Aughton and Malpass (1991) explained that there were two types of funding which were referred to as mixed funding: first, funding consisting of a proportion of HAG and private loans and public funding; and second, funding comprising a proportion of HAG from the Housing Corporation and other loans raised from local authorities or the Housing Corporation. The larger housing associations were expected to raise private loans to supplement the HAG. However, smaller housing associations that were not able to secure a private loan were able to receive public funding.

There are two ways available to housing associations to finance developments. The first is non-tariff mixed funding, which is now partly funded by social housing grants (SHGs; see below, p. 74) and partly by obtaining private loans. An assessment is made of individual schemes and some allowance is made for the costs of the development increasing while the works are in progress. Once the cost limit is set the association has to meet any excess cost above the set limit.

Non-tariff public funding is for those associations which are unlikely to attract private financing. The process is the same as for the non-tariff mixed funding except that the project approval stage is examined in terms of areas such as planning and land valuation.

In the November 1993 Budget, as part of a package of cuts aimed at reducing the size of the public sector borrowing requirement, the government announced a £300 million reduction in its funding of the Housing Corporation (for 1994/5), to be followed by a further cut almost as large in the following year – creating the conditions whereby housing associations would have to depend more and more on private finance if they were satisfactorily to perform their role as providers of social housing. The government (via the Housing Corporation) was, however, still the principal supplier of funds for housebuilding and renovation in the housing association sector. In addition, from 1990/91, the Housing Corporation was empowered to provide grants under the tenant's incentive scheme to help housing association tenants move into owner-occupation and thereby release subsidised rented accommodation for households in greater need of low-cost housing; and in 1990/91 and 1991/2 the government allocated £73 million to the housing associations to assist local authorities to provide housing for the homeless and to reduce the numbers of households in bed and breakfast accommodation.

The cutback in the size of HAGs and increased reliance on private finance (which, by necessity, requires a competitive rate of return) have had an inflationary effect on rents. The consequences of mixed funding schemes could become even more marked if public funding is eventually reduced to 50 per cent as was suggested in the White Paper, *Housing: The Government's Proposals* (Department of the Environment, 1987).

Whereas existing lettings are at fair rents as determined by rent officers, and are subject to rent increases every two years, under the Housing Act of 1988 all new lettings are at assured or assured shorthold tenure, with housing associations setting their own 'affordable rents'. In order to ensure that private capital is attracted into housing investment in this sector, however, average rents for assured tenancies rose markedly – in England, for example, from £24.50 per week in 1989 to £48.29 per week in 1995 (£10 per week higher than the average local authority rent in the same year) (Tables 3.9 and 3.11). With the government's intention to reduce its share of total investment in this sector (in 1994/5 it was reduced from 67 to 62 per cent and cut to 58 per cent in 1996 and 56 per cent in 1997), rents inevitably continued to escalate.

Even more private capital should be attracted into social housing under the Housing Act of 1996. This will permit local housing companies to be set up as non-profit-making social landlords to take over and manage an increasing proportion of the local authority housing stock. With boards consisting of councillors, tenants and representatives from the local community, they are to be funded partly by loans from a bank or building society, both of which would have an interest in ensuring that tenants are generally satisfied with management, notwithstanding the introduction of assured tenancy and the inevitability of higher rents.

All social landlords, under the 1996 Act, are in addition eligible for social housing grants (SHGs) allocated by the Housing Corporation – SHGs replacing HAGs as the principal mode of public funding allocated by the Housing Corporation.

Table 3.11 Housing association rents, 1980–95

Average rent per week (£)		*1980*	*1989*	*1995*
England	Fair rents	12.52	26.83	43.88
	Assured rents	—	24.50	48.29
Wales	Fair rents	13.53[1]	26.06	39.83
	Assured rents	—	26.00	42.16
Scotland	Fair rents	9.38[1]	23.37	27.88
	Assured rents	—	—	30.23

Source: Wilcox (1996)
Note: 1 1981

OWNER-OCCUPATION: INTERVENTION OR LIBERALISATION?

Although, in a market economy, the equilibrium of demand and supply determines both the price of a product and the quantity of that product purchased within the owner-occupied sector, it is often difficult to establish this equilibrium. The 'uniqueness' of each house, flat, maisonette or cottage; the imperfect knowledge of buyers and sellers; the expense, legal complexity and inconvenience of transferring property; and sometimes the slowness of planning authorities, mortgage institutions, the design professions and buyers and sellers to respond to market conditions – all are factors which make it difficult either to establish a market price based on the value of broadly comparable properties or to ensure that a transaction takes place at that price and equates demand with supply. Many households, from the beginning, weigh non-monetary factors (such as liking where they live) more heavily than monetary considerations, and therefore have no wish to sell even if a market price is offered. Thus, in Figure 3.8 although the market price (P) of a number of similar properties would equate demand with supply, transactions might not take place. Instead prices would either hover above the market price (at P_1) or slip below the market level (at P_2), resulting respectively in a glut of properties on the market (equal to $q_1 - q_2$) or a shortage of housing (again equal to $q_1 - q_2$).

Market failure is also apparent when, during a house price boom (and a substantial increase in demand) owners might be unwilling to sell because of expectations of further price increases or, conversely, during a house price slump (and a dramatic decrease in demand) buyers might be unwilling to buy due to anticipated further price reductions. Thus, although demand (in Figure 3.8b) could increase from DD to D_1D_1 in a boom, or fall from DD to D_2D_2 in a slump, and notional market prices might respectively rise to P_1 or fall to P_2, transactions would be postponed until more extreme price changes occurred.

Disequilibrium also occurs due to the slowness of

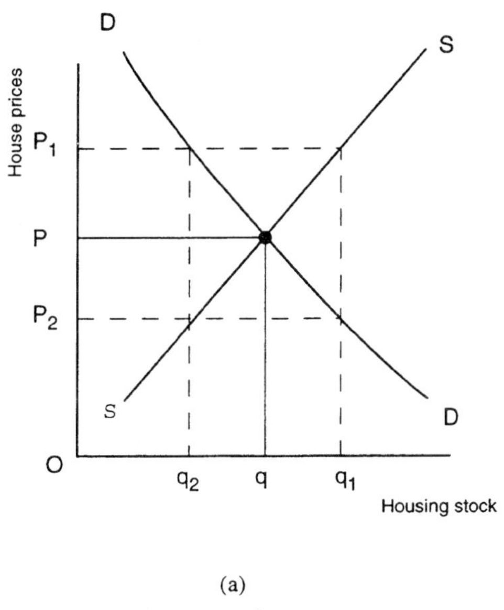

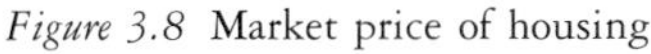
Figure 3.8 Market price of housing

the construction industry to respond to changes in demand, resulting either in a shortage of houses for sale when demand rises or a glut of houses on the market when demand falls. All too often there is also a stark inadequacy of supply in the case of low-cost affordable housing, while supply might exceed demand within the high price/luxury market. There are also very marked spatial disequilibria, most notably a shortage of housing in good condition in the inner cities, but a surplus of housing in many suburban and rural areas and, in general, a deficiency of supply in the South and an inadequacy of demand in much of the North and West.

Nevertheless, despite obvious inefficiencies, the market very broadly determines house prices and rents, resource allocation and the availability of housing. But in an attempt to reduce some of the inefficiencies and to achieve specific policy objectives, governments have for long, and in different ways, intervened in the market.

It is apparent that a high proportion of households favour owner-occupation rather than renting – a preference acknowledged and encouraged by governments in their attempt to create a 'property-owning democracy'. To increase demand (and in the hope of also increasing supply) governments for generations have therefore enabled mortgagors (house-buyers) to offset part of their cost of house purchase through tax relief and tax exemption, or have provided income support to facilitate mortgage repayment at times of unemployment – incentives which, in cash terms, soared throughout the 1970s and 1980s. Thus, largely as a result of substantial government support for owner-occupation, rented housing has become very much a second-best option for most households, and only sought by people either requiring temporary accommodation or unable to afford home ownership.

House prices

While owner-occupation began to expand in the 1920s and 1930s, since the 1980s the tenure has become consolidated as the most important numerical and therefore political factor in the housing market. Favourable government policy, difficulties of finding alternative accommodation, the development of specialist financial intermediaries (such as building societies) and an investment atmosphere favourable for property caused the number of owner-occupied dwellings to increase from 3.4 million in 1947 to 18 million in 1994, or from 26 to 68 per cent of the total stock of dwellings in the United Kingdom, with the most rapid growth occurring since 1979.

The price of owner-occupied housing in the post-war period increased steadily, at a rate marginally greater than the rate of inflation, as measured by the retail price index, but between 1971 and 1973 and 1986 and 1989 far exceeded this. This apparently inexorable increase encouraged early entry into the housing market as the investment aspect of house purchase was emphasised. Between 1974 and 1982, however, there was a realisation that rising house prices in real terms were not inevitable and in the early 1990s house prices fell in both real and absolute terms.

House price booms were the lagged outcome of low house price/earnings ratios and low mortgage interest rates, whereas house price slumps were the result of high house price/earnings ratios and high mortgage interest rates (Table 3.12).

Generally, the price of new houses will be determined by existing house prices. Because of the durability of buildings, the supply of houses is dominated by the existing stock. The net average

Table 3.12 Average house price and retail prices, 1970–96

	Average house prices £	*% increase*	*Retail price index increase (%)*	*Real increase in house prices (%)*	*House price/ earnings ratio*	*Mortgage interest rates (%)*
1970	4,975	7.2	6.4	0.9	3.14	8.50
1971	5,632	13.2	9.4	3.4	3.23	8.00
1972	7,374	30.9	7.1	22.3	3.78	8.50
1973	9,942	34.8	9.2	23.4	4.45	9.59
1974	10,990	10.5	16.1	−4.7	4.16	11.00
1975	11,787	7.3	24.2	−13.6	3.57	11.00
1976	12,704	7.8	16.6	−7.6	3.34	11.06
1977	13,650	7.4	15.9	−7.3	3.29	11.05
1978	15,594	14.2	8.2	5.6	3.30	9.55
1979	19,925	27.8	13.4	12.6	3.64	11.94
1980	23,596	18.4	18.0	0.4	3.56	14.92
1981	24,188	2.5	11.9	−8.4	3.23	14.01
1982	23,644	−2.2	8.6	−10.0	2.89	13.30
1983	26,471	12.0	4.6	7.0	2.98	11.03
1984	29,106	10.0	5.0	4.8	3.09	11.64
1985	31,103	6.9	6.1	0.7	3.05	13.47
1986	36,276	16.6	3.4	12.8	3.29	12.07
1987	40,391	11.3	4.1	7.0	3.40	11.43
1988	49,355	22.2	4.9	16.5	3.82	11.19
1989	54,846	11.1	7.8	3.1	3.89	13.66
1990	59,785	9.0	9.5	−0.5	3.87	15.10
1991	62,455	4.5	5.9	−1.4	3.74	12.47
1992	60,821	−2.6	3.7	−6.1	3.43	10.60
1993	61,233	0.7	1.6	−0.9	3.34	7.97
1994	64,762	3.9	2.5	1.4	3.40	8.14
1995	65,649	1.4	3.4	−2.0	3.33	7.98
1996	70,537	7.4	2.4	4.9	3.45	7.01

Source: Council of Mortgage Lenders (various) *Housing Finance*

addition to stock by new building is relatively insignificant: about 1.1 per cent per year between 1911 and 1991. Thus the supply of houses is relatively inelastic even over a long period; price changes are therefore normally caused by demand changes. House-builders will charge what the market will bear by reference to existing house prices, but they cannot always pass on increases in costs to the consumer unless the market is favourable. Consequently there is a broad correlation between new and second-hand prices (Table 3.13) and between housing starts and prices of second-hand houses. Generally, when second-hand house prices increase at a relatively faster rate, for example in 1987 and 1988, there will be an incentive for developers to increase the supply of new houses, other things being equal. The trend will need to be clearly established before the level of new starts is noticeably affected. But when the rate of increase in the price of second-hand houses decelerates, or when the price falls, the supply of new houses will decrease – for example between 1989 and the early 1990s.

The demand for owner-occupation

Population increase inevitably resulted in an increased demand for housing. Between 1951 and 1991, the population of the United Kingdom increased from 50.2 million to 57.2 million, and the number of households increased at an even faster rate due to young people leaving home earlier and a disproportionate increase in the number of single-person households. Because of a resulting high income elasticity of demand for housing, there was a need for a substantial volume of long-term finance to facilitate house purchase.

Building societies are still the dominant providers of long-term credit for buyers of residential properties. Building societies operate a loan system which is based on receiving money which must be repaid on demand or at short notice, such as investors, savings, but lending money for long periods of time, such as loans which are supported by a mortgage. The building society is able to foreclose on the loan if the borrower fails to comply with the terms of the mortgage, for example by falling into arrears with the payments. The building society can take

Table 3.13 Price indices of new and second-hand properties (quarter ending 31 March 1983 = 100), housing starts and second-hand house prices, 1982–94

	Prices of second-hand properties		*Price of new properties*	*Housing starts, private sector (000s)*	*% change*	*Average second-hand house prices £*	*% change*
	Modern	*Older*					
1982	90.4	91.4	94.8	140.8	19.9	25,167	3.3
1983	100.0	100.0	100.0	172.4	22.4	28,145	11.8
1984	112.7	113.7	112.5	158.3	−8.2	30,342	7.8
1985	125.5	127.8	125.0	165.7	4.7	32,673	7.7
1986	134.0	138.4	123.2	180.0	8.6	37,499	14.8
1987	153.7	160.9	150.3	196.8	9.3	43,427	15.8
1988	169.5	175.6	164.5	221.4	12.5	53,185	22.5
1989	225.8	233.7	211.1	170.1	−23.2	59,173	11.3
1990	224.2	239.7	205.1	135.2	−20.5	63,608	7.5
1991	212.7	217.9	190.1	134.9	−0.2	64,358	1.2
1992	204.6	209.4	187.0	121.0	−10.3	62,397	−3.0
1993	195.7	169.7	181.9	141.0	16.5	65,919	5.6
1994	194.6	209.4	189.2	157.7	11.8	67,729	2.7

Source: Council of Mortgage Lenders, *Housing Finance*; Nationwide Building Society, *House Prices*

the property into possession and sell it to clear the debt outstanding on the property.

Building society mortgages contain a variable interest clause so that when interest rates rise during the period of the loan the building society can advise borrowers that the interest they pay will also increase. Due to periods when interest rates regularly increase during the year, some building societies have adopted a system of increasing payments to reflect movements in interest payments once per year.

There has also been an increase in the number of building societies which offer mortgages with interest rates which are fixed for periods of up to five years, so that in times when interest rates are low many borrowers prefer to borrow on fixed-rate terms.

Until the 1980s, most borrowers took out repayment mortgages which meant that each payment made would include two elements: first, repayment of the principal element, and second, a proportion of the interest outstanding on the loan. During the early years of the loan, payments would include a high proportion of interest and a smaller proportion of principal. However, as the years passed then the proportion of interest paid would gradually decline and the principal part of the loan would gradually increase. The loan would be calculated to ensure that by the end of the loan period the full amount had been repaid.

During the 1980s, low-cost endowment mortgages became very popular. In this method the borrower takes out a life assurance policy which will at the end of the loan period yield enough to pay off the loan. The payments of this type of loan contain two elements, namely, the paying of interest during the period of the loan and an annual premium on the assurance policy. The low-cost endowment mortgage refers to an arrangement whereby the borrower reduces the cost of premiums on the life assurance policy by also taking out a with-profits endowment policy for a sum which is less than the amount required to repay the outstanding loan. The profits element of the policy is expected to cover the difference, although this is not guaranteed. There have been recent criticisms of this method because, as stated above, there is no guarantee that the profits element of the policy will yield enough profits to cover the difference and because lenders which provide loans on this basis receive substantial commission from the insurance company for the sold policies.

The effective demand for owner-occupation was thus overwhelmingly facilitated by mortgage loans. By 1993, building societies had over £221,000 million of outstanding loans, compared to £108,447 million owing to banks, £3,245 million to insurance companies and £651 million owing to local authorities. In 1988, a peak year, mortgage lenders financed 2,149,000 transactions, the number plummeting to 1,138,000 in 1992.

Normally, mortgage lenders are willing to provide finance up to two or three times the head of household's gross annual earnings. It could, there-

Table 3.14 Mortgages: main institutional sources, United Kingdom, 1980–93 (£m)

	Building societies	*Banks and miscellaneous institutions*	*Insurance companies, pension funds*	*Local authorities*	*Other sources*	*Total*
1980	5,722	500	263	456	341	7,282
1988	23,720	16,128	447	−329	144	40,111
1993	9,813	9,776	−399	−357	−2,389	16,461
Advances outstanding at end of 1993	221,142	108,447	3,245	651	24,203	357,688

Source: Council of Mortgage Lenders, *Housing Finance*

fore, be expected that average house prices would rarely exceed three or four times the level of average wages, taking into account that most borrowers would be unlikely to receive mortgages much in excess of 90 per cent of the transaction price of a house. However, house price/earnings ratios fluctuate from about 3.1:1 during a house price slump (for example in 1982) to over 5:1 during a boom (for example in 1989).

Public policy substantially distorted the market for home ownership and, in consequence, the investment market generally. For generations, mortgagors could claim relief on mortgage interest at the rate at which the claimant was liable for income tax. In the early 1990s, the owner-occupier was eligible for mortgage interest relief at source (MIRAS) at 40 and 25 per cent on mortgages up to £30,000 – the total amount of MIRAS having increased from £1,450 million in 1979/80 to £7,700 million in 1990/91. Markets were distorted in numerous ways:

1 MIRAS failed to help the poor since it pulled up demand and hence house prices beyond their means. With the supply of housing being comparatively inelastic, this boost to demand inevitably resulted in house prices being much higher than would otherwise have been the case (Figure 3.9).
2 MIRAS adversely affected household mobility, since during boom conditions it helped to widen North–South disparities (particularly inflating house prices in the South East) and was a disincentive to trade down, or buy a house in a cheaper location.
3 MIRAS distorted rural house prices, since commuters could easily outbid lower-paid local labour for a fairly fixed stock of housing.
4 To ensure that total government expenditure could be funded without too much reliance on public sector borrowing, tax revenue forgone as a result of MIRAS tended to be made up by higher rates of taxation generally (including higher rates of income tax).
5 MIRAS was inefficient, since the tax revenue forgone could have financed productive industry of more pressing social expenditure instead of lubricating the market for mainly second-hand houses. In addition, by helping to boost house prices, mortgage lenders – to facilitate house purchase – attracted a higher and higher proportion of personal savings (amounting to £414 billion in 1993), much of which, otherwise, could have been invested more productively.

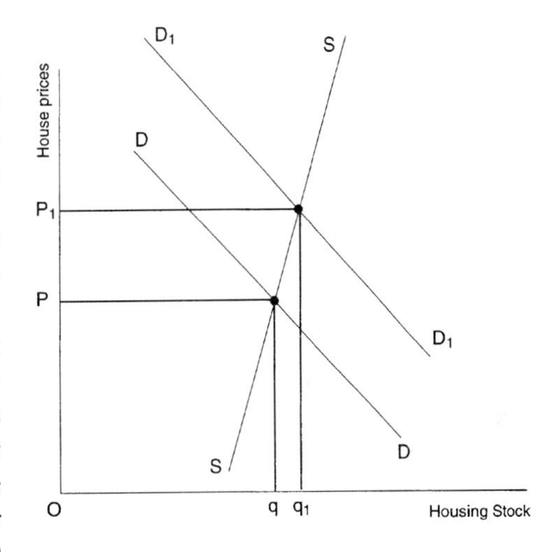

P = market price without tax exemption and relief

P_1 = market price with tax exemption and relief

Figure 3.9 The effect of mortgage interest relief on house prices

The house price boom of the late 1980s

In the period 1985–9 average house prices in the United Kingdom increased dramatically by 76 per cent in cash terms, considerably ahead of increases in the retail price index and average earnings. The causes of the boom were numerous, but essentially derived from an increase in the overall level of demand for home ownership:

1 Ahead of supply, there was a big rise in post-tax incomes, in part due to the reduction in the basic

rate of tax from 30 to 25 per cent and in the top rate from 60 to 45 per cent.

2 Liberal mortgage lending policies encouraged by financial deregulation (for example, by means of the Building Societies Act 1986) made it easier for house buyers to obtain larger and larger mortgages (Bank of England, 1991; Joseph Rowntree Foundation, 1991). The number of mortgages increased from 7.7 million to 9.1 million between 1985 and 1989, while net advances increased from £19.1 billion to £40.1 billion. Mortgage debt as a proportion of gross domestic product therefore increased from 21 per cent to 58 per cent, compared to only 20 per cent in France, Germany and Japan, and the house price/earnings ratio soared to 4.31:1 in 1989. In the United Kingdom in the late 1980s, 100 per cent mortgages were common, whereas in France mortgages were no higher than 80 per cent and in Germany and Japan rarely exceeded 60 per cent.

3 Since both realised and unrealised capital gains were greater than mortgage interest payments, the strong speculative motive in house purchase encouraged buyers to maximise their borrowing. With the end of multiple tax relief from 1 August 1988 (announced in the previous spring's Budget), there was a last-minute hike in demand between March and the end of July which substantially accounted for the 22 per cent increase in average house prices in 1988. Demographic change (notably high birth rates in the 1960s) also stimulated demand. In the late 1980s, there was a large increase in the number of young people wanting to buy.

4 The house price boom can also be attributed to a deficiency of supply. House-building was increasing more slowly than the number of new households – due to planning constraints imposed by county councils (in attempts to protect green belts, other areas of high landscape value and the alleged interests of existing communities), and possibly due to private owners withholding land from the market for speculative reasons.

Effects of the boom

One major attribute of the boom was that there were very wide regional variations in the rate of increase in average house prices between 1985 and 1988 and in house price/earnings ratios. Whereas in East Anglia, the South East and the South West prices soared by over 75 per cent and house price/earnings ratios were large (in excess of 3.5:1), in Northern Ireland, Scotland and the North the increase in prices was 33 per cent or less and ratios were comparatively small (less than 2.4:1). Regional disparities clearly restricted household mobility. Households wishing to move from areas of high to low unemployment, for example from the North to the South East, would not only have had to have paid considerably more for a house but would have had to secure higher-paid employment to afford their housing needs (see Balchin, 1995).

Conversely, few households would have been willing to move from high- to low-price regions (assuming jobs were available) since they would have had to forgo opportunities for substantial capital accumulation, and might also have feared that they would be priced out of the house market should they have wished to return.

A further effect of the boom was that much mortgage finance leaked out of the housing market into the rest of the economy. By 1988, at the peak of the house price boom, the mortgage institutions lent a total of £25 billion more in mortgages than was being spent on buying or improving homes – one of the reasons why general consumption grew 2 per cent faster than the rate of growth of the gross national product, with subsequent adverse effects on the rate of inflation and the balance of payments.

Building societies and other mortgage lenders, whose activities either caused or were in response to house price inflation, clearly helped to create a 'new class of wealth owners' (Hamnett, 1988). Where the total value of inherited residential property was £465 million in 1968/9, it increased to £7.2 billion in 1988/9, particularly benefiting households inheriting estates in the higher-price regions of the South and widening the North–

South divide still further. In 1989, for example, the average value of an estate in Greater London was £87,000 compared to only £45,000 in the North East (Durham, 1990).

The effects on the macro-economy in general were no less important. Since house prices were a major element in the cost of living of house-buyers, Muellbauer (1986) argued that soaring house prices in 1984–6 put pressure on the labour market and added 4 per cent to the level of real wages (he predicted even larger increases in nominal wages in the late 1980s). Because wage agreements are often nationally based, house prices in the South East thus influenced wage levels throughout Britain. The view that house prices during a house price boom effectively determine wage levels (rather than vice versa) could be supported by the notion that people buy the most expensive house they can afford with the largest mortgage they can raise, but then need to earn more to maintain their customary level of consumption and savings.

The house price slump

While average house prices in the United Kingdom increased by 15.6 per cent in 1988, the annual rate of increase subsequently decelerated and prices eventually fell, by 3.8 per cent in 1992. The slump was particularly severe in East Anglia, the South East, Greater London and the South West, where house prices fell continually from 1989.

A substantial decrease in demand was the principal cause of the slump:

1 In response to an increase in the bank rate, interest rates on new mortgages rose from 9.5 per cent in April 1988 to 15.4 per cent in February 1990 and remained about 10 per cent until September 1992, with severe effects on borrowing and the number of house-buying transactions, the number diminishing from 1.9 million in 1988 to 1 million in 1992.
2 Unemployment increased from 5.5 per cent in May 1990 to 10.9 per cent in January 1993, reducing the ability of many potential home-owners to buy and lowering the purchasing confidence of others.
3 Other would-be buyers awaited further falls in prices and by remaining outside of the housing market helped to bring about the price reductions expected.
4 In response to falling prices and less security, mortgage lenders were increasingly reluctant to provide 100 per cent mortgages, and advances fell from £40 million to £27 million between 1988 and 1991.
5 The ending of multiple mortgage interest relief in 1988 and the end of the 'stamp duty holiday' in August 1992 both decreased demand (from December 1991, stamp duty on transactions from £30,000 to £250,000 was suspended to encourage house-buying).
6 The high level of demand in the late 1980s was very unstable and it was probable that house prices had risen to an unrealistically high level in relation to incomes during the previous boom and were destined to fall (see Table 3.12). The house price/earnings ratio reached a peak of 5.03:1 in 1989 compared to an average of 3.7:1 throughout the 1980s and an average of only 3.3:1 throughout the period since the 1950s. In total, demand decreased from 1.9 million to 1 million transactions between 1988 and 1992.

The slump was also due to an excess supply of housing for sale. Many households inflicted with unemployment, failed businesses or bankruptcy were unable to sell their homes at an acceptable price (or at all) and were thus unable to repay their mortgage loans. The number of properties repossessed (and often put on to the market at greatly deflated prices) increased from 19,300 in 1985 to 75,540 in 1991 (Table 3.15). Although the number of repossessions fell slightly in 1992, there was a marked increase in the number of mortgages in arrears of more than 12 months with the possibility of resulting repossession and sale. There was also an increasing number of unsold newly built houses on the market in the early 1990s which helped to depress prices.

Table 3.15 Mortgaged properties taken into possession and mortgages in arrears, 1985–95

	Properties taken into possession		*Mortgages in arrears*			
		% of mortgaged properties	*6–12 months*	%	*More than 12 months*	%
1985	19,300	0.25	57,110	0.74	13,120	0.17
1991	75,540	0.77	183,610	1.87	91,740	0.93
1992	68,540	0.69	205,010	2.07	147,040	1.48
1993	58,540	0.58	164,620	1.62	151,810	1.50
1994	49,210	0.47	133,700	1.28	117,100	1.12
1995	49,410	0.47	126,670	1.20	85,200	0.81

Source: Council of Mortgage Lenders, *Housing Finance*

Effects of the slump

The regional impact of the slump was the reverse of that resulting from the former boom. There was a regional convergence (instead of a divergence) in the rate of change in average house prices, house price/earnings ratios and rates of unemployment. Whereas East Anglia, the South East and Greater London experienced decelerating house prices and lower house price/earnings ratios during the slump of 1988–92, a number of regions particularly Scotland and the North appeared to continue to enjoy a boom (Table 3.16). Despite convergence, however, house prices were still markedly higher in, for example, the South East than in much of the rest of the country (and unemployment in this region was now also high). Thus convergence, as such, did little to encourage any notable increase in household mobility.

The effects of the slump in house prices on owner-occupiers and the macro-economy were, however,

Table 3.16 North–South disparities in house prices and house price/earnings ratios, 1985–92

	Average house price	*Increase in average house price*	*House price/ earnings ratio*	*Average house price*	*Increase in average house price*	*House price earnings/ ratio*
	1988 £	*1985–8* %	*1988* *: 1*	*1992* £	*1988–92* %	*1992* *: 1*
East Anglia	57,295	81	3.60	64,610	13	3.41
South East (excl. Greater London)	72,561	79	3.60	75,189	4	3.16
South West	58,457	77	3.51	77,416	32	3.88
Greater London	77,697	75	3.81	77,446	−1	2.91
North	30,193	33	2.38	46,624	54	2.75
Northern Ireland	29,875	30	2.21	39,240	31	2.18
Scotland	31,479	17	2.28	52,274	66	2.62

Source: Council of Mortgage Lenders, *Housing Finance*

substantial. By 1992, over 1.5 million households (disproportionately in the South East, East Anglia and Greater London) were caught in the 'negative equity trap'. Since the value of their property was £6 billion below their mortgage debt (Bank of England, 1992) they were thus generally unable to sell their properties and buy elsewhere – the market, to an extent, ceasing to work.

The increase in the number of repossessions (referred to above) was clearly not only a cause of the house price slump, but also an effect. Had house prices been buoyant, mortgagors facing repayment difficulty could have sold their properties and traded downward. With the malfunctioning of the market, this was no longer possible and, at worst, dispossessed owner-occupiers found themselves homeless.

With the fall in house prices and higher unemployment in the South East, pressures on the labour market in the region eased in the early 1990s, resulting in a stabilising effect on national wage settlements. The house-building industry, in particular, responded to falling house prices by decreasing output and laying off labour. Falling house prices in the southern regions of Britain, moreover, deterred house-buyers from taking out second mortgages to purchase consumer goods and services, and nationally equity withdrawal was further constrained by the value of inherited properties rising less rapidly in the early 1990s than in the 1980s. Consumer demand was consequently reduced with disinflationary effects on consumer prices.

Plate 3.3 Starter homes completed for the owner-occupied market, 1997
The construction of affordable owner-occupied housing was increasing in the second half of the 1990s, particularly in Greater London and the South East, where average house prices were increasing by at least 10 per cent per annum

Recovery

By 1996 the house price recession was over. Average house prices increased by over 7.5 per cent but it was uncertain whether or not this heralded a return to the boom conditions of the late 1980s or to a period of gradual house price inflation characteristic of the 1960s.

A policy of contradiction

Although they were determined to increase owner-occupation in the 1980s-90s, Conservative governments adopted means to achieve this aim that were, to an extent, contradictory. On the one hand, an interventionist approach was adopted in response to the effects of the house price slump in the early 1990s, but on the other hand the government seemed wedded to a neo-liberal free market ideology as demonstrated by the conversion of much of the local authority stock into owner-occupied housing and the substantial reduction in mortgage interest relief at source (MIRAS).

In December 1991, the government introduced a £1 billion rescue package intended to help the victims or potential victims of repossession, many of whom had recently become unemployed. Cheap loans became available to housing associations to enable them to acquire repossessed homes in order to rent them back to their former mortgagors, and funds were allocated to building societies to facilitate the conversion of existing mortgages into rental agreements. In addition, income support for unemployed mortgagors (amounting to about £1.25 billion per annum by 1991) was paid directly to mortgage lenders to reduce the extent of mortgage arrears and the risk of foreclosure.

Further measures to protect home-owners from the slump were introduced in 1992. Thus building societies were permitted to raise the amount they could lend unsecured from £10,000 to £25,000 per person in an attempt to enable owner-occupiers (with negative equity in their homes) to sell and to subsequently buy elsewhere, and the Budget of March 1993 raised the transaction threshold from £30,000 to £60,000 on which stamp duty (at 1 per cent) would apply. The government, regarding low interest rates as central to monetary policy, hence withdrew the United Kingdom from the Exchange Rate Mechanism and brought down the base rate and interest rates on mortgages to the lowest levels they had been since the 1960s – other things being equal, a stimulus to housing demand. In an attempt to reduce excess housing supply, the government made £750 million available to the Housing Corporation (under the Autumn Statement of 1992) for allocation to housing associations to facilitate the purchase of unsold housing – mainly new stock and repossessions. By March 1993, over 21,000 empty houses were thereby removed from the market.

The privatisation of a sizeable proportion of the local authority housing stock (see above) and reductions in MIRAS were, however, indicative of a free market approach to housing. The government was concerned that MIRAS might be getting out of hand. It had greatly increased (from £1,450 million in 1979/80 to £7,700 million in 1990/91), it was one of the largest subsidies enjoyed by owner-occupiers and its distorting effects on the housing market were increasingly recognised. The Budget of 1992 therefore limited MIRAS to the basic rate of tax (25 per cent) and ended relief at the higher rate (40 per cent) – measures aimed at curbing further excessive price increases (after the house price boom of the late 1980s), and avoiding the repetition of over-borrowing and large-scale repossession (consequences of the boom and subsequent crash) towards the end of the decade. But, more importantly in the mid-1990s, it seemed likely that the Conservative government as part of its neo-liberal agenda was intent on abolishing MIRAS altogether in the long term, paving the way for a free market in owner-occupied housing, an approach broadly welcomed by housing reformers across the political spectrum. In the event of abolition, however, owner-occupiers would still benefit from not having to pay the full cost of acquiring an asset. Since 1962, they have not had to pay Schedule A tax on 'imputed rent income' (the free use of accommodation equivalent in value to rent), and are also normally exempt from capital gains tax, exemptions respectively worth £7,000 and £3,000 million in the early 1980s (Shelter, 1982) rising to at least £10,000 and £4,500 million by the end of the decade. In addition, council tenants buying their own homes received discounts of over £1,000 million per annum throughout much of the 1980s.

Leasehold enfranchisement

As early as 1884, a Royal Commission on the Housing of the Working Classes recommended 'leasehold enfranchisement', the right of a leaseholder to buy the freehold of the land on which his or her house stands. But proposed legislation to implement this right was continually blocked. The Landlord and Tenant Act 1954 left the leaseholder in a hopelessly weak position compared with the power of the ground landlord. The Leasehold Reform Act of 1967 ended insecurity by giving leaseholders the right to buy the freehold of their houses or to extend their lease for fifty-nine years, provided: the house had a rateable value not exceeding £400 (in Greater London) or £200 (elsewhere); the lease was originally granted for more than twenty years; the lease was of the whole house; the lease was at low rent (equivalent to less than two-thirds of the rateable value of the house); and the leaseholder was occupying the house as his or her only or main residence and had been doing so for five years. But a high proportion of leaseholders failed to qualify for these rights. By the early 1990s, there were up to 2 million owner-occupiers who were still leaseholders – much as before 1967. They mainly consisted of occupiers of flats bought on long leases in the boom development years of the late 1950s and early 1960s and holders of leases of less than twenty-one years who bought at a premium and paid peppercorn rents (of, say, £100–£2,000 per annum). All had to face the possibility that the value of the asset might decrease in real or even absolute terms in the future, or the certainty that they would lose their home when the lease expired. Long before a lease expired, however, the incentive to maintain the property and its saleability diminished.

Under the Leasehold Reform, Housing and Urban Development Act of 1993, however, 750,000 leasehold flat-owners either became able to buy the freehold of their homes (provided their leases were for more than twenty-one years and two-thirds of leaseholders of a freehold property agreed) or, where they were eligible to buy, were given the right to extend their leases by ninety years.

AFFORDABILITY AND SUBSIDISATION

In terms of income, households across each tenure were considerably more prosperous in Greater London and the rest of the South East than in, for example, the North or North West, while it was clear that, irrespective of region, owner-occupiers (and mortgagors rather more than outright owners) were better off than private sector tenants and these, in turn, received higher incomes than social sector tenants (Table 3.17). It is also notable that some 12–15 per cent of the income of working tenants in social housing was derived from social security benefits (Wilcox, 1995).

In recent years, however, a growing number of households have been unable to afford housing, regardless of region or tenure. Rising unemployment, an increasing precariousness of labour in low-wage part-time or temporary employment, and job losses in traditional production industries being only partly replaced by low-skilled service sector jobs, have constrained demand, while the supply of low-cost housing has been severely restricted by a decreased level of house-building in the social sectors, by privatisation, and by gentrification within the private rented stock (by high-income households displacing low-income tenants). The most obvious outcome of market dysfunction is that social problems, such as the increased number of people vulnerable to being homeless or poorly housed (notably single-parent families and the young), have become exacerbated (Bull, 1996).

Table 3.17 Gross household incomes by region and housing tenure

	£ per week			
	Owner-occupation			
	Mortgagors	Outright owners	Private renting	Social renting
Greater London	620	430	366	210
South East (excl. GL)	589	311	297	186
East Anglia	527	273	265	142
North West	496	259	200	152
South West	461	286	245	161
Yorkshire & Humberside	455	262	207	178
East Midlands	447	292	287	188
West Midlands	447	272	233	146
North	421	264	245	165
England	513	298	276	173
Scotland	507	333	299	173
Wales	426	277	202	165
Northern Ireland	517	245	220	140

Source: Department of the Environment (1993) *Family Expenditure Survey*, HMSO

Table 3.18 shows that homelessness in Great Britain increased by 155 per cent between 1979 and 1991, and the number of homeless households in temporary accommodation increased more than tenfold. Council waiting lists (also a manifestation of the need for affordable housing) also increased, to 1.4 million by 1992.

In all housing sectors, affordability was reduced throughout much of the period 1979–95 as a consequence of rising rents and house prices (see Case study 3.2, pp. 102–4). In the private and social rented sectors, rents increased substantially and represented a much higher proportion of household earnings at the end of the period than at the

Table 3.18 Local authority homeless acceptances and homeless households in temporary accommodation, 1979–94

Homeless households	*1979*	*1991*	*1992*	*1993*	*1994*
England	57,200	151,720	149,240	138,040	125,500
Scotland	8,126	17,304	19,176	17,289	15,700
Wales	4,676	9,843	10,207	11,125	9,897
Great Britain	70,038	178,867	178,686	166,454	148,057
Homeless households in temporary accommodation[1]					
Bed and breakfast	1,330	12,150	7,630	4,900	4,130
Hostels	3,380	9,990	10,840	10,210	9,730
Private sector leasing	—	23,740	27,910	23,270	15,800
Other	—	14,050	16,690	15,200	15,970
Total	4,710	59,930	63,070	53,580	45,630

Source: Department of the Environment; Scottish Office; and Welsh Office (various) *Housing and Construction Statistics*
Note: 1 England only

beginning, while in the owner-occupied sector, house price/earnings ratios and mortgage interest rates rose to record levels in 1979–80 and again in 1989–90.

An increase in the supply of affordable housing would, other things being equal, exert a downward pressure on rents and house prices, benefiting both tenants and first-time buyers. Subject to an adequate supply of building land with planning permission attached, and low-interest mortgage finance available to below average-income buyers, the market might be expected to provide low-cost housing such as starter homes. In respect of social housing, however, although Holmans (1995) predicted that there would be a need for 90,000–100,000 new social units per annum over the period 1991–2011, this would clearly require a considerable increase in mixed funding and in rates of housebuilding – which in the housing association sector amounted to only 32,900 starts in 1995, and in the local authority sector numbered a meagre 1,200 starts in the same year. The supply of local authority housing had moreover been severely depleted by the scale of RTB losses (1,529,000 in 1980–94).

To meet the future demand for affordable housing it can only be hoped that *Circular 7/91* (Department of the Environment, 1991) is encouraging planning authorities to show land in their local plans scheduled for the provision of social housing and successfully to negotiate with developers to ensure that an element of affordable housing is included in their schemes; and, more specifically, it is to be assumed that *Planning Policy Guidance Note 3* (Department of the Environment, 1993b) is encouraging planning authorities to target specific sites for the development of affordable housing 'based on evidence of need and suitability'.

But an adequate increase in the supply of affordable housing is unlikely to be forthcoming unless the current system of subsidies is radically changed. Under Conservative administrations in the 1980s and 1990s, an ideological opposition to maintaining the size of the local authority housing stock (within the context of large-scale privatisation) resulted in a very substantial reduction in bricks-and-mortar subsidies to that sector. Whereas Exchequer subsidies and rate fund transfers amounted to £2,130 million in 1980/81, by 1995/6 these had diminished to such an extent that most local authorities realised surpluses on their housing revenue accounts, producing an overall surplus of £361 million in 1995/6 (Table 3.19). But because it was government policy under the Housing Acts of 1980 and 1988 and the Local Government Housing Act of 1989 to raise private and social sector rents towards market levels, it was deemed expedient also to raise individual allowances. Housing benefits (rent allowance to private and housing association tenants and rent rebates to local authority tenants) therefore escalated by respectively 2,839 and 550 per cent, 1980/81 to 1995/6, rent rebates in the latter financial year becoming the largest housing subsidy of all.

Under the Social Security Act 1986, a new housing benefit scheme was introduced and was intended to be simpler to understand and to administer than hitherto. This was an attempt to ensure that people in similar circumstances were treated the same regardless of whether they were employed or not. It is an income-related benefit administered by local authorities and designed to help people who rent their homes and have difficulty in meeting their housing costs. All claimants are assessed on the basis of the income support means test and if their income is found to be equal to or less than the applicable amount, then the claimant is entitled to full housing benefit which would cover 100 per cent of the eligible housing costs. If the claimant's income is higher than the applicable amount, then the benefit tapers or is reduced accordingly. The current taper is 65p for every pound above the applicable amount.

Claimaints for housing benefit can be either public or private sector tenants. There are two avenues that a claimant can take when claiming housing benefits. First, when claiming income support an application is submitted at the same time for housing benefit. The second avenue is to apply for housing benefit if employed on a low income. In Great Britain in 1996, 3,006,000 local authority tenants were in receipt of an average weekly housing benefit of £30.70, while 544,000 housing association tenants were in receipt of an average weekly housing benefit of £39.50.

In assessing a claimant, the local authority has to consider their income level (if any), any capital available such as savings and shares, eligible housing costs such as rent and council tax, and contributions which non-dependants are expected to contribute to the household. Within the private rented sector, rent officers are responsible for determining market rents for housing benefit purposes. Where a claimant's accommodation is deemed to be too large or where the rent is seen as unreasonably high, the rent officer determines the eligible rent in order to ensure that the local authority restricts the amount of housing benefit which it will pay. This has been introduced to ensure that where private tenants receive full housing benefit the landlord is charging reasonable levels of rent, and to give tenants an incentive either to negotiate a lower rent or to choose a cheaper home. In Great Britain 1,141,000 private tenants are in receipt of an average weekly housing benefit of £51.10.

The Local Government and Housing Act 1989 introduced the housing revenue account subsidy (HRAS). The way this works (see p. 66) can result in tenants paying the full rent finding themselves paying for the rebates of their neighbours claiming housing benefit.

Income support on mortgage interest (paid to unemployed mortgagors) increased by at least fourteenfold over the period 1980/81 to 1995/6 – attributable largely to higher interest rates and rising unemployment in the early 1990s – and mortgage interest tax relief (MITR) escalated until 1990/91 before diminishing as a result of falling house prices and lower interest rates in the mid-1990s (Table 3.19). Clearly, governments were willing to subsidise people but not bricks and mortar.

As might be expected, the increase in total rent allowances paid to private and housing association tenants was associated with the substantial increase in the number of claimants (from 240,000 in 1980/81 to 1,776,000 in 1994/5) resulting from fair rents being superseded by rents at market levels. The

Table 3.19 Principal bricks-and-mortar subsidies and individual allowances, Great Britain 1980/81 to 1995/6

	(£m)			% change
	1980/81	*1990/91*	*1995/6*	*1980/81–1995/6*
Bricks-and-mortar subsidies				
Exchequer subsidy	1,719	1,221	(328)[1]	−119
Rate transfers	411	(9)	(33)[1]	−108
Total	2,130	1,212	(361)[1]	−117
Individual allowances				
Housing benefits:				
Rent allowances	183	1,779	5,378	+2,839
Rent rebates	841	3,368	5,470	+550
Income support on mortgage interest	71	553	1,035	+1,358
Mortgage interest tax relief	2,188	7,700	2,700	+23
Total	3,283	13,400	14,222	+333
Total subsidies	5,413	14,612	14,944	+176

Sources: Wilcox (1996)
Notes: 1 Surplus of rents over subsidies

increase in total rent rebates was also an outcome of the increase in the number of claimants, from 1.3 million in 1980/81 to almost 3 million in 1994/5 (Table 3.20). More and more local authority tenants became eligible for this form of subsidy as Exchequer subsidies and rate fund transfers were being cut back or abolished and, as a consequence, target and actual rents increased at a greater rate than average earnings and the retail price index.

The number of owner-occupiers eligible for income support on mortgage interest relief at source or MIRAS also increased dramatically in the 1980s and 1990s (Table 3.20). The increase in the number of claimants for income support (from 134,000 in 1980 to 529,000 in 1994) was due largely to the increase in the level of unemployment and the increase in rates of mortgage interest in the early 1990s, while the increase in the number of recipients of MIRAS (from 5.9 million in 1980/81 to 10.5 million in 1995–6) was an outcome of the increasing appeal of owner-occupation as a tenure (particularly during the speculative boom of the late 1980s) and the decreasing attraction of both private and social sector renting, notwithstanding the large increase in housing benefit payments.

Subsidies per dwelling or household have also diverged. Whereas Exchequer subsidies and rate fund transfers respectively averaged £179 and £67 per local authority dwelling in 1980/81, by 1995/6 rent hikes had, in net terms, eliminated the need for subsidies and produced surpluses of £61 and £7 per dwelling (Table 3.21). In contrast, average rent allowances, rent rebates and income support on mortgage interest all increased, respectively by 1,039, 527 and 278 per cent from 1980/81 to the mid-1990s. Private and housing association tenants were not only the largest recipients of subject subsidies but had received the largest increase in subsidies in the period under consideration. By comparison, mortgagors, on average, after enjoying an increasing amount of MIRAS in the 1980s, thereafter received a diminishing amount of subsidy (Table 3.21).

It is clear that the substantial increase in housing benefits did little or nothing to stimulate housebuilding or increase accessibility to affordable rented housing. In the local authority sector, rent rebates increased from £841 million in 1980 to £5,354 million in 1994, yet the number of housing starts plummeted from 34,599 to 1,300 over the same period, a decrease associated with the reduction in Exchequer subsidies and rate fund transfers from £2,130 million in 1980 to −£133 (in net terms) in 1994 (Figure 3.10). While the number of housing association starts, however, increased from 14,799 in 1980 to 39,814 in 1994, this was attributable mainly to housing association grants (HAGs) irrespective of the notable increase in rent allowances; and in respect of the private rented sector, while the supply of accommodation has

Table 3.20 Number of recipients of individual allowances, Great Britain, 1980–81 to 1994/5

	(000s)			*% increase*
	1980/81	*1990/91*	*1994/5*	*1980/81–1994/5*
Housing benefits				
Rent allowances	240	1,044	1,776	+641
Rent rebates	1,330	2,944	2,981	+124
Income support on mortgage interest	134[1]	310[2]	529[3]	+295[4]
Mortgage interest tax relief	5,850	9,600	10,500[5]	+79

Source: Wilcox (1995)
Note: 1 1980 2 1990 3 1994 4 1980–94 5 1995/6

Table 3.21 Average bricks-and-mortar subsidies and individual allowances, Great Britain, 1980/81 to 1995/6

	1980/81	*(£)* *1990/91*	*1995/6*	*% change 1980/1–1995/6*
Bricks-and-mortar subsidies for local authority dwellings				
Exchequer subsidy	279	140	(61)[1]	−122
Rate fund transfers	67	18	(7)[1]	−110
Individual allowances				
Housing benefits average payments:	199	1,095	2,268[2]	+1,039[3]
Rent allowances	240	903	1,505[2]	+527[3]
Rent rebates				
Income support on mortgage interest (average payments)	529[4]	1,785[5]	1,998[6]	+278[7]
Mortgage interest tax relief (average relief)	330	820	270	−18

Source: Wilcox (1995)
Note: 1 Surplus of rent over subsidy
2 1994/5
3 1980/81 to 1994/5
4 1980
5 1990
6 1994
7 1980–94

remained fairly constant in absolute terms (decreasing only slightly from 2,337,000 units in 1981 to 2,298,000 units in 1994), rent allowances in the sector have soared.

If housing benefits failed to have a significant influence on the supply of rented housing, they undoubtedly had an impact on demand. As Figure 3.11 shows, the substantial increase in rents in the local authority, housing association and private rented sectors in recent years required a simultaneous increase in demand fuelled by rent rebates and rent allowances. In respect of local authority housing, Figure 3.11a shows that a decrease in supply (resulting from privatisation) from SS to S_1S_1 was coupled with an increase in demand from DD to D_1D_1 to produce an increase in rents from r to r_1. Figure 3.11b shows that an increase in the supply of housing association dwellings (from SS to S_1S_1) is not sufficient to prevent rents rising substantially (from r to r_1) as a result of a large increase in demand supported by housing allowances; and Figure 3.11c shows how an increase in rents in the private rented sector derives solely from an increase in allowance-induced demand for a relatively fixed supply of housing.

The rise in house prices has similarly been fuelled by a subsidy – mortgage interest relief at source (MIRAS) (Figure 3.9) – particularly up to the time when MIRAS peaked at £7.7 billion in 1990/91, and consequently failed to help low-income households acquire affordable owner-occupied housing. But the reduction of MIRAS in the 1990s, from 40 and 25 per cent in 1991 to 15 per cent in 1995, has reduced demand for owner-occupied housing and depressed house prices. Figure 3.12 shows that an initial increase in demand from DD to D_1D_1, resulting from a high level of MIRAS, inflated prices from P to P_1 but that, following a cut in MIRAS, demand decreased to D_2D_2 and prices fell to P_2, rendering

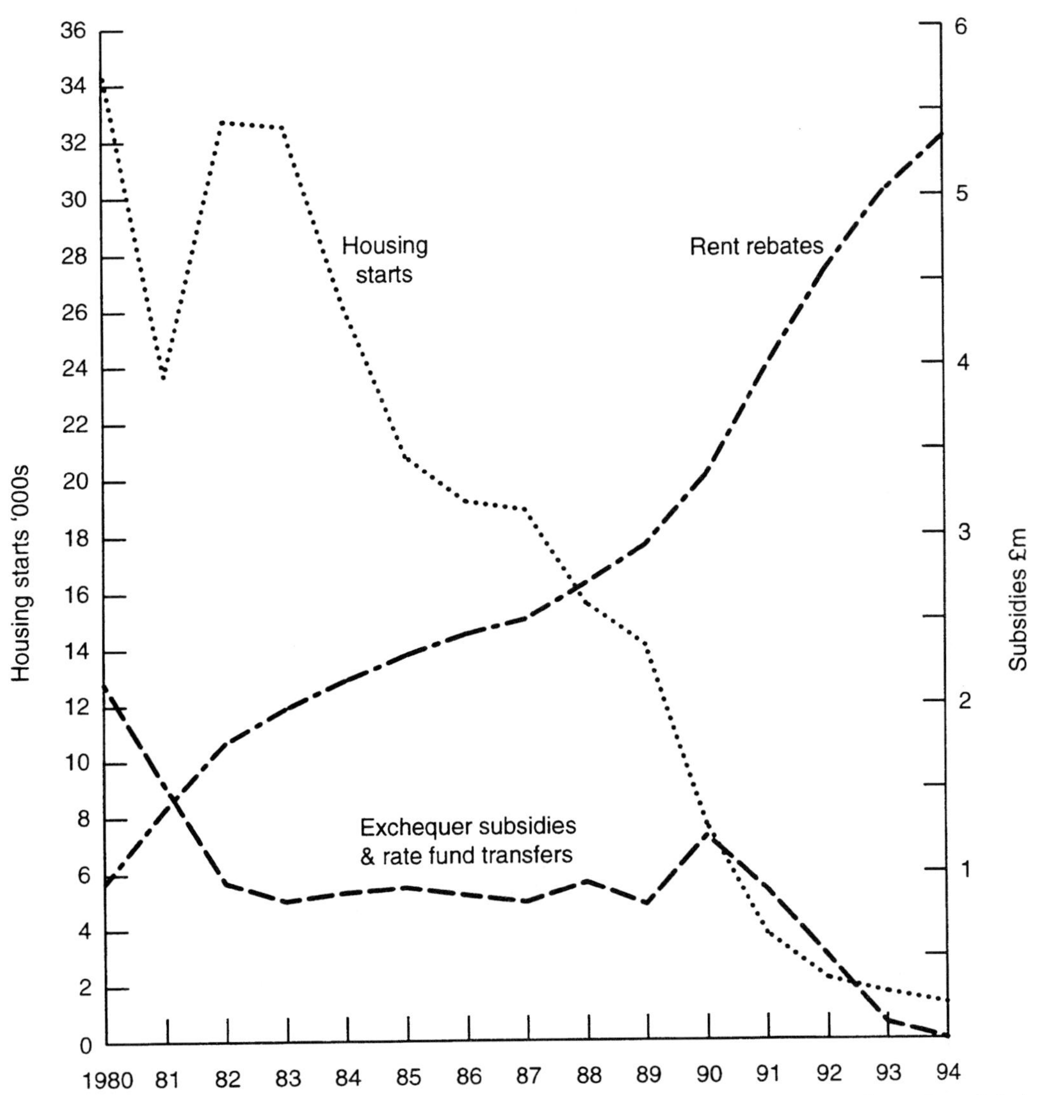

Figure 3.10 Local authoritiy housing starts, Exchequer subsidies, rate fund contributions and rent rebates, Great Britain, 1980–94

owner-occupation more affordable, other things being equal.

Within an increasingly market-dominated economy, it was becoming clearer that, in the virtual absence of bricks-and-mortar subsidies for new house-building, any increase in housing benefits would be translated into higher rents and vice versa. An increase in local authority and housing association rents was thus no longer resulting in net public expenditure (Wilcox, 1995). Higher rents were also inflationary and were extending the housing benefit poverty trap, deterring tenants from securing low or modestly paid employment. The White Paper, *Our Future Homes* (Department of the Environment,

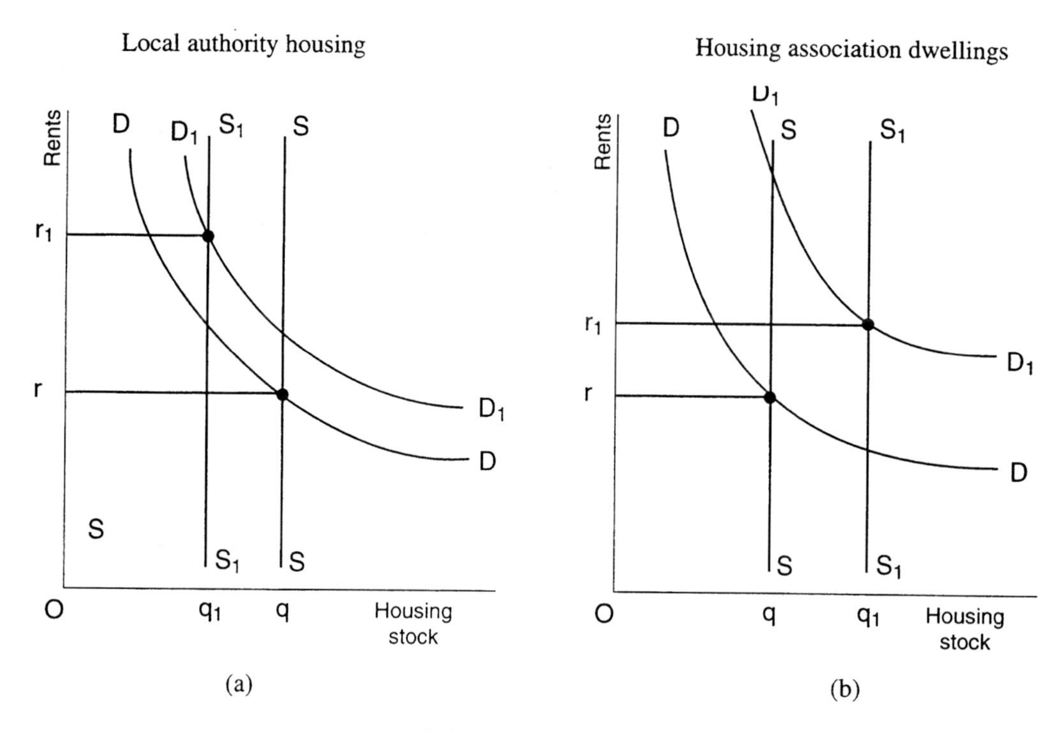

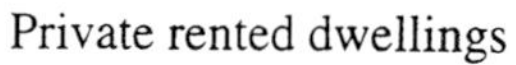

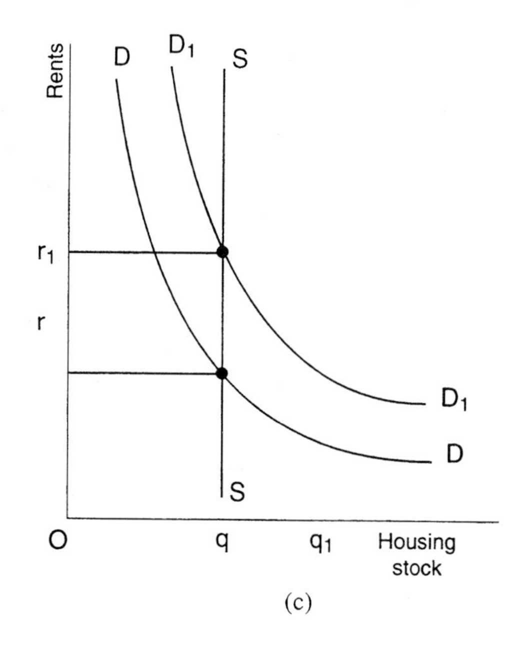

Figure 3.11 The effect of benefit-induced increases in demand upon rents in social and private housing sectors

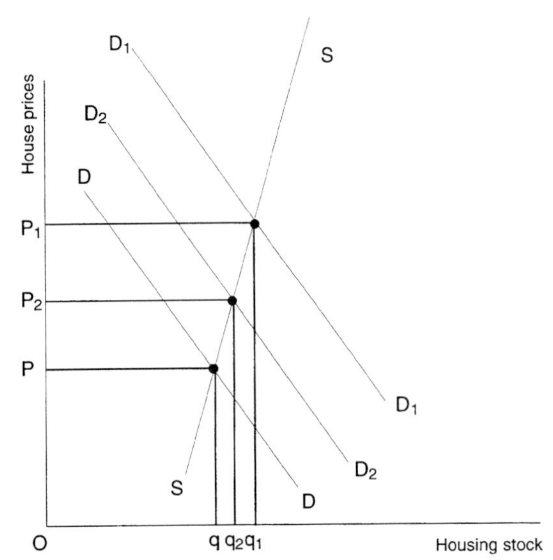

Figure 3.12 The effect of an increase and decrease in mortgage interest relief upon housing demand and house prices in the owener-occupied sector

1995b), therefore envisaged no further significant rises in social housing rents in the near future. There was also concern that in the private rented sector, housing benefits were being abused by both tenants and landlords, and therefore the 1994 Budget contained proposals to limit private rent levels eligible for 100 per cent housing benefit. Rents in excess of 'local reference rents' were to be only 50 per cent eligible for benefit from January 1996 (Wilcox, 1995).

In the owner-occupied sector, the decrease in MIRAS in 1991–5 may not only have had a depressing effect on house prices – potentially benefiting low-income first-time buyers – but, since rates of interest on mortgage loans plummeted over the same period, existing mortgagors also found house-buying more affordable. For example, a household with a £40,000 mortgage would have incurred net repayments of £431 per month in 1990/91 but only £289 in 1995/6. If rates of interest were to remain low, further decreases in MIRAS might be anticipated.

The increasing dependence on market rents (albeit with restricted levels of housing benefit) and less subsidised house prices is, however, unlikely generally to solve the principal manifestation of the affordability problem: homelessness and lengthy council waiting lists. Only a greater balance between individual allowances and bricks-and-mortar subsidies (rather than the virtual absence of the latter) is likely to ensure that there is an equilibrium between demand and supply at an affordable rent or price.

HOUSING TENURE

The principal outcome of Conservative housing policy in the 1980s and 1990s, as outlined in this chapter, is that there has been a marked increase in owner-occupation and a notable decrease in renting in the local authority sector (Table 3.22). Over the years 1981–94, the number of owner-occupied dwellings increased by 3.84 million, while the number of local authority dwellings decreased by 1.7 million, mostly as a result of privatisation under the Housing Act of 1980 and subsequent legislation. At the same time, there was an attempt to revive private renting and to expand the housing

Table 3.22 Housing tenure, United Kingdom, 1981–94

	Owner-occupied		*Private rented*		*Housing association*		*Local authority*	
	000s	*%*	*000s*	*%*	*000s*	*%*	*000s*	*%*
1981	12,169	56.4	2,375	11.0	472	2.2	6,570	30.4
1988	14,765	64.0	2,095	9.1	622	2.7	5,587	24.2
1994	16,004	66.5	2,310	9.6	881	3.7	4,868	20.2

Source: Department of the Environment, *Housing and Construction Statistics*

association sector, particularly from the late 1980s under the Housing Act of 1988.

Regionally there are clear disparities in the distribution of tenure. Whereas over 70 per cent of dwellings in the South East (excluding Greater London), the South West and East Midlands were owner-occupied in 1994, in the North, Greater London and Scotland owner-occupation accounted for fewer than 62 per cent of dwellings (Table 3.23). Local authority housing, on the other hand, was a disproportionately large tenure in Scotland and the North, where respectively 32 and 26 per cent of all dwellings were council owned, in contrast to the South East and South West where less than 13 per cent of dwellings fell into this sector. Private and housing association renting was particularly 'over-represented' in Greater London. Although, spatially, there are distinct economic, social and political disparities which largely explain the regional distribution of tenure, the housing policy of successive Conservative governments in the 1980s and early 1990s, which favoured the expansion of owner-occupation and the contraction of the local authority rented sector, arguably benefited the South East, the South West and the East Midlands (notwithstanding negative equity, repossessions and other effects of the housing slump in the early 1990s), but penalised Scotland, the North and Yorkshire and Humberside, where a disproportionate number of households relied on council housing as the only accessible affordable tenure.

Clearly, owner-occupation is more within the means of households with comparatively high incomes than those with low incomes. Table 3.24 shows that, on average, home-buyers in 1993 received gross weekly incomes of £508, whereas outright owners (many of whom were retired) received £297, private tenants £274 and tenants of social housing only £172. Economically active non-manual home-buyers were the most well-off, receiving £625 per week, while the unemployed and the retired among the tenants of social housing were the least well-off, receiving only £131 and £114 per week respectively.

By comparison, within Western Europe, there is little correlation between national prosperity and housing tenure. In respect of home ownership, there were ten countries with a higher gross national product (GNP) per capita than the United Kingdom in 1995, but in nine of them owner-occupation in the early 1990s represented a smaller proportion of total housing stock than in the United Kingdom.

Table 3.23 Regional distribution of housing tenure, 1994

	Owner-occupation	*Private renting*	*Housing association*	*Local authority*
South East excl. Greater London	73.7	9.9	5.0	11.4
South West	72.6	12.2	3.0	12.3
East Midlands	70.6	9.1	2.5	17.8
East Anglia	69.1	12.6	4.9	13.5
North West	67.8	7.6	4.4	20.2
West Midlands	67.6	7.4	4.1	20.9
Yorkshire and Humberside	65.7	9.3	3.1	22.0
North	61.5	8.4	4.0	26.0
Greater London	57.4	14.2	6.6	21.7
England	67.7	10.1	4.4	17.8
Scotland	57.0	7.6	3.1	32.3
Wales	72.1	6.7	3.3	33.1
Northern Ireland	72.3	—	2.2	25.5

Source: Department of the Environment, *Housing and Construction Statistics*

Table 3.24 Gross household income by tenure, United Kingdom, 1993

Tenure	(£ per week) Economically active			Economically inactive			All households
	Non-manual	Self-employed	Manual	Unoccupied	Unemployed	Retired	
Home-buyer	625	494	430	305	304	238	508
Outright owner	558	441	372	274	220	224	297
Private renting	500	358	285	160	156	157	274
Social renting	317	242	290	136	131	114	172

Source: Department of the Environment (1993) *Family Expenditure Survey*

Switzerland had the highest GDP per capita in Europe, but the smallest owner-occupied sector (Table 3.25).

Conversely, in no country poorer than the United Kingdom was there a larger social rented sector. Indeed, there was proportionately more social rented housing in the comparatively prosperous Netherlands and Germany. However, what really distinguishes social renting in the United Kingdom from the social sector elsewhere in Western Europe is that

Table 3.25 Gross domestic product per capita and housing tenure, Western Europe

	GDP $ per capita[2]	Tenure %[3]			
		Owner-occupation	Private renting	Social renting	Other tenure
Ireland	15,100	80	9	11	—
Spain	12,500	76	16	2	6
Finland	20,410	72	11	14	3
Greece	8,400	70	26	0	4
Italy	18,400	67	31	2	—
Luxembourg[1]	—	67	31	2	—
United Kingdom	18,950	66	10	24	—
Portugal	6,900	65	28	4	3
Belgium	22,260	62	30	7	—
Norway	26,590	60	18	4	18
France	23,550	54	21	17	8
Denmark	29,010	50	24	18	8
Netherlands	21,300	47	17	36	—
Sweden	23,270	43	16	22	19
Austria	25,010	41	22	23	14
Germany	26,000	38	36	26	—
Switzerland	36,430	31	60	3	6

Source: CECODHAS (European Liaison Committee for Social Housing) (1985) *Housing Tenure in Europe*; Economist Publications (1994) *The World in 1995*

Note: 1 Included in the GDP per capita for Belgium

2 Estimate for 1995

3 Estimate for 1993

the United Kingdom still has, by far, proportionately the largest stock of local authority or municipal housing, whereas housing associations, housing companies and other forms of social housing constitute most or all of this sector elsewhere.

Of the countries constituted, only two – Ireland and Italy – had lower proportions of private renting than the United Kingdom, with Switzerland, Germany and Belgium having the largest private rented sectors. In both social and private rented sectors, therefore, there appears to be no discernible relationship in Western Europe between the level of prosperity and the desire to rent.

Within the rest of Western Europe, economic, political and social disparities are on a greater scale than in the United Kingdom and thus, as a result of market forces, create greater spatial variations in tenure than those existing between the regions of the United Kingdom. Nevertheless, national dissimilarities in housing policy also play an important role. Whereas in the United Kingdom, private and social housing has been at best neglected and at worst run down in recent years, these sectors have generally been more actively promoted in much of Europe, while owner-occupation has not, to the same extent, enjoyed as much support as in the United Kingdom (see Balchin, 1996).

CASE STUDY 3.1

THE FINANCE OF PRIVATE RENTED HOUSING

A residential property or building can be owner-occupied or rented, the latter being an investment property. In the present market it may of course be vacant, resulting from being surplus to the owner's requirements or a poor investment! A large proportion of residential property is owner-occupied but most of the conventional property texts and theories are applied to an investment market. Property finance is money either raised on the back of existing properties or raised for the purpose of expenditure on properties. Whether the property is owner-occupied or an investment property may alter the criteria for the raising and application of the funds but the fundamental concepts of finance may well be the same. For instance, funds could be raised by an organisation internally or externally but the criteria for the internal loan or transfer of funds may well need to match those in the market. Here, the presentation is mainly about the application of funds to residential property in the investment market and funds are considered to be private rather than public sector monies. This is to make the analysis simpler but, as has already been said, the principles and concepts of public finance could well be similar.

The investment market for residential property cannot be seen in isolation from other investment markets. The application of funds to residential property has to reflect competition from other forms of investment. The decision to invest in a particular area will be a comparison of return and security, for instance, and thus knowledge of alternative investments and knowledge of the application of finance to other investments could be very important. This can clearly be seen in the securitisation and unitisation of property, which involves the breaking down of property assets into tradable units as, for example, through the former Business Expansion Scheme (BES) (see p. 63). Another important point concerns the nature of the lender and the property to which finance is applied. At its simplest the financial arrangement may deal with an individual purchasing a single property with a single loan, but it is rarely this basic. Finance might, for example, be raised by corporate entities such as large residential landlords using existing property and other assets as collateral for the purchase of a portfolio of assets which may include property assets but not comprise them exclusively.

The structure of the investment market

There are three major areas of traditional investment opportunity (ignoring gold, commodities and works of art): fixed interest securities, company stocks and shares, and real property (including residential property). The stock exchange provides a market for listed shares and certain fixed interest securities such as those issued by the government, local authorities and public bodies. The market in rented residential property contrasts with that of company shares and other securities. The property market is fragmented and dispersed while that of shares and other securities is highly centralised. The London Stock Market is an example of this centralisation. The centralisation of markets assists the transferability of investments, as does the fact that stock and shares can be traded in small units. Compared with other traditional investment opportunities, residential property investment has the distinguishing features of being heterogeneous, being generally indivisible and having inherent problems of management. The problems of managing property assets may include collecting rents, dealing with repairs and renewals, and lease negotiation. These problems may mean that real property is likely to be an unattractive proposition for the small investor. A decentralised market, such as exists for property, will tend to have high costs of transfer of investments and also there will be an imperfect knowledge of transactions in the market.

These factors make property difficult to value. There is no centralised market price to rely on and the value may be too difficult to assess unless a comparable transaction has recently taken place. The problems of valuation relate to difficulties of trying to relate comparable transactions to properties being valued or even to trying to assess what transactions could be considered comparable. Because of the nature of the real property market, individual investors have tended to withdraw from the market. This is also due to the channelling of savings into collective organisations, such as pension funds and insurance companies, rather than individuals using their savings for direct investment.

Property company shares

Property company shares provide a medium for indirect investment in residential property which deals with some of the disadvantages of direct investment previously outlined. Equities are available in smaller units and can be easily traded. Property shares, specifically, have been viewed as an effective protection against inflation because of the durability of property. Shares of a property investment company most of whose revenue is derived from rental income also provide the investor with a high degree of income security. Thus property shares traditionally were seen to provide both an element of protection against the effects of inflation and greater security to the investor.

Methods of finance

Methods of finance used in the private rented housing market, and the rest of the property market, are outlined below. This overview looks at debt finance, joint venture arrangements and mezzanine finance, at comparison of loan terms, and finance for small companies.

Debt finance

Clearing banks and, to a lesser extent, merchant banks have been prepared to provide loans for property development. Generally the loans have been secured on collateral beyond the property to be developed (recourse finance). Interest is charged on a fixed or variable basis. Non-recourse finance is now increasingly popular. This can be defined as loans on property unsupported by outside collateral. Finance is available from banks for up to 70 per cent loan-to-value ratio without outside security being provided. Banks take first charge over the site and advance monies during the development phase. Interest rates are a margin over the London Inter Bank Offered Rate (LIBOR), say between 1 and 2 per cent, but this is reduced with pre-let or pre-sale.

Lenders generally require less outside collateral (recourse) when the developer undertakes to:

- pay cost overruns;
- inject equity stakes either up-front or side-by-side;
- pay interest after completion;
- complete within certain time limits;
- complete within certain costs.

Banks will normally allow interest to be capitalised during the construction period and developers generally include this as a construction cost, thus not affecting profits during the development process. Some of the main types of debt finance available are as follows.

Commercial mortgages

These are straightforward loans in which the interest rate is either paid currently or capitalised. The principal is either amortised over the term of the loan or repaid by a single payment at the end. The interest rate can be fixed or variable and other capital market instruments can be used, including caps (compensating for interest rate rises over a certain rate) and floors (preventing interest rate falls below a certain level). These instruments are used to minimise interest rate fluctuations, thereby reducing risk and obtaining finer pricing of the loans. The length of loans will vary, but the maximum for many banks is five years, although insurance companies and building societies may lend for up to twenty-five years or more.

Equity participation or convertible mortgages

This structure allows the lender to share in the uplift of the value of the property in exchange for a reduction in the interest rate payable. The mortgage loan outstanding can be converted into the ordinary shares of the company lending the money.

Mortgage debenture issue

This is a traditional method of raising corporate finance. The debenture issue is secured against the property or other assets and yields either a fixed or an index-linked return.

Multi-option facility

In this structure a group of banks agree to provide cash advances at a predetermined margin for a certain period.

Deep discounted bonds

Deep discounted bonds are a method of raising long-term finance with a low initial interest rate. Interest payments can be stepped to accord with rent review and there are also tax advantages. Bonds can be placed with institutional investors and can be very finely priced. It is anticipated that bond issues will become increasingly used to finance major projects to overcome the need to refinance a completion of the development.

Joint ventures

There are a number of different types of joint venture in the private rented sector. The concept involves the coming together of two or more parties to undertake a transaction. Joint ventures are a useful means of bringing together parties with different interests in order to complete deals.

The reasons for increased use of joint ventures in property development include:

1. Bank of England pressure to reduce the level of bank debt in the property sector;
2. increased risk in property development;
3. lack of equity in the property market, with property companies less able to raise new funds in the stock market;
4. demand from overseas investors, who have a preference for joint venture arrangements.

Forms of joint venture structures are:

1. limited liability companies;
2. partnership, where one party must have unlimited liability;
3. profit participation.

The decision on creating a joint venture will depend on the purposes of the venture:

1 whether there is one property or one of a number of schemes;
2 the tax situation;
3 stamp duty considerations.

On negotiation of a joint venture agreement, important points to consider are:

1 the level of funding to be provided;
2 the development period/time;
3 who will control the decision-making process;
4 how the profit is going to be distributed;
5 what the provisions are for dissolution in the event of failure;
6 how disputes are to be settled.

Mezzanine finance

There is often a gap between the costs of development and available limited recourse loans and this can be filled by a mezzanine loan. The amount of mezzanine finance varies, but can take debt up to 95 per cent of cost. Mezzanine funders require rates of return between 30 and 40 per cent, secured on a subordinated loan (a loan raised to help repay an earlier loan) with a share of profit, normally side-by-side (simultaneously), with a cap at an agreed level. Should the project run into problems, the priority return that the mezzanine funder seeks can quickly diminish any profit. Interest payable on the mezzanine finance can be capitalised, producing cash-flow benefits.

Comparison of loan terms

There are four main areas to be considered prior to entering into a loan facility in order to ensure that a suitable structure related to the lender's requirements can be arranged. These are cost, flexibility, risk and accounting presentation.

Cost

The best rates are available for the best-quality covenants. Where the lending institution takes more risk, a higher return is required. For a non-recourse transaction the interest rate margin above LIBOR required by the lender is increased.

Flexibility

The greater the level of security and recourse, the less likely there is to be any restriction on management control. Flexibility depends on interest rate structure.

Risk

The greater the lender's risk, the greater the cost to the borrower.

Accounting

A number of creative packages are available which can remove debt from the balance sheet.

Finance for small companies

The finance for small property companies is extremely important and is a specialist area. Four stages in the financing needs of small companies can be distinguished.

Venture capital

This is the funds necessary to meet the start-up costs of a new company.

Development capital

This capital is needed to finance expansion once the initial phase of establishing the company has been completed.

Increasing the number of shareholders

This finance is needed when it becomes necessary to widen the equity base of the company. In such situations the loan capital may have reached an undesirable level for the stability of the company, and thus the gearing (ratio of loan capital to share capital) is too high.

Acquiring a stock exchange listing

Smaller firms suffer a number of disadvantages in the market. Small borrowers generally have to pay higher rates of interest. In addition, many small firms lack knowledge about potential sources of finance. Thus, because of problems related to credit and interest rates, the general condition of the economy will affect smaller firms more acutely than larger ones. The other problem that small firms face in raising finance is that they are very wary of entering into debt arrangements. There are a number of reasons for this, including fear of losing control over the company as well as fear of not being able to meet the conditions related to the loans. Possible sources of funding for small firms include merchant banks, which provide medium- and long-term loans and equity interest; specialist funders such as Investors in Industry, which provide loans and equity interest for venture and development capital; and finance houses and leasing companies, which provide finance for equipment and vehicle financing. There are also monies available from factoring houses, which provide cash for debts, and the clearing banks, which provide overdraft facilities. Insurance companies provide some financing for property and, finally, there is government funding through a number of intermediary bodies for financing technological innovation.

Institutional sources of finance

The main lenders in the market are:

- High street/clearing banks;
- Foreign banks;
- Building societies;
- Merchant banks;
- Insurance companies;
- Finance houses.

High street/clearing banks

The big four clearing banks are Lloyds, Barclays, National Westminster and Midland. These are probably the first port of call for people looking for loans, especially if they have established relationships as account-holders. However, these banks are conservative and may view new transactions related to property in a less than enthusiastic way. Smaller high street banks such as the Royal Bank of Scotland, the Bank of Scotland, Yorkshire Bank and the Clydesdale Bank may be more useful potential funders. The smaller banks are likely to have less of a bad debt problem and may want to increase market share. The larger banks are burdened at the moment by over-exposure to the property sector, and consequently the emphasis at present is on debt repayment rather than new lending.

Foreign banks

The foreign banks tend to be more aggressive sources of property finance, or were in the early 1990s, but by 1993 they were showing less interest. They are useful sources of funding particularly for quality and corporate transactions. The collapse of BCCI made borrowers wary of dealing with foreign banks. The collapse of a bank half-way through a development may mean that it could take years to unwind the legal problems and thus would put the borrower's own financial position at risk.

Building societies

The building societies fared badly in the 1990–95 slump in the property market. In the late 1980s, their inexperience and desire for a market share in rented property lending led to substantial bad debt. They are now putting their respective houses in order with rationalisation and more qualified staff and are likely to be an important source of investment finance in the future.

Merchant banks

Merchant banks rarely lend their own money but act as advisers and concentrate especially on large corporate transactions. They are unlikely to be interested in ordinary debt transactions because there would be little opportunity to use their expertise and add value to such a transaction.

Insurance companies

The insurance companies are limited providers of funds. They do, however, offer the attraction of long-term, fixed-rate funds priced over gilts, which can be useful in certain transactions.

Finance houses

In essence, the finance houses have been the principal providers of funding to the secondary leisure and retailing markets, providing finance for the purchase of freehold shops, pubs, restaurants and hotels. Their small trader exposure has made them particularly vulnerable to the latest recession, which has resulted in most of them leaving the market.

Lending criteria

The cost and availability of lending is a function of the value of any particular project and the amount of cost to be financed. Important attributes of the residential property are the design, mix, location and likely demand. The letting conditions are important, as are whether or not the investment is pre-let or speculative. The quality of the tenant, and who will be providing the cash-flow to the investment, will also be important. Other important criteria are the track record of the landlord and the strength of security. Finally, the duration of the loan will be important, as will the details of repayment; for instance, the anticipated regularity of repayments and the size/amount of repayments prior to redemption.

Most lenders look for the same aspects in a lending proposal, which in simplified terms are the 'four Cs':

- Character
- Cash stake
- Capability
- Collateral.

Character

This relates to the trading history or development experience of a borrower. In respect of a property developer or an investment company, the lender will want to know whether the borrower has the experience to complete the development, manage the investment or run the business, as appropriate. The lender will also be interested in whether the client is respectable and trustworthy. The lender may also cynically wonder why the borrower's own high street bank will not lend the money.

Cash stake

This relates to how much equity (the borrower's own money) is going into the transaction. In addition, the bank will want to know where the equity has come from:

- Is it lent from someone else?
- Is it from other profitable activities?
- Is it simply a surplus which has arisen on the revaluation of property?
- Is it already pledged as security?
- Is it legally acquired money – not laundered money?

Capability

Does the borrower have the capability to service the loan; that is, to pay the interest when it arises and the capital as and when repayment is required? Does the borrower have accounts or a business plan to show his/her present financial position and any estimate of future cash flow?

Collateral

The lender will want to know what security will be offered for the loan, its value and its saleability. The lender will want to know who valued the security and on what basis. The lender will need to assess the ratio of the loan to the value of that security. Finally, will personal guarantees be given by the borrower?

Government assistance

Government assistance is available to residential developers of sites in certain locations under certain criteria. The loans available include money for City Grants for projects in Urban Development Corporation (UDC) areas, set up under the Local Government Planning and Land Act 1980. Funding is also provided through the Simplified Planning Zones under the Housing and Planning Act 1986.

The City Grant came into operation on 3 May 1988. This replaced the Urban Development Grant and the Urban Regeneration Grant. To quality for a City Grant, it is necessary for the developer to prove that development project cannot proceed without such a grant and that the project will benefit a run-down city area. Development projects must be worth over £200,000 when completed to quality for grant aid. Apart from being available in the UDC areas, City Grants can be applied in the fifty-seven inner city regions which have been designated as target areas. In addition, to obtain grant a funding, there should also be substantial private sector funding for the development, normally at a leverage ratio (of public sector to private sector funding) of about 1:4. The amount of grant obtainable depends upon the difference between costs and value, with cost exceeding value in order to qualify.

In March 1993 it was announced that the Urban Programme of government funding would be phased out. The Urban Programme formed a major part of the government's urban renewal expenditure along with City Grants, Derelict Land Grants, Urban Development Corporations, City Action Teams and City Challenge (see Balchin *et al.*, 1995). Under the Leasehold Reform, Housing and Urban Development Act 1993 the government established English Partnerships (EPs) to take over policy decision-making, to complement regional policy, administer the English Industrial Estates Corporation, buy and develop inner city sites, and assume the responsibility of the Urban Programme to award City Grants and Derelict Land Grants and to administer City Challenge. A Single Regeneration Budget was created in 1994 to include expenditure on Urban Development Corporations, English Partnerships' activities, Housing Action Trusts and other programmes. The total planned expenditure for 1995/6 was £1,332 million and for 1996/7 £1,324 million (Balchin *et al.*, 1995: 273).

CASE STUDY 3.2

EARNINGS, RENTS AND HOUSE PRICES

In each of the housing sectors there are problems of affordability. In the private rented sector, as many as 40 per cent of households were non-earning in the early 1990s, only one-in-five contained more than one earner and as many as one-half of all households in the sector had disposable incomes of less than £8,500 per annum (Office of Population Statistics and Surveys, 1994), yet under the Housing Act of 1988, fair rent tenants were being increasingly brought into the assured tenancy or assured shorthold system (with minimum lettings of only six months), and tenants suspecting that their rents might be above the market level lost their right to refer the matter to the Rent Assessment Committee (Department of the Environment, 1995a). As Table 3.26 illustrates, private sector rents (as a percentage of earnings) increased substantially throughout the 1980s and early 1990s.

As is shown in Table 3.27, in the local authority sector, rents as a proportion of earnings have risen rapidly to reduce the scale of the Exchequer subsidies and to eliminate subsidisation from rates –

Table 3.26 Rents and earnings, private sector, 1980–95

		Rents as % of earnings[1]		
		1980	*1987*	*1995*
England	Fair rents	9.7	11.1	15.5
	Market rents	—	—	22.0
Scotland	Fair rents	6.5[2]	9.7	10.8[3]
Wales	Fair rents	8.4[2]	9.7	12.2
	Market rents	—	—	20.0

Source: Wilcox (1996)
Notes: 1 Average manual male earnings
2 1981
3 1993

Table 3.28 Rents and earnings, housing associations, 1980–95

		Rents as % of earnings[1]		
		1980	*1987*	*1995*
England	Fair rents	11.2	12.3	15.0
	Assured rents	—	—	16.5
Scotland	Fair rents	7.5[2]	11.4	9.8
	Assured rents	—	—	10.6
Wales	Fair rents	11.2[2]	12.3	14.0
	Assured rents	—	—	14.8

Source: Wilcox (1996)
Notes: 1 Average manual male earnings
2 1981

despite 60 per cent of local authority tenants having disposable incomes of less than £8,000 per annum (Office of Population Censuses and Surveys, 1994). The privatisation of much of the local authority housing stock has also inflated rents, since the loss of stock (particularly in older surplus-yielding estates) has reduced the extent to which newer high-cost council housing can be cross-subsidised through the process of rent pooling. The government's recent proposals to reduce local authority assistance to homeless households by greater use of temporary private accommodation (Department of the Environment, 1995b) will also put upward pressure on rents in areas of stress. It is not therefore surprising that rents as a proportion of earnings increased markedly in the 1980s and 1990–94.

Housing association tenants similarly experienced an increase in rents as a proportion of earnings (Table 3.28). Since, under the Housing Act of 1988, an increasing proportion of capital expenditure in this sector is funded by private financial institutions (56 per cent by 1997), rents have risen rapidly to ensure a competitive return on investment.

Table 3.27 Rents and earnings, local authority sector, 1980–95

	Rents as % of earnings[1]		
	1980	*1987*	*1995*
England	6.9	9.2	13.1
Scotland	6.1[2]	8.1	10.1
Wales	9.5[2]	9.9	12.5

Source: Wilcox (1996)
Notes: 1 Average manual male earnings
2 1981

It is clear from the above tables that whereas private sector rents were the least affordable, local authority rents would have been the most affordable. It must be taken into account, however, that the tables show only rents as a proportion of average male manual earnings, and therefore affordability would have been significantly less in both sectors in respect of households with lower-paid jobs or none.

With regard to owner-occupation, the degree of affordability is normally indicated by the house price/earnings ratio and the mortgage interest rate. When ratios and/or mortgage interest rates are high, there is a relatively low level of affordability, but when ratios and/or interest rates are low, affordability is relatively high. But ratios and interest rates do not indicate the varying degree of affordability among different sorts of households. When ratios were high, for example 4.95:1 in 1973 or 4.36:1 in 1989 (and in the latter year when the mortgage interest rate was over 13 per cent), households, with above-average incomes were willing to afford a higher level of housing expenditure in order to trade up for speculative motives, believing that house prices would continue to rise in the foreseeable future. But when ratios were comparatively low (for example in 1992 and 1993), households with below-average incomes (including many first-time buyers) and those with insecure incomes would have found house prices unaffordable, even though prices might have fallen dramatically from previous peaks in real and even in cash terms. National house price/earnings ratios and mortgage interest rates also fail to show substantial regional variations in affordability.

Research undertaken by Bramley (1990) suggested that during the boom conditions of 1989, affordability among first-time buyers under thirty years old was even less than hitherto acknowledged. He showed that, nationally, only 22 per cent of potential new home-owners earned enough to buy a new three-bedroom house, while in London the proportion was as low as 10 per cent (compared with 38 per cent in the North and North West). Bramley (1991) subsequently showed that, despite the onset of the slump in house prices, affordability among young first-time buyers was even less than during the boom since mortgage interest rates remained at a comparatively high level (at over 11.5 per cent throughout 1991).

Notwithstanding the above problems of affordability, 400,000 households became owner-occupiers for the first time in 1991, and over 350,000 in 1992 (Boleat, 1992). The Conservative government was undoubtedly achieving its aim of increasing property ownership. By 1993, the house price/earnings ratio and mortgage interest rates had fallen to their lowest levels since the 1950s and 1960s and houses were thus more affordable than at any time over the past three or four decades. Towards the end of 1993, the *TSB Affordability Index* (Trustee Savings Bank, 1993) showed that typical first-time buyers needed to spend only 26 per cent of their take-home pay on mortgage payments as against 67 per cent in the first quarter of 1990, and even in the most expensive region, the South East, the proportion was barely higher at 30 per cent. Only unemployment – or the fear of unemployment – and a general lack of confidence in the market prevented a noticeable increase in house-buying and an upturn in house prices in 1993 and 1994.

By the end of 1996, it was evident, however, that house prices were rising by at least 7.5 per cent per annum nation-wide, and it was predicted that in the South East house price inflation would be at a rate of 10 per cent in 1997, but it was uncertain whether or not the boom conditions of the late 1980s would return. Affordability was still constrained by the tightening up of building society and bank lending, by many first-time buyers still having negative equity (a mortgage debt greater than the value of the property), and by lower-income households failing to gain from economic growth in the 1980s and 1990s, and, in the near future, affordability might be constrained by single-person households making up a higher proportion of total households – with possibly a greater preference for renting rather than buying (Bull, 1996).

CONCLUSIONS

With the return of a Conservative government in 1979 under the premiership of Margaret Thatcher, the uneasy consensus of the previous half century or more broke down. Thatcherism, involving as far as possible a lurch back to the free market, dominated housing policy for a generation – in latter years under the nostrums of the Major administrations, 1991–7. A neo-liberal welfare regime was created, whereby housing investment in the local authority sector plummeted, rents were permitted to rise to market levels within both the private and social rented sectors, local authority and housing association dwellings were sold off, rehabilitation was forced to rely more and more on private means, housing association investment became increasingly dependent upon bank and building society funding, and subsidisation of the owner-occupied sector was gradually reduced, with home ownership becoming dependent more and more on private funding.

In general, subsidies were either dramatically

reduced (particularly in respect of house-building) or increasingly available only to the poor on a means-tested basis – state funding providing little more than a safety net for those most in need, a welfare service for the disadvantaged. But without an adequate and affordable supply of social housing, many low-income families were forced into short-term private rented accommodation or, as marginal buyers, into owner-occupation – with the more seriously disadvantaged becoming homeless.

This review of housing markets and finance provides an essential background to an appreciation of subsequent chapters, particularly Chapter 4 on equal opportunities, Chapter 5 on planning and housing development, and Chapter 7 on environmental health and housing.

QUESTIONS FOR DISCUSSION

1 Examine the causes of the house price and house-building cycles, and discuss the impact of these cycles upon the macro-economy.

2 Recently there have been calls to expand the private rented sector. Examine some of the ways this sector could be expanded, and whether such expansion would be economically beneficial.

3 Examine the economic advantages and disadvantages of the privatisation of council housing.

4 Within the context of affordable housing, examine some of the economic consequences of local authorities changing their role from that of 'provider' to 'enabler'.

5 'Housing needs will only be satisfied if there is a shift of emphasis from subsidising households to subsidising "bricks and mortar".' Discuss.

NOTE

1 In 1980, the Housing Corporation's responsibilities in both Scotland and Wales ceased and were taken over respectively by Scottish Homes and Tai Cymru. In Northern Ireland, registered associations are directly controlled by the Department of the Environment for Northern Ireland.

RECOMMENDED READING

Balchin, P. (1995) *Housing Policy: An Introduction*, 3rd edn, Routledge, London.

—— (ed.) (1996) *Housing Policy in Europe,* Routledge, London.

Maclennan, D. and Gibb, K. (eds) (1993) *Housing Finance and Subsidies in Britain*, Joseph Rowntree Foundation, York.

Malpass, P. and Means, R. (1993) *Implementing Housing Policy*, Open University Press, Buckingham.

Malpass, P. and Murie, A. (1995) *Housing Policy and Practice*, 4th edn, Macmillan, London.

REFERENCES

Ambrose, P. (1986) *Whatever Happened to Planning?*, Methuen, London.

Aughton, H. and Malpass, P. (1991) *Housing Finance: A Basic Guide*, Shelter Publications, London.

Balchin, P. (1995) *Housing Policy: An Introduction*, Routledge, London.

—— (ed.) (1996) *Housing Policy in Europe*, Routledge, London.

Balchin, P., Bull, G. and Kieve, J. (1995) *Urban Land Economics and Public Policy*, 5th edn, Macmillan, Basingstoke.

Ball, M. (1983) *Housing Policy and Economic Power*, Methuen, London.

Bank of England (1991) *Quarterly Bulletin 1*, Bank of England, London.

—— (1992) *Quarterly Bulletin 3*, Bank of England, London.

Barlow, K. and Chambers, D. (1992) 'Planning agreements and social housing quotas', *Town and Country Planning*, 61 (May).

Bassett, K. and Short, J. (1980) *Housing and Residential Structure: Alternative Approaches*, Routledge & Kegan Paul, London.

Boleat, M. (1992) 'Britons still dream of their own home', *Observer*, 13 September.

Bramley, G. (1990) *Bridging the Affordability Gap in 1990*, Association of District Councils and House Builders' Federation, London.

—— (1991) *Bridging the Affordability Gap in 1991*, Association of District Councils and House Builders' Federation, London.

—— (1993) 'The enabling role for local housing authorities: a preliminary evaluation', in Malpass, P. and Means, R. (eds) *Implementing Housing Policy*, Open University Press, Buckingham.

Bull, G. (1996) 'Implications of the changing social/private housing mix on housing provision, affordability and social exclusion', paper presented at the European Network for Housing Research Conference, *Housing and European Integration*, Helsingar, 26–31 August.

Department of the Environment (1977) Green Paper, *Housing Policy: A Consultative Document*, Cmnd 6851, HMSO, London.

—— (1985) Green Paper, *Housing Improvements: A New Approach*, Cm 9513, HMSO, London.
—— (1987) White Paper, *Housing: The Government's Proposals*, Cm 214, HMSO, London.
—— (1988) *English House Condition Survey 1986*, HMSO, London.
—— (1989) *Local Authorities' Housing Role; 1989 HIP Round*, HMSO, London.
—— (1991) *Circular 7/91*, HMSO, London.
—— (1993a) *English House Condition Survey 1991*, HMSO, London.
—— (1993b) *Planning Policy Guidance Note 3*, HMSO, London.
—— (1995a) *Annual Report*, HMSO, London.
—— (1995b) White Paper, *Our Future Homes*, Cm 2901, HMSO, London.
—— (1996) Green Paper, *Household Growth: Where Shall We Live?*, Cm 3471, HMSO, London.
Durham, M. (1990) 'Families growing rich in inheritance bonanza', *Sunday Times*, 29 July.
Edwards, S. (1992) 'A long term risk for all of society', *Observer*, 13 September.
Evans, A.W. (1987) *House Prices and Land Prices in the South East: A Review*, House Builders' Federation, London.
Frankena, M. (1975) 'Alternative models of rent control', *Urban Studies*, 12.
Gibb and Munro (1991) *Housing Finance in the UK: An Introduction*, Macmillan, Basingstoke.
Hall, P., Gradey, H., Drewitt, R. and Thomas, R. (1973) *The Containment of Urban England*, vol. 2, Allen & Unwin, London.
Hamnett, C. (1988) 'Housing the new rich', *New Society*, 22 April.
Harloe, M., Issarcharoff, R. and Minns, R. (1974) *The Organisation of Housing: Public and Private Enterprise in London*, Heinemann, London.
Holmans, A. (1995) 'Housing demand and need in England 1991–2001', *Housing Research*, 157, Joseph Rowntree Foundation, York.
Joseph Rowntree Foundation (1991) *Inquiry into British Housing: Second Report*, JRF, York.
Karn, V. (1993) 'Remodelling a HAT: the implementation of the Housing Action Trust legislation 1987–92', in Malpass, P. and Means, R. (eds) *Implementing Housing Policy*, Open University Press, Buckingham.
Maclennan, D., Gibb, K. and More, A. (1991) *Fairer Subsidies, Faster Growth: Housing, Government and the Economy*, Joseph Rowntree Foundation, York.
Merrett, S. and Gray, F. (1982) *Owner-occupation in Britain*, Routledge & Kegan Paul, London.
Moorhouse, J.C. (1972) 'Optimal housing maintenance under rent control', *Southern Economic Journal*, 39.
Muellbauer, J. (1986) 'How house prices fuel wage increases', *Financial Times*, 23 October.
Northern Ireland Housing Executive (1993) *Northern Ireland House Condition Survey 1991*, NIHE.
Office of Population Censuses and Surveys (1994) *Family Expenditure Survey*, HMSO, London.
Ricardo, D. (1971) *Political Economy and Taxation*, Penguin, Harmondsworth. (First published 1817.)
Scottish Homes (1993) *Scottish House Condition Survey 1991*, Scottish Homes, Edinburgh.
Shelter (1982) *Housing and the Economy: A Priority for Reform*, Shelter, London.
Town and Country Planning Association (1996) *The People: Where Will They Go?*, TCPA, London.
Trustee Savings Bank (1993) *TSB Affordability Index*, 4, TSB, London.
Welsh Office (1988) *Welsh House Condition Survey 1989*, Welsh Office, Cardiff.
Wilcox, S. (1995) *Housing Finance Review 1995/96*, Joseph Rowntree Foundation, York.
—— (1996) *Housing Review 1996/97*, Joseph Rowntree Foundation, York.
Wolmar, C. (1985) 'Shrinking the Green Belt', *New Society*, 21 June.

4

EQUAL OPPORTUNITIES AND HOUSING

Maureen Rhoden

The elderly do not have any legislation to protect them but have to endure the disadvantages which can restrict their access to good-quality services such as housing and health. Since the introduction of the Sex Discrimination Act 1975 and the Race Relations Act 1976, there has been a range of studies (see Commission for Racial Equality, 1984, 1988, 1989a, 1989b; Greater London Council, 1986; Rao, 1990; and Sexty, 1990) which have served to highlight the fact that women and black and ethnic minority people still encounter discrimination in terms of access to areas of provision such as housing. There is also evidence to suggest that there is a strong link between ill-health and poor housing.

The purpose of this chapter, therefore, is to focus on how the elderly, single women, black and ethnic minority households and the sick are often disadvantaged in housing markets in terms of choice, access and affordability, and to review public policy response. Specifically, the chapter examines:

- The elderly in relation to:
  - Household formation.
  - Housing tenure and housing condition.
  - Household incomes and wealth.
  - Staying put or moving into alternative accommodation.
- Women in relation to:
  - Home ownership.
  - The family home after divorce or separation.
- Black and ethnic minority households in relation to:
  - Home ownership.
  - Housing association tenure.
  - The local authority rented sector.
  - Racism.
- Health and housing:
  - Health and housing tenure.
  - Health and poor living conditions.

HOUSING AND ELDERLY PEOPLE

The housing circumstances, incomes and opportunities available to elderly people have resulted in the aged becoming increasingly marginalised. During the 1980s and 1990s, the living standards of elderly people declined significantly and for the majority the goal of real housing choice and income security remains elusive. Many find themselves dependent on state benefits and many are severely disadvantaged (Sykes, 1994).

Household formation

The future provision of housing is affected by the major changes which are taking place in the demographic and family structure of this sector of the population. There is a growing number of elderly households and there is also a growing trend for elderly people to live alone. It has been estimated by the Institute of Actuaries (1994) that by the year 2031 there will be 6.8 million people aged over sixty in Britain, 3.2 million of whom will require regular or continuous care.

The number of single elderly person households

in Britain is expected to grow to 620,000 by 2001. Since the turn of the century the number of elderly people living alone has increased from 10 per cent in 1901 to 27 per cent of people aged 65–74 and 50 per cent of people aged over seventy-four (Table 4.1). Elderly people have since 1945 preferred to live alone or with their spouse rather than with friends and relatives as in the past. Balchin (1995) states that there are two reasons for this trend: the increase in the rate of divorce in recent years, and the low remarriage rate of divorced people who are elderly.

Housing tenure and housing condition

The housing tenure of the elderly is roughly divided between half residing in rented accommodation, of whom the largest number are aged over eighty, and half in the owner-occupied sector. Elderly people find themselves to be in poorer housing generally than other age groups and this pattern is common across all tenures. Sykes (1994) states that elderly people living alone are more likely to live in poor-quality housing. In addition, the length of residency is usually an indicator of poor housing and so those people who have occupied a property for more than twenty years are more likely to be elderly and are more likely to be residing in poor housing conditions.

Although there has been some significant improvement in the provision of basic amenities for elderly people, there are still significant numbers living in poor conditions. If we look more closely at the over-seventy-five age group, there are 112,000 homes lacking basic amenities, 116,000 homes which are unfit for human habitation and 275,000 properties in poor repair. It is therefore estimated that over 1.3 million elderly people are living in poor housing conditions (Balchin, 1995).

Table 4.1 People living alone by age and sex (% figures)

	Men	*Women*	*Total*
16–24	4	4	4
25–44	8	5	6
45–64	10	13	11
65–74	17	36	27
75 or over	29	61	50

Source: Gilroy (1994)

Household incomes and wealth

During the 1980s, the government attempted to portray the elderly population as 'Woopies', well-off older persons with greater spending power, greater housing choice and generally greater opportunity during their retirement. However, the debate on the new prosperity of elderly people does not reflect the true picture of this sector of the population.

Bull and Poole (1989) and Bosanquet *et al.* (1989) identified three groups of elderly people (Table 4.2). The first group represent 20 per cent of the elderly population and are seen as the 'Woopies' mentioned earlier. They have adequate financial resources to cover them as their health declines and they become less mobile. They have the ability to obtain the housing which best suits their changing needs and can buy in services which they may require. This group are usually people who were in well-paid employment and so enjoy enhanced pensions, investments, high-value properties and significant savings.

The second group represents approximately 40 per cent of the elderly population and are seen as 'not rich, not poor'. This group were on middle incomes during their employment and have usually paid off their mortgages. They are often referred to as asset rich but income poor. Many will have savings which may be just above the benefit thresholds, thus resulting in their not claiming any state support. While this group of elderly people may be better off materially, they may not have sufficient funds to make real choices in the housing options.

The third group are the large number of elderly people who are impoverished and extremely disadvantaged. They are to be found on low incomes which have to be supplemented by state benefits.

Table 4.2 Typology of not rich, not poor

	Well-off 20%	*Not rich, not poor* 40%	*Poor* 40%
Income	High	Middle to Low	Low
Savings/investments	Substantial	Limited	Limited
Tenure	Owner-occupiers	Mainly owner-occupiers	Local authority or private tenants
State benefits	None	Partially dependent	Highly dependent
Occupational pensions	High levels	Small amounts	Low

Source: Bull and Poole (1989)

The elderly people in this group are often found in rented accommodation in either the social rented or the private sector. They are more likely to suffer from poor housing conditions and live at poverty levels.

Bull and Poole (1989) stated that approximately 80 per cent of the elderly population have limited means and so are restricted in the choice available to them in terms of housing options.

Sykes (1994) states that elderly people derive their income from four main areas: the retirement pension and other related state benefits; occupational pensions; savings and investments; and wages. The income levels of elderly people are likely to change as the income of a surviving partner after the death of a spouse is likely to fall while expenditure remains stable. Furthermore, expenditure increases with age in terms of the additional care which is often required, while income levels are often reduced at this time.

The income level of the elderly population has gradually reduced since the 1970s. During the 1980s, pension increases were linked to the retail price index rather than being linked to average increase in incomes as in the past. Whereas the average weekly earnings of the population as a whole amounted to £343 (gross) in 1993 (*Employment Gazette*, 1993), the basic weekly pension in 1993/4 was £57.60 for a single person and £92.10 for a married couple.

Furthermore, in households whose head is aged sixty or more, a disproportionate amount of income is spent weekly on housing, fuel and food. Pensioners living alone spent 62 per cent of their income on these items, compared to 39 per cent for all households (Central Statistical Office, 1993). A National Opinion Poll survey (British Gas, 1991) also discovered that 21 per cent of retired people who owned their own homes had incomes so low that they often had difficulty in paying for essential household items mentioned above. Elderly people are therefore at a disadvantage when compared to those members of the population in employment, and so many are income poor.

There has been much discussion about the growing affluence of elderly people as a direct result of increasing levels of occupational pensions. However, manual workers are less likely to have an occupational pension and if they do receive an occupational income this is likely to be at a low level. Those elderly people who earned high wages during their period of employment generally receive a high level of occupational pension. However, in general the level of savings which elderly people have is comparatively low.

Staying put or moving into accommodation for the elderly?

Sykes (1994) indicates that approximately 90 per cent of elderly people reside in general housing, with the remainder living in other types of housing such as sheltered housing, residential care and nursing homes. Many elderly people prefer to remain where they live rather than move elsewhere as they become older. Those who decide to move have choices to make between the different housing options, which are dependent on a number of factors, including their existing housing tenure, their income and savings and the family support which is available to them (see Table 4.3).

However, although it could be argued that, in any case, wherever possible elderly people prefer to remain in their own accommodation, cuts in public expenditure have led the government to support schemes which could help retired people to stay put. The public sector found that with capital expenditure being cut back by central government during the 1980s, it was cheaper to shift to the option of adapting, modernising and repairing an elderly person's existing property than try to find them an alternative. Tinker (1984) revealed that whereas the public sector spent on average £5,000 per annum to keep an elderly person in sheltered accommodation, the comparative cost of staying-put schemes could be as low as £3,000 per annum.

Both the Griffiths Report (1988) and government policy as set out in *Caring for People: Community Care in the Next Decade and Beyond* (Secretary of State, 1989) actually encouraged elderly people to remain in their own homes for as long as possible. In addition, the subsequent National Health Service and Community Care Act 1990 gave local authorities the responsibility for local community care for elderly people. In fact Mackintosh and Leather (1992) found through their research that these schemes were able to help elderly owners to remain for longer in their own properties.

Elderly people who are owner-occupiers are seen as having the widest choice from the housing options which are available. However, their choice is also restricted by the value of their property and, given this, the elderly owner-occupier may have to trade down to a smaller property, and so release some surplus cash. They can then decide whether to move into general housing or specialist housing for elderly people. Elderly people in the rented sector are limited in their ability to choose alternative housing options due to the factors referred to above and their move is generally to specialist rented sheltered accommodation.

The 1970s saw the introduction of housing built specifically for the elderly. In previous years the only option available for the elderly person looking for suitable housing to meet their growing needs was a residential home. The general private housing market and social rented sector was able to cater for

Table 4.3 Housing options for the elderly

Existing housing	*Option for future*
Local authority tenant	Sheltered housing to rent (local authority; housing association) Other, non-specialist rented accommodation
Private sector tenant	Sheltered housing to rent (local authority; housing association) Other non-specialist rented accommodation
Owner-occupation	Sheltered housing to rent or buy Smaller owner-occupied property Retirement housing to buy Staying put assistance

Source: Rolfe, S. *et al.* (1993) *Available Options*, Anchor Housing Trust, Oxford

those retired persons who required low-dependency housing such as ground-floor flats near to shops and services. There was also a need to develop specific forms of housing more suited to those requiring medium-dependency housing, such as sheltered housing, and high-dependency housing, such as very sheltered housing and residential and nursing care (Table 4.4).

Sheltered housing is particularly popular with elderly people who have a medium dependency level. Sheltered housing generally refers to accommodation with a resident warden, alarm system and some communal facilities such as a residents' lounge, guest room and laundry (Age Concern, 1990), although some sheltered housing only has an alarm system.

The Ministry of Housing and Local Government *Circular 82/69* (1969) introduced certain categories for different types of sheltered accommodation. The Category One sheltered accommodation had communal facilities and wardens as options and the bungalows or flats usually without a lift were intended for low-dependency residents. The Category Two schemes are flats under one roof with communal facilities, resident wardens and lifts and are intended for medium-dependent residents.

During the 1980s, very sheltered housing (Category 2½) schemes were introduced for those elderly people with high dependency. They provide all the services of a Category 2 scheme but also include the provision of meals and home help domestic assistance.

Henwood (1992) indicates that in the ten years between 1976 and 1986 the number of residential places in the private sector grew by 260 per cent. During the 1980s there was a rapid increase in private sheltered accommodation built by developers such as McCarthy and Stone. It is estimated that by the end of the period 90,000 private sheltered properties were available, a further growth of 400 per cent. In addition to the private developers building during this period, many local authorities moved to transfer their homes into the private sector. The availability of residential homes which are run by local authorities varies. Furthermore, the growing use of local policies and eligibility criteria has also affected the access to and provision of care which is available for elderly people throughout the country. However, as the property market slump began to take effect at the end of the 1980s private developers were beginning to find it difficult to sell these properties. They were predominantly bought by purchasers aged seventy-six or more and 80 per cent were women who had traded down on the death of their husbands. Concerns have been voiced about the high costs of this private sheltered accommodation, as much was seen as overpriced with high service charges. Also, many leases in this sector contained a clause which compelled the resident to vacate the property at the onset of a serious disability.

Table 4.4 The tenure of sheltered housing, England, 1989

Local authority	303,061
Housing association	120,911
Other social sector	2,905
Private	38,798
Total	465,675

Source: Department of the Environment, *Housing Investment Programme Statistics*

During the 1980s, a range of schemes intended to encourage elderly people to remain in their existing accommodation were introduced. These included the use of mobile local authority wardens, the distribution of dispersed alarm systems, the provision of home equity release schemes by insurance companies, and the Care and Repair and Staying-put schemes. Under the Care and Repair and Staying-put schemes, the whole repair, improvement or adaptation process was undertaken on behalf of the elderly owner-occupier or private tenant by the local authority. An aspect of the schemes was that not only the housing needs of the occupier were addressed but also any other problems, such as income maximisation. The funding for the schemes is obtained on the owner's behalf and the building work is supervised on their behalf by the local authority.

Sykes (1994) states that for elderly people there

are two important factors. First, any elderly person who moves, whether wealthy or impoverished, is looking to improve the quality of their life and their property. Second, a move is likely to reduce their housing options in the future as they would generally be trading down or moving into specialist housing for elderly people.

Halpern *et al.* (1996) state that government policy on long-term care of elderly people has not developed in a co-ordinated manner and as a result the care available to elderly people is not equitable. Thus, some elderly people may still be able to acquire NHS facilities, such as community care beds or stay in community care hospitals, and while these do vary in the quality of service which is available the elderly patients are not charged for the service provided. However, there are also elderly people with the same medical need who may find that they do have to pay for the service they receive if they are residing in a private nursing home. The security of tenure for residents in such accommodation is not guaranteed and if they reach a stage where they are no longer able to afford the fees or become too frail, then they can be asked to leave. Those elderly people who may prefer to reside in a local authority funded home may find it difficult to locate and gain access to such homes as there are often waiting lists in operation. Furthermore, Henwood (1992) states that they may receive a poorer quality of service compared with those who are self-financing.

According to Laing (1993), the National Assistance Act 1948 stated that the provision for long-term care of elderly people who were 'sick' or 'infirm' was the responsibility of the NHS, which was free to the user, while the care of those who were 'frail' or 'old' should be dealt with through the local authorities and was means tested. However, the boundaries between what is free to the user and that which is not has been blurred in recent years. Partly as a result of this and coupled with the financial restraints endured by many hospitals, hospitals are tending to discharge their elderly patients earlier to nursing and residential homes than occurred in the past.

Until 1993 residential and nursing homes were completely funded by social security subsidies. The introduction of community care has, however, resulted in the transfer of financial responsibility to cash-limited local authorities. As a consequence, all long-term care which is provided by the local authority is means tested.

Despite the number of residential homes which are available, a number of elderly people prefer to remain at home and are cared for by relatives. Research has found that of the estimated 7 million carers in Britain caring for an elderly person many are elderly themselves. The NHS and Community Care Act 1990 entitles carers to have their needs assessed. Those who reside with the elderly person receive less assistance, and assistance is concentrated on the elderly people who live alone (Laing, 1993).

WOMEN AND HOUSING

The 1991 census revealed that in England and Wales there were 19,877,272 households of which almost 31 per cent were headed by women (Table 4.5). Gilroy (1994) states that the normal practice is to identify the male as the head of household where there are a man and woman residing together in the same property. Only when the woman lives alone is she identified as head of household. The 1991 census attempted to overcome this by suggesting that the person responsible for the budget within the home should be identified as the head of household, but to

Table 4.5 Tenure of male and female heads of households (as % of each tenure category)

	Women %	*Men* %
Owner-occupation	23.81	76.19
Private rented	30.37	69.63
Housing association	53.78	46.22
Local authority	47.46	52.54
All tenures	30.54	69.46

Source: Census (1991)

what extent this was followed by those completing the census details is difficult to measure. In addition, it should be noted that where women are identified as head of household they are often single women, single mothers, women in lesbian households and women caring for dependent adults.

Women and home ownership

The General Household Survey 1991 (Office of Population Censuses and Surveys, 1992) (see Table 4.6) revealed that while 77 per cent of married couples are home-owners, relatively few women are able to enter the sector. Women experience great difficulty in their ability to enter the owner-occupation sector, as their ability to purchase is more restricted than that of a man in terms of their income and capability to afford a mortgage.

In view of the reduction of government capital and revenue for the social rented sector, the inability of women fully to participate within this sector is a matter of concern, given the marginalisation of rented accommodation in the social rented sector and the image of owner-occupation as the preferred tenure.

Furthermore, the system of mortgages in Great Britain, with heavy repayments during the early years of a mortgage, is structured to suit the double-income household or the single high earner, and so generally disadvantages many women. The structure of owner occupation in this country is based on an idealised picture of the typical family as:

Table 4.6 Home-owners by sex and marital status (%)

Married couples		77
Women	single	44
	widowed	51
	divorced/separated	46
Men	single	54
	widowed	50
	divorced/separated	54

Source: Office of Population Statistics and Surveys (1992), table 3.33b

> the young, married, heterosexual, white middle-class couple with two children, a boy and a girl, all of whom live together in their own house. The husband is the main breadwinner and the wife is full-time housewife/mother who may, however, work part-time.
>
> (Gittens, 1993)

Women generally acquire home ownership through a number of means, principally marriage, separated and divorced women retaining the family home and staying there to raise the children, widows remaining in their married location and single women who may inherit a home. Women's home ownership is therefore not generally derived from purchase through their own means.

The family home after divorce or separation

With the increase in divorce and separation for many families, the ownership of the family home becomes a major issue. It is widely believed that in such situations women automatically continue to reside in the marital home. According to the *General Household Survey 1991*, women are more likely to remain in the marital home in the short and long term than men, while men are more likely to improve their tenure status from that of tenant to owner-occupier in the short term (Table 4.7)

However, research by McCarthy and Simpson (1991) identified that the men with custody of their children were more likely to remain in the owner-occupied marital home after divorce than women with custodial responsibility. However, for women another important issue was their income level in deciding whether they would be able to afford to remain in the marital home after divorce. Those in the lower income bracket were more likely to lose their owner-occupied tenure and move into council housing.

McCarthy and Simpson (1991) also found in their research that of the men and women who were unemployed at the time of divorce, 17 per cent of women dropped out of owner-occupation compared to 9 per cent of men. For those who divorced during the first ten years of marriage, it was found that

Table 4.7 Summary of tenure status: of the marital home; one year after divorce; and at the time of interview

Tenure of the former marital home	*Tenure one year after divorce ('short term')*	*Current tenure ('long term')*	%		
			Men	*Women*	*Total*
Owner-occupied	Owner-occupied	Owner-occupied	33	36	35
Owner-occupied	Rented/not a householder	Owner-occupied	6	7	7
Owner-occupied	Owner-occupied	Rented/not a householder	3	3	3
Owner-occupied	Rented/not a householder	Rented/not a householder	17	13	14
Rented	Rented	Rented	18	27	23
Rented	Owner-occupied/not a householder	Rented	3	3	3
Rented	Rented	Owner-occupied/not a householder	8	6	7
Rented	Owner-occupied/not a householder	Owner-occupied/not a householder	12	6	8
Base = 100%			204	339	543

Source: Gilroy (1994)

women in this situation would invariably have young children and be likely to remain in the home during these years, given the high cost of good-quality child care. Furthermore, the home is likely to be at the bottom end of the property market and to have very little equity. A survey in 1992 (OPCS) identified that 29 per cent of female adults were economically inactive, compared to 12 per cent of male adults.

BLACK AND ETHNIC MINORITY HOUSEHOLDS

There are currently approximately 3 million black and ethnic minority people residing in Britain, which is 5.5 per cent of the population. The 1991 census collected data on the ethnic origin of respondents for the first time. The information collected indicated that black and ethnic minority people are more likely to live in urban areas, with 70.5 per cent in the South East and West Midlands. Greater London contains 44.8 per cent of the South East region's black and ethnic minority people, constituting 20 per cent of the capital's population.

Home ownership

The 1991 census identified that 67 per cent of white households were owner-occupiers compared to 48 per cent of black Caribbean households. Among Asian households, 82 per cent of Indian households were in owner-occupied properties, the comparable figures for Bangladeshi households being 45 per cent (Table 4.8).

The house price booms of the 1970s and 1980s have benefited black and ethnic minority households, especially in London and environs, where they are more often to be found. However, in areas outside London with high concentrations of black and ethnic minority households, it has been found that the value of their properties has been difficult to maintain as they are often found in declining areas.

In addition, mortgage interest tax relief is regressive and so does not benefit those black and ethnic minority families on low incomes. Many families are also unable to benefit from mortgage interest tax relief, as many purchased their homes with loans which are informal and short term and so do not attract mortgage interest tax relief (Karn *et al.*, 1985).

Ginsburg (1992) states that estate agents, surveyors, solicitors and vendors encourage residential segregation through their practices. It is the sign of a successful estate agent that they are able to match the potential buyer with the most appropriate property in terms of aspects such as price, location and desirability. Thus, for example, the guiding of black and ethnic minority buyers to particular properties which may be in less desirable locations is difficult to investigate and measure.

Table 4.8 Percentage of households in each tenure by ethnic group, Great Britain, 1991

	Owner-occupiers	*Private renters*	*Social renters*
Indian	82	8	10
Pakistani	77	11	13
White	67	9	24
Chinese	62	21	17
Head born in Ireland	55	13	31
Black Caribbean	48	7	45
Bangladeshi	45	12	43
Black African	28	20	52

Source: Office of Population Censuses and Surveys, Census Report for Great Britain, Crown copyright
Note: 'Private renters' includes renting with a job or business

Housing association tenure

Housing associations house a larger proportion of black and ethnic minority households compared to the aggregate for all tenures – social and private (Table 4.9). An investigation into nominations to housing associations by Liverpool City Council found that white households were twice as likely to be offered a house, four times more likely to be offered a newly built property with a garden, and twice as likely to be offered a property containing central heating, as black and ethnic minority households (CRE, 1989b).

The Commission for Racial Equality found in its research into forty housing associations conducted in 1993 that, while some housing associations operated non-discriminatory practices, many did not. The research found that although a large number of housing associations monitored areas of service delivery such as lettings, several housing associations did not analyse the information collected and so consequently did not change or improve their practices. The possibility of discrimination was identified with regard to their allocations policies. It was found that, as a result of officer discretion, new tenants nominated to housing associations were more likely to receive offers of flats or maisonettes than of houses. The study found that only twenty-six of the forty housing associations studied maintained records of nominations and referrals from local authorities. The CRE also found that two-thirds of housing association management committees did not have black and ethnic minority members. In addition, where there were black and ethnic minority members they were more likely to be found in subcommittees and regional committees than in the main decision-making committee/s or boards.

Despite the growth of housing associations which cater solely for black and ethnic minority residents, their numbers are small in comparison to housing associations which cater for both white and black and ethnic minorities. It is therefore important that these housing associations monitor the service they provide and ensure that the staff and committee members are representative of the range of residents they serve.

The local authority rented sector

Within the social rented sector black and ethnic minority people are over-represented in the most unpopular estates, which are often in tower blocks, older properties and estates and properties of poorer physical quality (MacEwan, 1991). Ginsburg (1992) states that the process of residential segregation has continued since the 1970s, when black and ethnic minorities were allocated inferior accommodation.

Racism

Ginsburg (1992) identified three forms of institutional racism:

1 Relative inadequacy in the physical standards of dwellings in relation to black applicants' needs. particularly the numbers of bedrooms, due to a failure to take into specific account the needs of black applicants in home construction and acquisition programmes of the past.

Table 4.9 Ethnicity and tenure, 1991

Ethnicity and migrant status of household head	*Housing association tenure (% of total)*	*Local authority tenure % of total*	*All tenures (% of total)*
Non-white total	8.4	5.0	4.5
Black	5.5	3.2	1.7
South Asian	1.5	1.0	1.8

Source: Office of Population Censuses and Surveys, Census Report for Great Britain, Crown copyright

2 Formal local policies creating differential access for black applicants, such as dispersal policies, residency requirements, and exclusion of owner-occupiers, co-habitees, joint families and single people from housing waiting lists.
3 Managerial landlordism involving racialised assessment of respectability and deserving status of applicants, and assumptions about preferred areas of residence for different racial groups and about the threat of racial harassment by whites causing avoidable trouble for managers.

(Ginsburg, 1992: 143)

Ginsburg (1992) explains that the processes of institutional racism are not conducted overtly by staff or politicians but, instead, are invariably covert and that they are possible within an organisation without any direct contribution from individuals. The procedures and rules which govern how an organisation will function can include covert discrimination which works to the detriment of black and ethnic minorities using their services.

Structural racism is identified by Ginsburg (1992) within the social rented sector as including a number of features. The reduction of public expenditure and the changing role of local authorities as enablers rather than providers of housing have occurred at a time when growing numbers of black and ethnic minority households have had to turn to the social rented sector for accommodation.

Second, the right to buy introduced in the Housing Act 1980 (now Housing Act 1985) has been extremely successful in terms of the number of council properties which have been sold under the scheme. However, the sales have been to residents who benefited from the discriminatory allocation practices of the past. Thus, white residents are more often able to buy the desirable properties, the house with a garden, so removing these from the local authority housing stock and squeezing black and ethnic minority residents into the less desirable and poorer-quality properties and estates.

Racial inequalities in housing are maintained through the threat of racial harassment, which is used against many black and ethnic minority groups. The possibility of the use of violence against black and ethnic minority people has resulted in the allocation of people to areas where they are less likely to encounter such behaviour. In addition, many tenants may choose to reside in those areas where they feel that there is less chance of their having to suffer any racial harassment. These areas are more often to be found on the least desirable estates where, in the past, black and ethnic minority people were allocated. Often the prospective resident is not confident that, if they were to encounter racial harassment, there would be any effective support from the local authority. There still exist housing officers who see racial harassment as an issue which should not be a part of their work and can thus render any policies to remedy the situation useless by their inertia or tardiness in dealing with cases which may arise in their patch or area of work.

Many local authorities have amended their tenancy agreements to ensure that legal action can be taken against perpetrators of racial harassment. However, while the use of legal action has been clarified, the actual number of prosecutions has tended to remain very small (Forbes, 1988).

The situation is similar in both the private rented and owner-occupied sectors where residents and owners from black and ethnic minority groups are concentrated in particular areas in order to reduce the possibility of encountering racial harassment. However, the issue of racial harassment of black and ethnic minority people was not recognised as a problem until the 1980s (CRE, 1987). Access to good housing is seen as the gateway to other important areas such as education and health. Thus, if black and ethnic minority people are consistently allocated poorer-quality accommodation in the social rented sector, or rent or buy properties in the least desirable areas, then other provisions such as education will tend also to be of a poorer quality.

HEALTH AND HOUSING

Leather and Wheeler (1988) state that the options for home-owners with health needs are dependent on the resources they may have available through the equity which may exist within their home. Those owner-occupiers whose health begins to decline in later age are often fortunate in that they may have paid off their mortgage and may be able to take early retirement with the benefits which may be available with such packages from their employer. Smith (1991) indicates that the home becomes a source of housing services and makes equity available to owners if they need to pay for social services or medical care.

However, where access to housing in the private sector is governed by the consumer's ability to pay, those individuals with health problems may be prevented from entering the private sector or may be marginalised into poor-quality housing. Smith (1991) states that owner-occupiers may find that an illness which increases the cost of living could result in a higher proportion of their disposable income being used to obtain special items and/or facilities which may not be available through the NHS. This could result in the need to move to smaller and/or cheaper accommodation or to move into the private rented sector.

The need to maintain a property in a condition which is warm, well ventilated and free of damp, cold, mould and other problems is an important element of owner-occupation which is often overlooked by owner-occupiers. People with health problems may find that they are not able to maintain the property due to their own poor health and the reduced opportunity of obtaining grants for repairs, maintenance and improvements.

Smith (1991) states that although the private rented sector may be viewed as an alternative means of accommodation for people suffering with health problems, given the market rents which are now set it would still be necessary for the renter to have some means of regular income. In addition, the reduced security of tenure which is available under the Housing Act 1988 would also militate against this sector being suitable for someone with health problems.

Connelly and Roderick (1991) have found that there are difficulties in the separation of the independent contribution of housing from the effects of environmental and socio-economic factors on health, especially with regard to illnesses such as anxiety and mild depression. The World Health Organization (WHO) has, however, acknowledged the effect which poor living standards has on health.

Health and poor living conditions

Shanks (1991) states that the homeless form a subgroup different from the rest of the population in terms of patterns of morbidity. Homeless people are more likely to suffer from health problems such as alcoholism, psychiatric illness and respiratory diseases, which are often the result of their poor living environments. They are further disadvantaged in obtaining adequate health care by factors such as poverty, poor education and unemployment. Homeless families are often forced to reside in overcrowded temporary accommodation where they face health risks (especially with regard to the children) such as poor nourishment, chest infections and diarrhoeal illnesses.

There is growing evidence that there are increasing numbers of people who are disadvantaged in terms of their living conditions. Fox and Benseval (1995) state that disadvantaged groups are more likely to experience higher levels of illness and disability compared with more advantaged groups in the population. In addition, their living conditions often result in a higher probability of premature death. Jacobson *et al.* (1991) found that there would be approximately 42,000 fewer deaths per annum among people aged 16–74 if manual workers experienced the level of health care which non-manual workers enjoy.

According to a study by Woodroffe *et al.* (1993), premature death among unskilled workers was three times as common among men and twice as common among women as it was in professional workers. Children from an unskilled family are twice as likely to die before the age of fifteen compared to children in professional families.

It has been estimated that if half the mothers of children under five living on low incomes were able to obtain good-quality cheap childcare enabling them to gain employment, then this would move them out of poverty. Lack of adequate heating and consequential dampness or condensation can result in poor health for both children and adults: asthma, bronchitis, tuberculosis and psychiatric problems (Bines, 1994). This, coupled with unemployment, can mean a great decline in the living standards and health of the family.

The links between family poverty, poor housing conditions and long-term unemployment and health can be seen to be closely related.

In terms of housing, investment encourages construction and results in increased employment in the construction and related industries. Meen (1992) indicates that an extra £100 million invested in housing would create 1,000 jobs in the construction industry in the first year. In addition a further 540 jobs would be created in response to the increased spending power of these workers and so it is estimated that over a two-year period a total of approximately 3,000 extra jobs would be generated. The increased employment would have a direct impact on central government as it would receive taxes on the income and spending of the new workers, and there would be a consequent reduction in welfare payments.

Research by Noble *et al.* (1994) and Green (1994) has found that one effect of the drive by the Conservative government to increase the number of owner-occupiers is that those who have been unable to purchase their own home have been marginalised and can usually be found in council or housing association accommodation and in the less desirable properties in the private rented sector. Deprivation is now concentrated in particular areas, such as inner city areas. The concentration of households suffering deprivation can strain the resources available to them, such as health services.

CONCLUSIONS

Three groups of households are particularly disadvantaged in the housing market in terms of choice, access and affordability, while a fourth group, sufferers of ill-health, often have to tolerate inadequate housing conditions. A significant proportion of the elderly, single women and black and ethnic minority households undoubtedly do not enjoy equal opportunities in satisfying their housing needs. A high proportion of elderly people have insufficient means to secure the type of housing they require – a situation compounded by cuts in public expenditure which have inhibited realistic choices from being made between 'staying put' and moving to alternative accommodation. Single women similarly find themselves disadvantaged in the housing market, in part due to problems of access to social rented housing and in part due to comparatively low incomes and the problems of securing and retaining a mortgage. Low incomes also restrict black and ethnic minority households from gaining access to owner-occupied housing, while different forms of racism limit choice of housing in both the private and the social sectors. Possibly most seriously of all, there is undoubtedly a strong causal relationship between poor health and the ability to secure reasonable housing. An analysis of social policy in Britain provides the opportunity to examine the social context of housing provision while considering the role and consequent impact of government policies. Government policies have done nothing to remedy the housing choices which are available for those individuals who are on a low income or are unemployed, especially traditionally disadvantaged groups such as women, elderly people and black and ethnic minorities. Public investment in housing is essential in order to stimulate the sector in terms of construction and rehabilitation and to provide assistance to people who find it difficult to satisfy their housing needs at a price they can afford and maintain.

While Chapters 2 and 3 also shed light on housing exclusion in terms of economics and finance, the problem of unequal opportunities and housing is further examined in Chapter 9 on management and Chapter 10 on policy-making and politics.

QUESTIONS FOR DISCUSSION

1 Consider and discuss the social and economic inequalities which are experienced by women and black and ethnic minority groups in Britain.

2 Consider whether it can be said that housing is sorting people into two groups: (a) the deprived; and (b) the privileged.

3 Discuss the social and economic changes experienced by elderly people since 1979.

4 Discuss whether anti-discrimination legislation has had any impact in changing the housing situation of black and ethnic minority groups.

5 Consider whether inequality is a natural feature of housing policies.

RECOMMENDED READING

Balchin, P. (1995) *Housing Policy: An Introduction*, 3rd edn, Routledge, London.

Bull, J. and Poole, L. (1989) *Not Rich, Not Poor: A Study of Housing Options for Elderly People on Middle Incomes*, Anchor Housing Trust, London.

Benzeval, M., Judge, K. and Whitehead, M. (eds) (1995) *Tackling Inequalities in Health: An Agenda for Action*, King's Fund, London.

Commission for Racial Equality (1989) *Racial Discrimination in Liverpool City Council: Report of a Formal Investigation into the Housing Department*, CRE, London.

Gilroy, R. & Woods, R. (eds) (1994) *Housing Women*, Routledge, London.

REFERENCES

Age Concern (1990) *Sheltered Housing for Sale*, Fact Sheet, London.

Balchin, P. (1995) *Housing Policy: An Introduction*, 3rd edn, Routledge, London.

Bines, W. (1994) *The Health of Single Homeless People*, Centre for Housing Policy, University of York.

Bosanquet, N., Laing, W. and Propper, C. (1989) *Elderly People in Britain: Europe's Poor Relations?*, Laing & Buisson, London.

British Gas (1991) *The British Gas Report on Attitudes to Ageing 1991*, Burston Marsteller, London.

Bull, J. and Poole, L. (1989) *Not Rich, Not Poor: A Study of Housing Options for Elderly People on Middle Incomes*, Anchor Housing Trust, London.

Central Statistical Office (1993) *Social Trends, 23*, HMSO, London.

Commission for Racial Equality (1983) *Collingwood Housing Association Ltd: Report of a Formal Investigation*, CRE, London.

—— (1984) *Race and Council Housing in Hackney: Report of a Formal Investigation*, CRE, London.

—— (1987) *Living in Terror: A Report on Racial Violence and Harassment in Housing*, Commission for Racial Equality, London.

— (1988) *Homelessness and Discrimination: Tower Hamlets*, CRE, London.

— (1989a) *Race and Housing in Glasgow: The Role of Housing Associations*, CRE, London.

—— (1989b) *Racial Discrimination in Liverpool City Council: Report of a Formal Investigation into the Housing Department*, CRE, London.

Connelly, J. and Roderick., B. (1991) 'Medical Priority for Rehousing: An Audit', in Smith, S.J., Knill-Jones, R. and McGuckin, A. (eds) *Housing for Health*, Longman, London.

Employment Gazette (1993), December, Department of Education and Employment, London.

Forbes, D. (1988) *Action on Racial Harassment: Legal Remedies and Local Authorities*, Legal Action Group, London.

Fox, J. and Benseval, M. (1995) 'Perspectives on Social Variations in Health', in Benseval, M., Judge, K. and Whitehead, M. (eds) *Tackling Inequalities in Health: An Agenda for Action*, King's Fund, London.

Gibb, K. and Munro, M. (1991) *Housing Finance in the United Kingdom: An Introduction*, Macmillan, London.

Gilroy, R. (1994) 'Women and Owner Occupation in Britain: First the Prince, Then the Palace?', in Gilroy, R. and Woods, R. (eds) *Housing Women*, Routledge, London.

Ginsburg, N. (1992) 'Black People and Housing Policies', in Birchall, J. (ed.) *Housing Policies in the 1990s*, Routledge, London.

Gittens, D. (1993) *The Family in Question*, 2nd edn, Macmillan, London.

Greater London Council (1986) *Private Tenants in London: The Greater London Council Survey 1983–1984*, Greater London Council Housing Research and Policy Report no. 5, London.

Green, A. (1994) *The Geography of Poverty and Wealth*, Institute of Employment and Research, University of Warwick.

Griffiths Report (1988) *Community Care: Agenda for Action*, HMSO, London.

Halpern, D., Wood, S., White, S. and Cameron, G. (eds) (1996) *Options for Britain: A Strategic Policy Review*, Dartmouth, Aldershot.

Henwood, M. (1992) *Through a Glass Darkly: Community Care and Elderly People*, Research Report 14, King's Fund, London.

Institute of Actuaries (1994) *Elderly Care*, IA, London.

Jacobson, B., Smith, A. and Whitehead, M. (eds) (1991) *The Nation's Health: A Strategy for the 1990s*, King's Fund, London.

Karn, V., Kemeny, J. and Williams, P. (1985) *Home Ownership in the Inner City*, Gower, Aldershot.

Laing, W. (1993) *The Ageing Popluation*, Macmillan, Basingstoke.

Leather, P. and Wheeler, P. (1988) *Making Use of Home Equity in Old Age*, Building Societies Association, London.

McCarthy, P. and Simpson, R. (1991) *Issues in Post Divorce Housing*, Gower, Aldershot.
MacEwan, M. (1991) *Housing, Race and Law*, Routledge, London.
Mackintosh, S. and Leather, P. (1992) *Housing and Later Life*, SAUS. University of Bristol, Bristol.
Meen, G. (1992) *Housing and the Macro Economy: A Model for the Analysis of Housing Policy Options*, Institute of Housing Economics and Statistics, University of Oxford.
Ministry of Housing and Local Government (1969) 'Housing Standards and Costs: Accommodation Specially Designed for Old People', *Circular 82/69*, HMSO, London.
Noble, M., Smith, G., Avenell, D., Smith, T. and Sharland, E. (1994) *Changing Patterns of Income and Wealth in Oxford and Oldham*, Department of Applied Social Studies and Social Research.
Office of Population Censuses and Surveys (1992) *General Household Survey 1991*, HMSO, London.
Rao, N. (1990) *Black Women in Public Sector Housing*, Commission for Racial Equality, London.
Secretaries of State (1989) *Caring for People: Community Care in the Next Decade and Beyond*, HMSO, London.
Sexty, C. (1990) *Women Losing Out: Access to Housing in Britain Today*, Shelter, London.
Shanks, N. (1991) 'Homeless Women and Pregnancy', in Smith, S.J., Knill-Jones, R. and McGuckin, A. (eds) *Housing for Health*, Longman, London.
Smith, S.J. (1991) 'Housing Opportunities for People with Health Needs: An Overview', in Smith, S.J., Knill-Jones, R. and McGuckin, A. (eds), *Housing for Health*, Longman, London.
Sykes, R. (1994) 'Older Women and Housing Prospects for the 1990s', in Gilroy, R. and Woods, R. (eds) *Housing Women*, Routledge, London.
Tinker, A. (1984) *Staying Home: Helping Elderly People*, HMSO, London.
Woodroffe, C., Glickman, M., Barker, M. and Power, C. (1993) *Children, Teenagers and Health: The Key Data*, Open University Press, Milton Keynes.

5

TOWN PLANNING AND HOUSING DEVELOPMENT

John O'Leary

The purpose of this chapter is to introduce readers to the history of town planning in the United Kingdom from the nineteenth century to the 1980s and to examine its relationship to housing, then to examine the planning system and housing in the 1990s and to consider some future causes for concern. The chapter specifically focuses on:

- Legislative developments, 1875 to the 1930s.
- Post-war legislation.
- Planning, housing and design in the post-war years.
- Planning and private sector house-building in the 1960s and 1970s.
- Planning in the 1980s.
- The planning system and housing in the 1990s.
- Environmental assessment.
- Planning and housing: the future agenda.

INTRODUCTION

Housing and town and country planning in Britain are inextricably linked, both in the history that the two disciplines share and in some areas of overlapping contemporary policy and legislation. The latter governs the volume, type and (to some extent) the affordability of new housing, as well as controlling demolition, alteration and changes of use to dwellings. Indeed there is an active pressure group entitled the National Housing and Town Planning Council which campaigns for, among other things, higher standards of housing design. There is another pressure group, the Town and Country Planning Association, that campaigns for the development of stand-alone new settlements beyond the confines of existing urban areas. Both these pressure groups illustrate the joint concerns of those involved in housing and planning. The pressure groups are, however, separate and distinct from the professional bodies that represent those employed in housing (the Chartered Institute of Housing) and planning (The Royal Town Planning Institute). These professional bodies do of course consult one another from time to time, while keeping a respectful distance, and this perhaps best illustrates the relationship that currently exists between the two disciplines in our occupationally specialised world.

This chapter will begin by briefly examining something of the history of planning in relation to housing, before going on to discuss the present-day situation. The treatment of planning in this chapter is deliberately selective and tailored towards the needs of those readers with a housing perspective. Those readers who wish to discover more about planning should refer to the various texts on the subject. For example Cullingworth and Nadin (1997) provide probably the best all-round coverage, while Rydin (1993) provides more of an analytical and theoretical approach and Adams (1994) discusses how urban planning interacts with the development process.

PLANNING AND HOUSING: COMMON HISTORY AND COMMON CONCERNS

The planning of settlements can be traced back through ancient history, although a more relevant starting-point for this particular discussion is the middle of the last century, when Victorian reformers began to raise the issue of housing for the working classes. The chaos and disease of the Victorian city, driven by unfettered capitalism, is often cited as the raison d'être for planning. The industrial revolution in Britain brought with it not only technological innovation but rapid population growth. The country's population was 10.5 million in 1800, rising to 21 million in 1850 and 37 million in 1900 (Telling and Duxbury, 1993: 1). As described by Dyos and Wolff (1973), the problems in the Victorian city centred on overcrowding, which led to unsanitary conditions, disease, low life expectancy and political unrest. In the middle of the nineteenth century reformist politicians such as Lord Shaftesbury were able to generate ad hoc legislative changes in response to some of the problems, for example through sponsoring the Labouring Classes Lodging Houses Act 1851 (Greed, 1993: 82). During the same period, pioneering urban local authorities such as Birmingham, embarked on schemes to provide much-needed infrastructure in their cities by way of levying rates. These initiatives were worthy attempts to remedy difficult circumstances, but they were not enough to prevent sporadic outbreaks of typhoid and cholera which threatened not only the poorer strata of society in the city.

While government was beginning to recognise that an unregulated private sector could not be relied upon to build the 'city beautiful', there were examples of enlightened industrialists building complete new suburbs and in some cases whole towns. Titus Salt at Saltaire (1853), the Cadbury family at Bourneville (1878) and the Lever family at Port Sunlight (1887) were the outstanding examples of the genre. As Darley notes (1991: 5–24), the model industrial villages and towns were driven by a curious mixture of idealism, benevolence, paternalism and enlightened self-interest. The latter related to the fact that a healthy, well-housed and well-motivated workforce was more reliable and better for business. The industrial sponsors of the model communities also sought a financial return on their investment, although this was usually set below what could be achieved on other investments in the open market. For example, as Darley points out (1991: 10), Joseph Rowntree's trust for the establishment of the model village New Earswick near York (1904) settled for between 3.5 and 4 per cent return on capital. Similar trust arrangements operated for the other model towns and villages.

It was evident that the design, layout and quality of housing units, as well as the range of community facilities within the planned settlements, were far superior to those of the cramped estates which were being hastily developed on a speculative basis in the cities. It took a campaigning Victorian with an appropriately Victorian name – Ebenezer Howard – to give direction to the flow of ideas and model settlements. Howard is seen by many as the father of the town planning movement in this country; his energy and charisma brought together like-minded individuals in the Garden City Association founded in 1899 (now known as the Town and Country Planning Association). Howard was also the author in 1898 of *Tomorrow – a Peaceful Path to Real Reform*, which was re-issued in 1902 as *Garden Cities of Tomorrow* (see Ward, 1992: 4). This book, reprints of which are still available today (Howard, 1985), explains Howard's vision as well as the principles for establishing garden cities. In an age before computer-generated imagery, Howard's use of graphic icons in the book, such as the three magnets diagram (see Figure 5.1), to explain those principles is quite impressive. The three magnets diagram suggests that a combination of town and country could provide a more civilised life for ordinary people. The short and accessible book went on to consider population thresholds and management systems for a

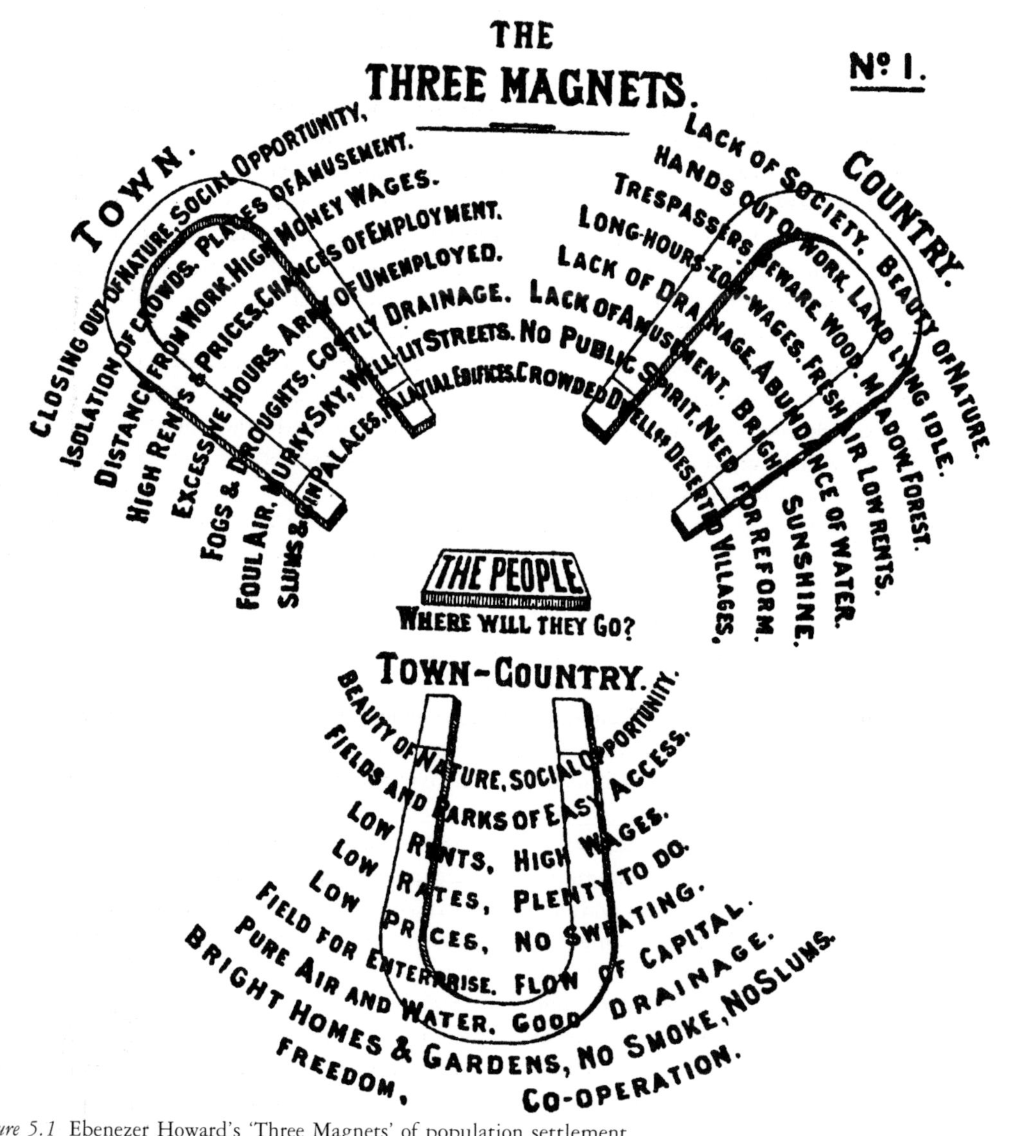

Figure 5.1 Ebenezer Howard's 'Three Magnets' of population settlement

garden city. Howard's discussion on how the garden city community could collect and retain the increases in land values, as the settlement developed, is, as Darley notes (1991: 24), perhaps as relevant today as at the turn of the last century.

Howard was not just an armchair theorist and sought to put theory into practice at Letchworth, the first garden city, begun in 1903. Today Letchworth may seem 'ordinary' and suburban to the visitor, but at the turn of the century it marked a significant moment in the raising of standards in housing layout and design, land management and overall environmental quality.

At Letchworth, Howard's team, who had formed

themselves into a private company, secretly purchased agricultural land before beginning to implement plans drawn up by the designers Barry Parker and Raymond Unwin. Parker and Unwin were themselves heavily influenced by the Arts and Crafts movement and the design principles advocated by the Austrian artist Camillo Sitte. Sitte's work, which sought to demonstrate the need to incorporate artistic principles into town planning (see for example Collins and Collins, 1965), still features in the reading lists of undergraduate planners and architects. Parker and Unwin's design for Letchworth provided groups and short terraces of generously proportioned houses set in gardens within a framework of tree-lined avenues at an overall density of 10.5 dwellings to the acre. The design principles established at Letchworth and in the model villages such as Port Sunlight were to form the template for the better examples of local authority and garden suburb development which followed in the inter-war years. It should also be noted that both at Letchworth and in some of the industrial model villages, there was an attempt to create a social mix in the resident population. There was a view prevalent at the time, that great benefit could be derived from encouraging the social classes to mix. Historians and sociologists are divided on the extent to which this actually happened in the early communities and whether there were any tangible benefits from this aspect of town planning. The principle of social mixing may seem less relevant today in what some see as a less socially stratified society. However, the ideal of social mixing still pervades contemporary discussions on tenure, dwelling mix and the proportion of affordable dwellings that can be achieved in new housing schemes.

LEGISLATIVE DEVELOPMENTS, 1875–1930s

While Howard, Lever and Cadbury were providing practical and theoretical models of how the country's growing population could be housed, the march of legislation had been much less spectacular. The Public Health Act of 1875 enabled local authorities to make by-laws which exerted some control over the construction and layout of housing. By the 1880s most urban local authorities were operating model by-laws. 'By-law housing' originates from this era: its uniform appearance can be seen in virtually any inner-urban area of Britain (or in the television programme *Coronation Street*).

Although some local authorities had begun to develop public housing by the beginning of the twentieth century, the majority of house-building activity continued to be carried out by private builders on a speculative basis. There was control, as Booth (1996: 16) describes, over the form that housing units might take, as well as street widths and distances between the backs of dwellings. There was no control, however, over the scale and location of housing. The first notion that local authorities could control both the form and layout of housing and its location and volume (through the release of land) came in the curiously named Housing, Town Planning Etc. Act 1909. The word 'notion' is used intentionally, because the Act provided only a discretionary power to local authorities to prepare planning schemes for land being or likely to be developed for housing. In the event this discretionary power was only taken up in thirteen instances throughout England and Wales.

In 1919 another Housing, Town Planning Etc. Act came into force, providing stronger powers to local authorities. This particular Act gave rise to a spate of inter-war council house building, as it placed a duty on local authorities to assess and provide for their housing needs with recourse to central government grants. The Act also required that local authorities with a population of 20,000 or more should prepare a planning scheme for land that was 'ripe' for development.

The Town and Country Planning Act 1932 extended the ability of local authorities to plan not just for ripe development land, but for the whole of their areas. However, the requirement for the larger authorities to produce plans contained in the 1919 Act was dropped (Booth, 1996: 19). The majority of local authorities did, however, make a resolution to produce a planning scheme for their area, although, in the event, few local authorities followed through and produced a plan. Given the stop-start support for planning by central and local government, it was hardly surprising that a considerable amount of speculative house-building went on in the open countryside during the inter-war years. The phenomenon of ribbon development became particularly evident, as semi-detached housing threaded out into the rural areas along transport routes. Because the rather chaotic planning system could not prevent this wasteful and invasive form of development, Parliament resorted to specialised legislation by way of the Restriction of Ribbon Development Act 1935.

In his engaging book *Semi-Detached London*, which chronicles the period 1900–39, Jackson (1991) points out that the planning system which existed during the period was insufficient to prevent private house-builders from developing a whole swathe around London. Jackson refers to the resultant development as a wasted opportunity, while the collective product he sees as ugly, monotonous and appalling. Jackson does, though, acknowledge that thousands of ordinary people were able to find a home within their financial reach and that the sum of human happiness was immeasurably increased (1991: 258–9). Jackson believes that a more sophisticated system of planning (which was to emerge after the war) could have prevented the worst excesses:

> That the outer layer of London which emerged from all this activity was neither well balanced in its constituents or visually and psychologically satisfying, is common ground. It happened largely because there was no positive planning on a regional scale, and worse than it might have been because such planning powers as did exist were not properly used by most of the local authorities.
>
> (1991: 257)

POST-WAR LEGISLATION

It was not until after the Second World War, with the passing of the Town and Country Planning Act 1947, that local authorities were given a comprehensive set of powers with regard to planning and land use. The 1947 Act reflected political and social change that had occurred during the war, with a more widespread acceptance of state intervention and 'planning' in its widest sense. Cullingworth describes the climate of opinion and new legislation as follows:

> There was a widely shared vision of a rebuilding of the physical fabric of Britain which would parallel the social and economic reformation to be brought about by full employment, social security, a national health service and similar changes to the societal structure of pre-war Britain. The new positive planning powers would enable the building of new towns and the rebuilding of older cities; development would give priority to the public interest; speculation in land would be brought to an end by the nationalisation of land values; buildings and landscapes of community value would be protected.
>
> (Cullingworth, 1996: 172)

The provisions of the 1947 Act were detailed and comprehensive, especially with regard to compensation and betterment issues. The key features were that local authorities acquired new development control powers which effectively nationalised development rights. Thus anything that fell within the broad definition of development (building works or changes of use to land or buildings) required express consent from the local authority. Developers now required planning permission from the local authority before they could embark upon their schemes. Local authorities also had a new responsibility to produce development plans which were to look ahead over twenty years. The new plans were to identify not just future housing sites, but were required to allocate land for all other land uses as well as designating land for green belts, transport corridors, mineral extraction sites, and so on (Rydin, 1993: 26).

As well as creating a new system of town and country planning, the government of the day also passed the New Towns Act 1946. The Act enabled the establishment, by central government, of publicly funded new town development corporations to build whole new towns on greenfield sites surrounded by green belts. Many of Howard's original ideas for garden cities were incorporated into the new towns machinery, although as Ward (1992: 14–16) points out, there were significant departures from the Howard model. For a start the new town populations at around 80,000 were considerably larger than envisaged by Howard and the soft Arcadian feel of the Arts and Crafts movement was to be replaced by a harder 'modernistic' style of architecture (which in the later new towns such as Harlow included high-rise housing blocks). The principles of returning the rise in land values to the community were institutionalised as the development corporations acted as developer and land-trader on behalf of the government. The first post-war new town at Stevenage, though, does bear some of the hallmarks established by the earlier pioneers at Letchworth and Welwyn, and generally the early new towns sought to strike a balance between the garden city concept and the use of the motor car. New town designers sought inspiration from Radburn, an inter-war garden city in the United States (Fishman, 1992: 149–52) which provided a template for separating pedestrian and vehicular access to dwellings (see Figure 5.2). At Radburn the front of the dwelling faced on to pedestrian routes and greens, while culs-de-sac provided access to the rear of dwellings. Variations on the Radburn theme were experimented with in the new towns in an attempt to achieve an efficient use of land while providing public space, on to which dwellings could face without severance by a vehicle route.

PLANNING, HOUSING AND DESIGN IN THE POST-WAR YEARS

Leaving aside the new towns programme, there was a need in the 1950s and 1960s to repair large areas within bomb-damaged cities, as well as to provide additional housing to meet population growth. Parts of the remaining housing stock had become unfit for human habitation and needed to be replaced. As a response, central government set ambitious annual housing targets for the construction of new dwellings in both public and private sectors. The local authorities' role as housing provider was considerably expanded under the Ministry of Housing and Local Government and considerable sums of central government money were channelled into local authorities in the form of capital grant money to build public housing. The emphasis was on speed and completion rates and quite staggering figures were achieved during the late 1950s and early 1960s when compared with the rates achieved in the early 1990s. However, the quality of some of the housing provided, particularly in the public sector, was, with the benefit of hindsight, less than desirable.

The rise and fall of high-rise housing

Between the 1930s and the 1950s there was a near revolution in the culture and theories of architectural design. In particular modern movement thinking, inspired by architects such as Le Corbusier, swept through architectural schools, design offices and local authority departments responsible for housing. Le Corbusier's vision of high-density 'towers in the park' was seductive, both in the potent imagery (see for example Guiton, 1981) and in the apparent efficiency with which these towers could house an urban population (Figure 5.3). The volume house-builders were quick to realise the possibilities of system-built housing contracts with local authorities. Indeed central government had been so won over by the idea of high-rise housing that local authorities were offered additional grants for high-rise projects, geared to the number of storeys. Volume builders were able to supply system-built packages with incentives and discounts for those local authorities that signed multiple contracts (Dunleavy, 1981).

At the outset the professions, as well as local and central government politicians, supported high-rise housing, and even the early tenants were pleased with their new homes in the slab and tower blocks. Councillors even had blocks named after them, as had been the municipal tradition with local street

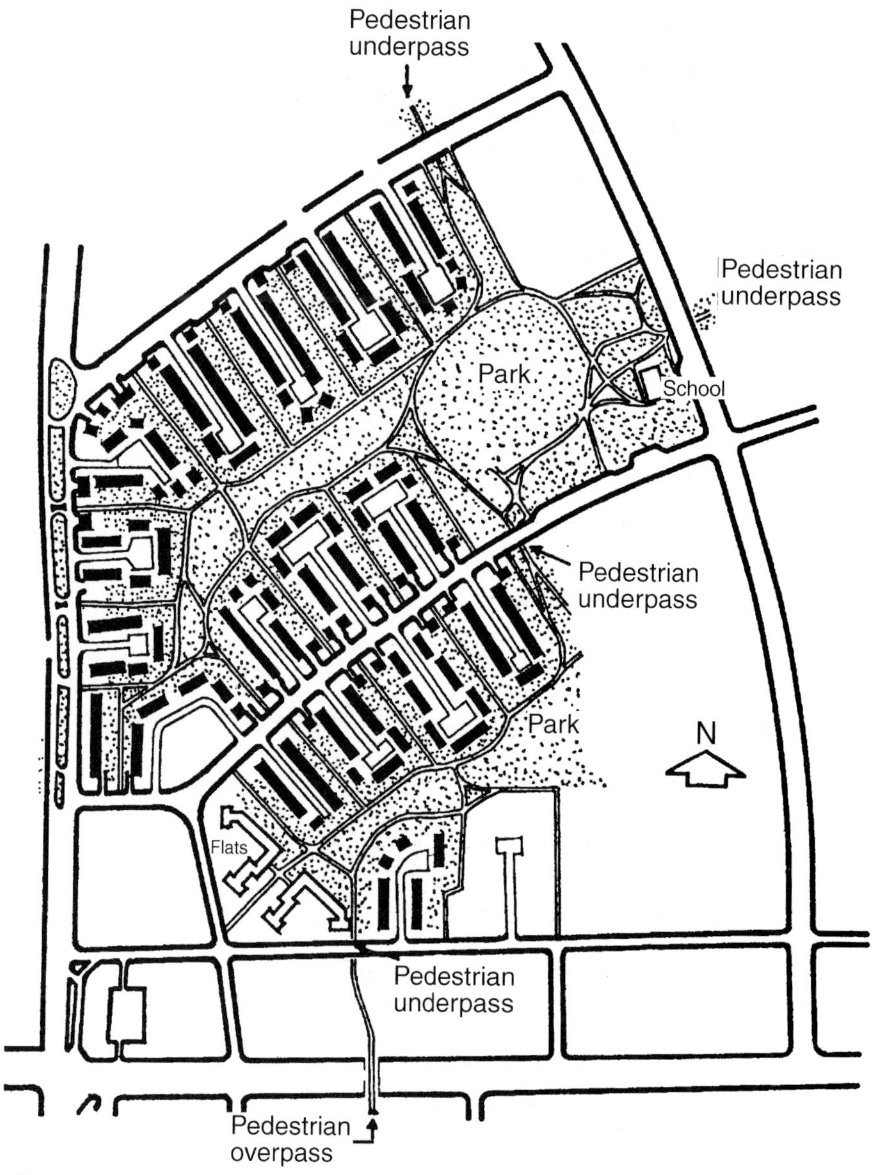

Figure 5.2 Radburn: a residential area of the town as developed, showing culs-de-sac, footpath system and open space

names. Indeed the now infamous Ronan Point, which was part of a larger contract to supply point (high-rise) blocks between the London Borough of Newham and Taylor Woodrow Anglian, was named after the local councillor who was the vice-chairman of the housing committee (Dunleavy, 1981: 242–54). It was the partial collapse of Ronan Point in 1969, following a gas explosion, which focused the nation's attention on the inadequacies of high rise. By the late 1960s, the honeymoon period for high-rise accommodation was over, as growing tenant dissatisfaction began to make itself felt. Housing

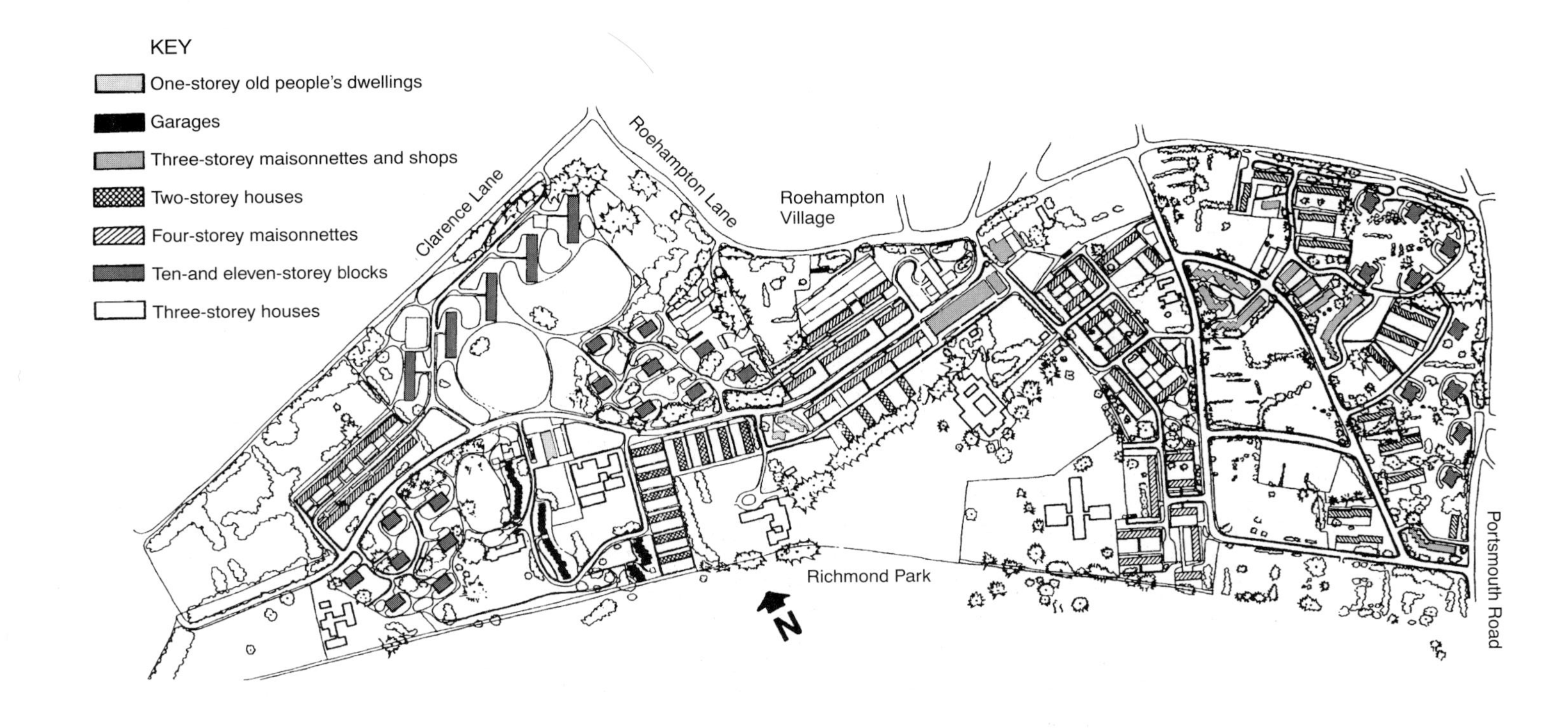

Figure 5.3 The Roehampton Estate, developed by the London County Council, 1952–9, incorporating the Le Corbusier concept of 'towers in the park'
Source: Reproduced courtesy of the *Architectural Review*

Plate 5.1 Ten-storey maisonette blocks, Roehampton Estate, London Borough of Wandsworth
When it was completed in the late 1950s, the Roehampton Estate was the largest housing scheme in Europe. Inspired by Le Corbusier, the London County Council provided housing for 9,500 people in five ten-storey maisonette slab blocks, twenty-five eleven-storey tower blocks and terraces of four-storey maisonettes and terrace houses. The buildings are set in mature parkland, but there is comparatively little segregation of pedestrians and vehicles

authorities began to face complaints from tenants about poor thermal and sound insulation in the flats, lifts were constantly breaking down and the lack of supervised play space for children became an issue.

Of course not all of the housing provided during the era under discussion was unsatisfactory and as early as 1969 there was the beginning of a switch away from comprehensive redevelopment to gradual renewal with the introduction of General Improvement Areas under the Housing Act of that year. GIAs provided grants on a blanket basis for between 300 and 800 older dwellings which could be improved, thus avoiding the necessity for major clearance where deterioration was allowed to become irreversible. As Cullingworth and Nadin (1997) point out, the Housing Act 1974 was subsequently to go further by explicitly recognising the relationship between the decaying fabric in the older areas of housing and areas of social stress. There was also an in-built requirement in the Act for Housing Action Areas (HAAs) to include meaningful resident involvement in the scale and nature of improvement works.

However, there remains a legacy of high-rise blocks, which are often part of larger local authority estates containing lower-rise flat developments also built in the modernist style. The exposed concrete surfaces in the 'brutalist' style weathered poorly from the outset, with graffiti soon adorning the grey, stained external surfaces. Parts of some estates became the focus for crime and delinquency, which gave rise to a heated debate on 'architectural determinism'. One school of thought believed that the poor design of the estates corrupted the tenants and made them behave badly, while others explained the phenomenon in terms of wider structural factors such as unemployment and poverty. According to the latter and more widely held view, the estates just happened to be where the urban poor lived and so society should not be surprised if these areas generated a greater than average incidence of crime and anti-social behaviour. The relationship between architectural design, crime and behaviour has been explored by a number of academics and researchers. A summary of important contributions by two authors, Oscar Newman and Alice Coleman, is relevant at this point.

The work of Oscar Newman

Oscar Newman conducted research on the relationship between crime, vandalism and the design of housing projects in the United States in the early 1970s (Newman, 1972). Newman defined the concept of 'defensible space' after showing a correlation between higher crime and vandalism rates in housing projects which lacked defensible space. The concept is simple to grasp in that the space outside buildings should have clear demarcations, whether they be physical or symbolic, to demonstrate ownership. For example, there is no ambiguity associated with the front garden of a house which has a fence or wall and a gate. It is quite clear that the garden 'belongs' to the house and there are physical and symbolic signs of that ownership in the form of the closed gate and garden wall, which prevent strangers or passers-by from using the space inappropriately. Of course the symbols of ownership act as much upon the psyche as they do in any physical sense. A determined wrong-doer will bypass these symbols, but they are robust enough to make the casual or opportunistic vandal think twice. There is also a sense of surveillance in that the windows of a house look out on to the garden. The street beyond is quite clearly public space, although it too will benefit from collective surveillance if all houses front on to the street, so that there is potential surveillance from both sides for the entire length of the street. There is thus the potential for any wrong-doing to be observed, reported or possibly challenged.

At present in the United Kingdom, the presence of Neighbourhood Watch signs at the entry to a street reinforces the sense of surveillance, which may be actual or potential. Either way it is a deterrent to casual crime and vandalism. Where space is not well defined and its ownership is ambiguous, the defensible space concept breaks down, as Newman found in his work. The phenomenon can also be observed with regard to the undefined space around the base of many urban tower blocks or spaces within large post-war estates which are often uncared for, or in a poor condition. The rationale for these spaces was never clearly articulated and often it is simply space left over between buildings, or SLOBB, to give it its acronym. Similarly the raised walkways, corridors and lower ground car parks which characterised many estates are not defensible in terms of their clear ownership or surveillance. It is, of course, these very spaces which attract vandalism and graffiti and provide the potential setting for other more serious crimes. In recent years the police force has been providing advice on planning out crime and is consulted on planning applications where crime might be an issue. In any event, Newman's defensible space principles have long since filtered into the training of architects and planners.

The work of Alice Coleman

Alice Coleman is a professor of geography at King's College, London, who has studied inner-city housing estates over a number of years with a team of researchers. Coleman has drawn some of the inspiration for her work from Oscar Newman. In terms of the architectural determinism debate, Coleman's research has led her to believe that bad design in housing is like a sliding scale that can tip the balance in favour of socially inadequate behaviour, where ordinarily (i.e. given adequate housing) the situation is finely balanced. As Coleman states:

> First, let us stress that we are not dealing with determinism. Even in the best housing there may be people who choose to behave badly, and even in the worst there are those who maintain impeccable standards. Bad design does not determine everything, but it increases the odds against which people have to struggle to preserve civilised standards.
>
> (1990: 83)

In order to arrive at these conclusions, Coleman's prosaically named Design Disadvantagement Team spent several years collecting data on the design attributes and day-to-day functioning of post-war housing estates in the London Boroughs of Southwark and Tower Hamlets. The team scrupulously recorded the incidence of litter, graffiti, vandalism and the proportion of children in care, as well as examining whether parts of the estates, such as poorly lit stairwells, had been used as public toilets. The data were then correlated against the design attributes of each estate in the study. The results have been published in Coleman's provocative book *Utopia on Trial* (1990).

To begin with, Coleman is particularly critical in

her book of utopian thinking in the field of housing provision, which she traces back to Ebenezer Howard and Le Corbusier (it is hard to think of a more unlikely pair of bedfellows). Coleman believes that the utopian ideas inspired by these thinkers were unscientific and lacked a basis in reality. According to Coleman, what actually happened in the post-war years was the development of a managerialist 'top-down' mentality which believed it could deliver utopia for the citizens. This process, Coleman argues, bypassed tried and trusted methods of supply and demand, in that housing 'customers' could no longer make signals to housing suppliers about what would and would not work in housing terms. Coleman is critical of the paternalistic attitudes that prevailed when many of the problem estates were being planned and designed. Coleman also cites Jane Jacobs (1961), whose book *The Death and Life of Great American Cities* contains a similar, if more emotive, attack on planning and housing renewal policy in the United States in the 1950s and 1960s (discussed more fully below under 'Public participation', p. 136).

Coleman's research has led her to advocate, among other things, design changes to estates in an attempt to make them more client friendly and therefore reduce the observable incidence of anti-social behaviour. Coleman's work and subsequent proposals have not gone without criticism from the academic and professional community. Critics would argue that Coleman's proposals are largely cosmetic and do nothing about the underlying causes of poverty, the physical and social manifestations of which have in any case, the critics argue, been recorded in an unscientific way by Coleman's team. Critics have also pointed out that high-rise, or apparently poorly designed, estates such as the Barbican or Chelsea Harbour function perfectly well because the inhabitants are from high-income groups and have chosen to live in the particular development.

In any event, local authorities have long since ceased to construct high-rise or modernist estates and are already demolishing their 1960s estates and rebuilding with more conventional housing where they are financially able to do so. There has also been a whole professional and political rethink about how housing should be designed, constructed and managed. While not all participants in the debate agree with Alice Coleman's prognosis, there is at least in the work a series of reiterations (and therefore the possibility of dissemination) of what has already been learnt by those working in the field. For example, Coleman advocates reductions in the number of families with children in the remaining high-rise blocks. Coleman also highlights the erroneous assumption that high rise can provide higher residential densities at a cheaper cost per dwelling and there are reminders for architects and planners to think about 'defensible space' in new housing design.

PLANNING AND PRIVATE SECTOR HOUSE-BUILDING IN THE 1960s AND 1970s

Returning to more conventional types of housing provision by the private sector, the planning system which operated during the 1960s and 1970s sought urban containment and the protection of the countryside (Hall *et al.*, 1973). Under a system of process planning, introduced by the Town and Country Planning Act 1968, the spatial distribution of economic and social trends (rather than physical standards) became the principal concern of planners. This effectively meant that the development control system operated by local authorities 'policed' new housing demand, allowing new development on infill sites or on the edge of larger towns and villages, but preventing development in the open countryside and in designated areas such as green belts and Areas of Outstanding Natural Beauty. Most local authorities in rural areas operated a strict village envelope policy which restricted new house-building to land within a tight line drawn around the village in the relevant development plan. Proposals to build housing outside the envelope were routinely refused by local authorities and subsequent appeals dismissed by the planning inspectorate.

Even the proposal to construct individual dwellings in the countryside for farm workers had to undergo a strict 'agricultural need' test as laid down in DoE *Circular 24/73* before planning consent would be granted. The agricultural need referred to above was verified by an officer of the Ministry of Agriculture, Fisheries and Food. If the agricultural need was not proven, the local authority would almost certainly refuse the planning application.

House-builders during this period became increasingly adept at identifying under-used or vacant sites within urban areas, or negotiating with local authorities for planned releases of land in the form of town or village expansion schemes. Very often the latter would coincide with infrastructure improvements, particularly roads. In the early New Towns the private house-builder had, through the workings of the New Towns Act, been almost totally excluded by the publicly funded New Town Development Corporations. A change in the legislation in the late 1960s resulted in increased involvement of private developers in the new towns during the 1970s and 1980s, as described by Hebbert (1992: 173–6). Outside the new towns and particularly in the South East and around London, the available 'windfall sites' for house-builders became increasingly scarce due to almost insatiable market demand. The latter was due to the fact that during the 1960s and 1970s there was considerable migration into the South East from other regions, primarily related to the search for employment. There had also been a 'baby boom' during the 1960s when the population of the country was experiencing growth rates of 700,000 per year according to Collins (1994: 94). The working population had also become more willing to commute longer distances from towns and villages beyond the green belt. While the population of Greater London stabilised at around 7 million, the population of the wider city region continued to grow at an alarming rate, which the new and expanded towns in the South East could barely keep pace with.

Given that the competition for housing was pushing up housing land values considerably, house-builders began to consider not just the obvious greenfield sites. Attention turned to large back gardens which could be 'rolled up' to provide land for a cul-de-sac with detached and semi-detached housing. Older houses in large grounds were also targeted because they could be redeveloped to provide low-rise flat developments with car parking within the curtilage. Given demographic change with more smaller households, private low-rise flats within commuting distance of major employment centres were easily sold on the open market. The era of backland development had arrived, which effectively meant a more intensive use of existing residential land for housing, or, looked at another way, the gradual raising of residential densities. Whitehand (1990) and Whitehand, Larkham and Jones (1992) have conducted research on this matter by examining case studies on the edges of the London and Midlands conurbations. The case studies show that house-builders would seek ownership or contractual control over backland sites, often over a number of years, before submitting planning applications to the relevant district council. Very often there would follow a series of adjusted and resubmitted planning applications, which were responses both to changes in the market and to objections from residents, planning officers and local councillors. The objections would typically relate to allegations of over-development and inadequate access and parking. Local authorities would often seek to operate density and parking policies which would effectively limit the amount of development on a particular site, while the developers would seek to push these limits as far as possible in order to maximise the number of dwellings on a site. The outcome, as Whitehand notes, was often a poorly co-ordinated compromise, with issues such as the design and appearance of the new dwellings relegated to secondary or minor considerations.

'Anywhere housing', design guides and development briefs

It is widely agreed that much of the housing development that went on the 1960s and 1970s was lacking in character and architectural merit. Part of the problem stemmed from the observations made above, in that the density and location of housing were the key considerations, while the design and appearance of the product were somehow forgotten in the process. The volume house-builders had also developed a limited range of 'no-frills' styles which used cheaper materials and a lowest-common-denominator design. These standard designs were deployed throughout the country, wherever the particular firm had an operative site. All that changed was the shape of the site and the road layout to service the units. The term 'anywhere housing' was coined to describe these characterless additions to the urban landscape.

It was not long before there was a cultural and professional backlash against this uniformity. An alternative, termed 'vernacular architecture', began to be explored which paid attention to context, regional styles and materials. Writers such as Cullen (1971), Sharp (1968) and Worskett (1969) had also re-awakened an interest in the whole subject of 'townscape'. The emergence of the vernacular and townscape ideas led local authorities to think seriously about the loss of quality and character which many areas were experiencing at the hands of the volume house-builders.

The local authority response came in the form of design guides, produced to enhance the quality of development control decision-making with regard to the layout, design and appearance of new housing. One of the earliest and certainly the most influential in the 1970s and early 1980s was the *Design Guide for Residential Areas* (1973) produced by Essex County Council for use by the district councils in an attempt to curb the spread of bland housing design and layout which was eroding the character of the county. The guide illustrated some of the shortcomings of the anywhere housing estate and advocated instead the use of vernacular styles and materials. Advice was also given in the guide on how to design-in privacy by careful positioning and dimensioning of widows, as well as integrating garages into the design of dwellings as outbuildings, rather than allowing them to dominate the street scene. The guide also sought to curb 'prairie-planning', which involved excessively low densities on endless estates and which was wasteful of land. The guide advocated traffic calming through shared surfaces and the conscious design of a road hierarchy ending in contained courtyards.

House-builders were for the most part supportive of the guide, as they saw potential returns from using sites more intensively and through increased market prices that could be achieved for a higher-quality product. Of course in practice the guide was 'interpreted' to meet local site conditions and the market niche aimed at by the particular developer, although there is agreement that the guide was instrumental in raising standards in the county. Other local authorities followed the example set by Essex, although they made appropriate adjustments to match the particular vernacular form appropriate for the area. Urban authorities such as Westminster also began to produce local guides governing alterations and extensions to housing, particularly in conservation areas.

Design control, urban design and Prince Charles

While local authorities are under no statutory obligation to produce design guides, they are still produced and updated and are accepted as 'material considerations' in development control decisions taken by authorities and inspectors. The extent to which local authorities can influence design is determined by central government policy advice contained in *Planning Policy Guidance Note 1: General Policies and Principles* (DoE, 1997). A summary of the advice is that local authorities must reject obviously bad designs; that is, those that are out of scale with their neighbours or are inappropriate to their context. Good design everywhere, so the advice states, should be the aim of all those involved in the development process. In practice, many local authorities already have design specialists within their planning departments who are able to discuss design matters with those submitting planning applications.

For large or important sites which are about to be developed, local authorities will commonly issue 'development briefs'. These are short documents which indicate the expectations which the authority has with regard to the development of the particular site. Design issues will normally feature in development briefs, giving an indication of acceptable surface materials, parking standards, and the general height, mass and configuration of buildings in relation to open space, building lines and site lines. The briefs do not seek to fetter the architect and are often open to negotiation and interpretation. The overall intention is to raise awareness of a site's potential in planning and design terms and to try to prevent conflict between the developer and local authority which could lead to unnecessary delays.

Since 1978 there has also existed an organisation called the Urban Design Group whose membership is made up of architects, planners, landscape architects and other professionals who are interested in raising the quality of the built environment. This group has been active in sponsoring debates through its quarterly journal *Urban Design* and through joint conferences with the various professions to raise the agenda of design and to encourage inter-professional co-operation, which is so often lacking in a professionally fragmented world. The UDG is closely related to a number of postgraduate courses in urban design which address many of the missing ingredients in the design of urban areas identified by Newman and Coleman. The overall contribution aims to raise the level of awareness and knowledge about design issues, and to seek to avoid many of the design mistakes made in the past.

Of course no discussion about architecture and urban design would be complete without a mention of HRH the Prince of Wales who, during the late 1980s and early 1990s, made a number of controversial statements regarding the design quality of much of our built environment. Prince Charles's well-publicised views, particularly on post-war additions to London and modern movement architecture, have been very critical, to the point of upsetting a large swathe of the architectural community. Many architects feel that the Prince's views are entrenched in a romanticised view of the past, wherein only classical architecture embodies quality and modern architecture by definition lacks quality and good taste. Taken to its logical conclusion, critics argue, such a view will stifle innovation and contemporary artistic and cultural expression. Many architects would support the principle of experimentation, recognising that some architectural concepts will inevitably fail, while others will succeed and possibly bring forth new masterpieces. There is also frustration in that the majority of architects would claim that they have not undertaken a long training just to end up copying historical styles. The 'battle of styles' continues to rage, with the Prince finding some supporters within the architectural community and among the lay public, many of whom prefer the romanticism and emotional association with heritage that classically designed buildings can inspire.

The Prince has made a less well known, but potentially more significant, contribution by playing an active part in the Urban Villages Forum, an amalgamation of various property professionals, developers and community activists who explored new ways of building urban areas. Publication of the Urban Villages Forum blueprint in 1992 was heralded with much critical acclaim, and recent government planning policy (DoE, 1997) has followed up by endorsing the principle of mixed-use urban villages. The Prince has also made a serious attempt to lead by example. On part of the Duchy of Cornwall estate, which Charles owns and which abuts the town of Dorchester, the Prince has promoted the development of an extension to the town. The scheme, which is called Poundbury, is nearing completion and it is apparent that it embodies many of the principles established in the work of the Urban Villages Forum. The Prince employed architect Leon Krier, who was sympathetic to Charles's views on town planning and architecture, which embrace the concepts of sustainability, mixed uses, neighbourhood, and human scale as well as attention to the design of buildings and public spaces. The results are far superior to many of the town and village extensions that litter much of our countryside and in this respect Charles must take credit for raising standards by example.

Public participation in planning

One of the themes touched upon by Coleman (see above, pp. 131–2) in the discussions on post-war planning and housing development was an alleged breakdown between the producers and users of the built environment. Here Coleman was drawing upon the work of an earlier American author, Jane Jacobs, who found that in the United States town hall officials had become detached from what was actually happening in urban neighbourhoods. The officials had become so remote from the public opinion that they were identifying whole areas for redevelopment, often against the wishes of the residents. Jacobs discovered that the officials believed they were acting in the best interests of all concerned, a utopian ideal which she traces all the way back to Ebenezer Howard and the garden cities ideal:

> His [Howard's] aim was the creation of self-sufficient small towns, really very nice towns if you were docile and had no plans of your own and did not mind spending your life among others with no plans of their own. As in all utopias the right to have plans of any significance belonged only to the planners in charge.
>
> (Jacobs, 1961: 27)

Jacobs felt that because there was inadequate public participation, whole neighbourhoods were subject to what she termed 'cataclysmic money' (heavy-handed modernist redevelopment) rather than 'gradual money' (sensitive regeneration of existing areas) (1961: 305–34). It is against this background that we can consider the development of public participation in planning in this country.

During the 1950s there was very little written on the subject of public participation in planning and we could deduce from this that locally elected councillors were relied upon to adequately represent their local electorate on planning matters. The theory and practice of a representative democracy thus held sway in most areas of local government activity, including planning. However, by the mid-1960s a process had begun of wider social change in which individuals were seeking greater involvement in decision-making that was affecting their lives. The Skeffington Report (1969) was in many ways a reflection of a popular clamour for greater control over the pace and volume of change being experienced, particularly in urban areas, in the 1960s. The report was the outcome of a government-appointed committee which sought to explore the actual and possible extent of public participation in plan-making. The Skeffington Report also publicised good practice, as the committee saw it, that was already being undertaken by local authorities in the field of participation. The report was very supportive of extending the principle of public participation, linking the benefits to mature and responsible citizenship and improved decision-making. The timing of the Skeffington Report appears a little curious in hindsight, in that it postdates the Town and Country Planning Act 1968, which made public participation in the preparation of development plans mandatory.

Planning was one of the first local government activities to 'go public' and the principles of participation spread to development control decision-making, beginning with statutory requirements to publicise bad-neighbour developments and finally all developments by way of site notices, press notices and letters to neighbours. There are statutory windows of opportunity for members of the public to scrutinise and make comments on all planning applications. Members of the public may also speak at planning committee meetings and attend public inquiries. They may also take part in public inquiries as third party witnesses if they can demonstrate 'standing' in relation to the particular case. Third party observations on development proposals are material considerations which must be weighed in the balance in decision-making under the planning Acts. Development projects can be altered significantly or refused following the consultation exercises, and developers are sensitive to cultivating public opinion, especially where a large or controversial project is envisaged. Local authorities frequently go beyond the statutory minimum requirements and mount exhibitions and/or conduct leaflet drops encouraging participation on particular local planning matters, although responses are typically poor unless a development proposal is in some way perceived to be a threat. Case histories, committee reports and decisions are all open to public scrutiny, often from a 'one-stop shop' counter at the local authority. The Ombudsman will also investigate planning decisions where there are complaints from members of the public about particular decisions or the handling of a particular case. Planning has thus been and remains a relatively open branch of government in comparison to other branches of public sector activity.

Of course one of the key issues with regard to participation is the extent to which participation actually affects the decision taken. Arnstein (1969) was one of the first to note in the United States that there were degrees of citizen participation. Her 'ladder of participation' (reproduced in Thomas, 1996: 173) remains one of the most helpful conceptualisations of this issue. On the bottom rungs of Arnstein's 'ladder' the decision-makers simply provide information about an issue or decision which affects the public, but no account of public views is actually taken in arriving at the decision. There is little genuine participation at this end of Arnstein's ladder, but rather publicity is provided about a decision that has already been taken. Genuine consultation would involve a potential change in the decision, or some degree of compromise depending on the results of the consultation. These middle rungs on Arnstein's ladder are thought to approximate the degree to which the public can influence decisions taken by government in this country at present. At the top end of Arnstein's participation ladder, citizens are fully empowered with decision-making responsibilities and power is widely diffused throughout society.

As well as degrees of meaningful participation to consider, Adams (1994: 198–208) notes that even within the systems which currently exist, there are marked differences in the extent to which different groups in society actually become involved in the opportunities to participate in decision-making. Well-organised development and land-owning organisations do not miss the opportunity to participate in planning and do so to some effect by influencing decisions and outcomes. The articulate middle classes are also adept at participating in the planning system when they need to. Less well-organised or motivated groups and individuals may be less aware of their ability to participate, or are less able to devote time to participation exercises. For this reason they may be less able to influence development outcomes. The extent to which genuine participation actually occurs within the present planning system is constantly debated, although many feel that a realistic balance has been reached between allowing individuals to 'have their say' while facilitating decision-making which allows needed developments to go ahead without undue delay. The more strident members of the development industry would probably say that we have too much participation in planning and it is a luxury we can ill afford. Protestors against development projects would arrive at a different conclusion. For a more in-depth discussion on this subject readers should consult Thomas (1996: 168–88).

PLANNING IN THE 1980s

Writers such as Rydin (1993: 59–84) and Thornley (1991) on the subject of planning in the 1980s refer to it as an anti-planning era during which market forces were allowed greater sway over environmental issues or the views of third party objectors. It is certainly true that planning became more market orientated and concerned with economic regeneration, as planners turned their hands to job creation and regenerating local economies. They had already begun to do this in the 1970s, and many in the profession were irritated by the allegations implicit in central government circulars which suggested that they were delaying decisions on planning applications and thus delaying job creation. The Conservative government did streamline some of the planning procedures and took a more relaxed view in policy guidance; for example, allowing more out-of-town retail development. Developers also found that their success rate at appeal increased as the planning inspectorate interpreted the new relaxed policy stance towards business development. However, the degree of change that took place in the 1980s should be kept in perspective, as the government still supported many of the central planks of the planning system. For example, it was still impossible for a consortium of house-builders to obtain planning consent for private new settlements of between 3,000 and 5,000 dwellings in London's green belt and surrounding countryside (Ratcliffe and Stubbs, 1996: 486–90). These latter decisions show how political planning policy can be, since there is widespread agreement that the refusal of these new settlement proposals was largely to avoid alienating the traditionally Conservative vote in the shire counties.

At the beginning of the 1980s, the government introduced mechanisms aimed at regenerating the inner city. Perhaps the best-known initiatives are the Enterprise Zones (EZs) and Urban Development Corporations (UDCs) which were created under the Local Government, Planning and Land Act 1980. The EZs and UDCs sought to stimulate private sector investment within a simplified planning framework. Brownill (1990) notes that in the London Docklands Development Corporation area developers and investors sought the certainty that planning provides and the LDDC now produces planning and design guidelines for remaining sites within its area. Both the EZs and UDCs have stimulated considerable volumes of development activity, although there is still a debate about whether this development would have occurred in any case and just happens to be concentrated in the EZs (where there are tax breaks for capital projects) and in the UDC areas. In the latter, land and infrastructure were prepared by the UDC and sites often disposed of cheaply to the private sector to create a development momentum. There is also a view that the regeneration that has occurred may not be self-sustaining, nor match the employment needs of the very areas it was meant to regenerate. The debate, as Thornley (1996: 189–204) notes, has now moved on in the 1990s to issues of 'capacity-building' (something which Jane Jacobs would have referred to with approval as 'gradual money') in the realms of urban regeneration and sustainability in the wake of the Rio Earth Summit (referred to as Agenda 21). How then does the modern planning system of the 1990s engage with housing?

THE PLANNING SYSTEM AND HOUSING IN THE 1990s

In order to understand the present relationship between planning and housing, we must first consider the statutory and policy framework which gives us our modern planning system. This has developed through gradual legislative changes, discussed above in this chapter, and now has two distinct elements. The first involves a hierarchy of 'development plans' which ultimately allocate sites for land uses, including new housing. The second element is termed 'development control' and has to do with the grant or refusal of planning consent for individual schemes. Both of these elements are discussed in turn below.

Development plans

The system of development plans that has evolved since the 1947 Act is essentially hierarchical. Policy effectively cascades down from central government to the local borough or district level where policies are given local expression and land use allocations are made in local plans. At the top of the planning hierarchy is the Secretary of State, who is assisted by middle-ranking and junior ministers and a permanent staff at the Department of the Environment, Transport and Regions (DETR). It should be noted at this point that since May 1992, the Department of National Heritage (renamed Department of Culture, Media and Sport in 1997) has been made responsible for a number of heritage-related planning functions, such as the 500,000 listed buildings on the national list and 13,500 ancient monuments.

The Secretary of State at the DETR and his ministers are elected politicians and it is they who give direction to the planning system as they see fit (and as agreed in Cabinet). For example, there may have been a pre-election pledge in the governing party's manifesto to increase the scale of green belts around cities and so, once elected, the ministers will instruct officials at the DETR to draft policy (or if necessary legislation) to implement the pledge. All major pieces of draft policy and legislation which emanate from the DETR are subject to at least one round of consultation with local authorities, professional institutions, business and amenity groups before they are adopted as government policy (or legislation). In general terms legislation is altered only in exceptional cases where substantial change or new powers are required or where a consolidating Act is required. In the field of planning, the government has shown a willingness to use Circulars and more latterly Planning Policy Guidance Notes (PPGs) and Regional Planning Guidance Notes (RPGs) to provide direction to local authorities and the development industry on how central government expects the planning legislation to work in practice. The DETR has now produced more than twenty PPGs ranging across various topics such as *PPG2: Green Belts* (1995), *PPG17: Sport and Recreation* (1991) and *PPG6: Town Centres and Retail Developments* (1996b). The most relevant to this discussion is *PPG3 Housing* (1992a), which indicates how the planning system should interact with housing. Frequent reference will be made to *PPG3* below, as the discussion develops.

Regional planning guidance is also important to this discussion in that it provides dwelling targets for each of the standard regions and metropolitan areas in the country. For example, *RPG9* (Department of the Environment, 1994) provides regional planning guidance for the South East of England in which there are, among other things, dwelling targets for all county councils in the region. In 1995 the government issued a supplement to *RPG9* which outlines a development strategy for the Thames Gateway on the eastern side of London. In the regional guidance, Kent for example is expected to make land available to accommodate an annual average of 5,800 new dwellings between 1991 and 2006. These figures are derived partly from statistical projections made by the DoE and partly in consultation with SERPLAN, which is the regional standing conference representing the planning authorities in the South East. The agreed figures form the basis for the county councils to produce their Structure Plans which, among other things, identify the broad geographical spread for the new dwellings throughout the respective county. The Secretary of State issues separate strategic guidance for London and the other metropolitan areas in which there are also housing targets for individual boroughs and districts. A similar consultative process is undergone, this time with organisations such as the London Planning Advisory Committee (LPAC), before strategic guidance for London is issued.

County Structure Plans

In the shire counties the Structure Plan and its policies provide the framework within which individual boroughs and districts produce their more detailed Local Plans. During the preparation of a Structure Plan, there is an opportunity for organised pressure groups, such as the House Builders' Federation (HBF) and the Council for the Protection of Rural England (CPRE), to air their views at an Examination in Public (EiP), which is arranged by the relevant county council. At this stage in the Structure Plan's preparation, its policies are to some extent in a state of flux and organisations such as the HBF will seek alteration to policies in the plan before they become established county policy. The HBF invariably argues for increases to the new housing allocations in the plans and may also seek to influence the geographical spread of the housing. Conservation-driven organisations, such as the CPRE, often argue for reductions in proposed dwelling figures for rural shire counties, arguing that new housing should be targeted at brownfield sites or recycled land in cities. The individual county councils may also have strong views about what their 'carrying capacity' is for accommodating new dwellings.

There is a growing feeling on the part of some of the shire counties in the South East that their environmental capacity is limited, and at the relevant EiP for their Structure Plan they have made strong arguments to provide less housing than is suggested in the Secretary of State's regional guidance. The latter figures in regional guidance are thus guidelines and not mathematical absolutes, which allows some scope for policy interpretation at the county level. County councils such as that of Berkshire have in the early 1990s taken a much publicised protectionist view and sought to limit new house-building in the county. Such an approach indirectly seeks to restrict the in-migration to the county which additional housing would encourage. In simple terms, Berkshire is claiming that it is 'full up' and can only meet internally generated housing demand in future. The impetus for these arguments is as much political as rational, since county councillors in counties such as Berkshire take the view that they are protecting their county from over-development. This stance taken by councillors, in instructing their planning officers to resist higher housing targets, is very popular with the majority of the county's electorate, who may feel that their environment is under threat from more house-building. Of course the HBF sees this as unduly alarmist and a manifestation of the 'NIMBY' (not in my back yard) factor which in aggregate, the HBF claims, produces a shortage of land for housing and increases the cost of the final product beyond the means of low-income households.

Discussions at Structure Plan EiPs can thus become very polarised between development-driven and conservation-driven arguments. Ministers, through the regional offices of the DoE, also make representations during the Structure Plan preparation process and have the power to 'call in' any part (or all) of a plan which, it is considered, requires amendment. However, it is unusual for a minister to intervene in such a direct manner as, in the case of housing figures, a compromise is usually arrived at. The key point to bear in mind is that new dwelling figures adopted by each county are as much the outcome of compromise, argument and 'expert' opinion as they are of mathematical forecasting based on population projections.

Local Plans

In considering what actually happens to the agreed housing targets and the strategic location for new housing in a county in an adopted Structure Plan, it is necessary to bring back into the discussion the metropolitan districts and London boroughs. These are 'unitary authorities' since they carry out both county and local government functions in the absence of a higher-tier authority (the former Metropolitan Counties and GLC were abolished in 1985). Unitary authorities found favour with the last government and, following a recent local government review, a number of new unitary authorities were created.

The unitary authorities take their housing targets directly from the Secretary of State's Regional Planning Guidance Notes. *RPG3* (1996c), for example, provides strategic guidance for London. Whether the source is a Structure Plan or an RPG, the local authority is required under the 1990 Act to produce a Local Plan in which housing allocations are discussed. For example, the borough of Dartford is one of the twelve local authorities in the county of Kent. Dartford borough is required under the planning Acts to produce a Local Plan which is in general conformity with the Kent county Structure Plan. Dartford's Local Plan thus makes detailed allocations for housing land on its proposals map, linked to written policies in the plan governing, among other things, density, access and design. In London and the Metropolitan areas the unitary authorities produce a Unitary Development Plan (UDP), which has strategic and local elements (in effect a Structure and Local Plan rolled into one). UDPs also comprise detailed policies and a proposals map which shows, along with all other uses, future housing sites. The UDPs and Local Plans go through a number of stages of preparation during which there are statutory windows of opportunity for public consultation before the plan is adopted by the local authority. The exact sequence of events and procedures need not detain us here, although readers might wish to consult Telling and Duxbury for detailed coverage on this matter (1993: 45–77). At the point of adoption by the local authority, the plan will achieve its full statutory status, after which it becomes the first point of reference for local authorities in making development control decisions, as Section 54A of the 1990 Act states:

> where, in making any determination under the planning Acts, regard is to be had to the development plan, the determination shall be made in accordance with the plan unless material considerations indicate otherwise.

The above clause in the 1990 Act was inserted retrospectively by a clause in the 1991 Act, which heralded what has become known as the 'plan-led' system.

The effect of what appears to be a minor legislative change to an Act has been to stimulate considerable interest on the part of the development industry, landowners and conservation groups in influencing the content of Local Plans. The forum at which this is attempted is the Local Plan inquiry where various organisations and the general public can make verbal and/or written representations to a plan inspector on the contents of draft plans. From the point of view of a landowner, there may be considerable importance in making a case that the Local Plan should allocate a tract of land for housing development rather than green belt. If the landowner is successful in his or her arguments, the ensuing rise in land value may well run into millions of pounds. The stakes are thus high in convincing a plan inspector that a parcel of land could more suitably be used for new housing than its existing use as part of the green belt.

Local Plans vary considerably from locality to locality, depending on the local geography and demographic make-up of the area. The myriad of local circumstances cannot be anticipated by the DETR, although guidance is provided to local authorities on how to draft robust policy which is able to achieve the aims and objectives of the particular planning authority (DoE, 1992b).

Housing policies in Local Plans: the example of the Greenwich UDP

Housing policies in a Local Plan usually form one chapter in a comprehensive document which considers many other planning topics. For example, the London Borough of Greenwich's Unitary Development Plan (1994) contains fourteen topic chapters which include jobs and the local economy, design and conservation, shopping, movement (transport) and tourism. Housing is therefore an important, but by no means the only, topic which the planning authority has to consider. Development plans such as the UDP look forward over ten to fifteen years, although they are all periodically reviewed – usually after five years.

The thirty-two housing policies in the Greenwich UDP are obviously specific to that London borough, although they are not in any way unusual in terms of 'typical' housing policies for an urban area. A brief examination of some of the housing policies in the Greenwich UDP may thus serve to illustrate the extent of the current relationship between Local Plans and housing in an urban context.

The Greenwich UDP provides a housing target of 11,000 new dwellings over the plan period 1987 to 2001, which follows precisely the borough's expected contribution towards London's overall housing allocation in the Strategic Guidance for London (*RPG3*). The plan's proposals map identifies the main housing sites which, when developed, will contribute to the 11,000 dwellings. There is also, in common with other urban Local Plans, an implicit recognition that approximately 50 per cent of all new housing will result from the re-use of brownfield sites. Thus the UDP includes a catch-all policy for such sites, which must be environmentally suitable for housing and consistent with other policies in the plan on affordability, design, privacy, and so on.

The overall volume of housing provision at 11,000 new dwellings over fourteen years is also pitched at a realistic level, since Greenwich, like other London boroughs and metropolitan districts, is already densely developed and there are few if any greenfield sites available for housing. Green land in boroughs such as Greenwich is usually protected in the Local Plan by designations such as Metropolitan Open Land or Green Chains which supplement parks and commons.

The UDP also discusses the growing number of smaller households and a growing affordability gap between households on low incomes and average dwelling prices in the borough. The UDP recognises the need for affordable housing and there is a policy to the effect that the council will seek to ensure that 30 per cent of new housing provided will be affordable to those on low incomes. Affordability policies are now common in most urban and rural Local Plans following recognition of this issue by the government in the late 1980s. In addition, as Bramley, Bartlett and Lambert point out (1995: 81–4), affordability can be pursued by local authorities by way of planning agreements under Section 106 of the 1990 Act and *PPG3* acknowledges that a community's need for affordable housing is a material consideration. *PPG3* advises that the best use of legal agreements to achieve affordable housing objectives is to involve housing associations or trusts to provide housing for rent (or shared ownership) so that the benefits of affordability are experienced by both initial and subsequent occupiers. Local need would thus have to be clearly demonstrated by the local authority before it embarks on this course of action. Demonstrating local need in a borough like Greenwich is not particularly difficult since, as the UDP points out:

> In the third quarter of 1994 the average price of a dwelling in the borough was £69,000 and the average price of a one bedroom dwelling was £47,000. As over half (53 per cent) of the households in Greenwich have incomes of less than £10,000 per year this level of house prices puts home ownership out of the reach of the majority of households, including those currently lacking their own separate dwelling.
>
> (London Borough of Greenwich, 1994: 63)

Guidance and the thresholds for affordable housing have been adopted by the government in *Circular 13/96* entitled *Planning and Affordable Housing* (DoE, 1996a).

With regard to the existing stock of dwellings in Greenwich, the UDP indicates that the council will normally refuse planning permission for demolition or changes of use of residential accommodation which would result in a net loss of dwellings in the borough. Again this is a conventional approach taken in Local Plans to protect and foster the existing stock of housing.

The density of new housing development is also covered in the UDP, which provides density ranges for different types of housing. For example, a density range of 60–75 habitable rooms per acre (HRA) is prescribed for family dwellings and households with special needs. Higher densities are permitted for non-family dwellings such as those on sites with good access to public transport (up to 100 HRA) and riverside sites (120 HRA). The UDP policy states that criteria also have to be met in conjunction with the density policies which seek to strike a balance between the efficient use of scarce housing land and the ability to meet car parking standards, landscaping and adequate design generally. The use of habitable rooms per acre (or hectare) is a standard way of controlling density in local plans. A 'habitable room' in this context includes all separate living rooms including bedrooms, but excluding bathrooms, toilets, landings, halls and lobbies. Kitchens are counted as habitable rooms if they have a floor area greater than 13 square metres.

Like all Local Plans, the Greenwich UDP contains car parking standards which, depending on the context, are applied as maxima or minima. For example, in relatively isolated locations or those that are not particularly well served by public transport, there is a recognition that most movement will be by car. In these circumstances car parking standards are applied as minima; for example, a minimum number of parking spaces must be provided on the site in relation to the volume of development measured in the number of dwellings (or gross floorspace for commercial developments). The reverse is true in town and city centres, where local authorities try to discourage the use of the private car in favour of public transport. In such locations the parking provision on development sites is restricted.

In a borough such as Greenwich, which has a traffic congestion problem and double-parking in most residential streets, the council tries to strike a balance between further on-street parking and restriction of the use of the car for private commuting into the borough's commercial and employment centres. There are thus two separate parking standards. For a typical residential development in a residential part of the borough, the UDP seeks a minimum provision of one car parking space in the curtilage (i.e. off-street) of a one- or two-bedroom dwelling. For a three- or four-bedroom dwelling, the requirement is a minimum of two car parking spaces within the curtilage. However, in Greenwich town centre a maximum of one parking space is allowed for every 750 square metres of office floorspace in a development

The Greenwich UDP policy on the sub-division of existing housing into flats is cautious and guarded, indicating that a series of criteria such as adequate parking and minimum room sizes need to be achieved before consent for sub-division is granted. While there is no commonly adopted standard for minimum internal room sizes in new and converted dwellings, these are material considerations which the local authority can take into account when granting or refusing planning consent. Greenwich's UDP encourages changes of use to residential from other uses where an acceptable standard of accommodation can be met.

In terms of backland and infill development, the UDP policy states that this will only be allowed in exceptional circumstances where specific criteria are met. The latter include consideration of privacy, character and amenity space; that is, where these are threatened by backland development, consent will not be granted by the council. Design and conservation policies are covered in another chapter in the Greenwich UDP, although references to design do permeate virtually all of the housing policies in the UDP ranging from landscaping, noise attenuation and internal space standards to storey height. Interestingly, and following experiences similar to those of other boroughs with high-rise housing (discussed elsewhere in this chapter), the council indicates in the UDP that it will not normally grant consent for blocks of flats higher than three storeys. It will though make exceptions for particular client groups (small households on riverside sites, students' housing). The UDP also contains policies which encourage the provision of sheltered housing and special needs housing, and there are mobility policies for households with disabilities.

Local Plans in rural areas

In rural areas Local Plans will seek to make land available for new housing in, or on the edge of, larger settlements which already have the social and physical infrastructure to support an increase in population. Thus in general terms, the larger villages, which are sometimes referred to as 'key settlements', tend to get larger, while the smaller villages remain on the same scale within their village envelopes. Land is, however, allocated for new housing and the sites are shown on the district's or borough's proposals map in much the same way as discussed above for the UDP. However, outside the metropolitan areas, *PPG3* requires that local authorities are able to identify the availability of a five-year supply of housing land judged against the target set in the appropriate Structure Plan. Local authorities are encouraged to produce with housebuilders joint land availability studies to identify genuinely available sites.

As in the Greenwich example, there is in rural areas also an issue of housing need. In rural areas there is often fierce competition in housing markets, which stems from those seeking second homes, retirement homes or commuter homes in the countryside, competing with local people seeking to get on to the housing ladder. Incomes for those employed in agriculture and related industries are generally considerably lower than those achieved by the urban middle classes who are seeking to move into the countryside. The result is that very often local people in rural areas simply cannot compete in the open market for housing in their own villages.

As in many urban areas, council housing is in increasingly short supply due to right-to-buy policies and a dearth of council house-building during the 1980s and 1990s. This is especially true of the countryside around London and also in the scenically attractive parts of the country, such as the

National Parks. In the Lake District (a designated National Park), for example, the first statutory responsibility for the local authority (the Lake District Special Planning Board) is to preserve the scenic beauty of the area. All other considerations follow this responsibility, which, as Shucksmith points out (1990: 107–37), causes some frustration for the authority, as the authority also recognises the need to provide homes for local people to underpin a working countryside. Throughout the 1980s the Lake District planners sought to restrict the occupancy of new homes for local people by way of legal agreements and conditions limiting occupancy to those who were employed locally. However, Shucksmith argues that the occupancy controls created difficulties for local people in raising mortgages, due to the restricted resale opportunities. It was also felt that non-locals were being explicitly discriminated against and that existing housing (without occupancy restrictions) became such a scarce resource that it experienced inflation way above the norm.

The government's response in the 1980s to local needs policies was opposition, as it was felt that planning should concern itself only with matters of land use and not occupancy characteristics or the price or tenure of housing. However, current central government advice to local authorities is contained in *PPG3* and *Circular 13/96*, which together broadly endorse the use of legal agreements to secure affordable housing. In rural areas the same approach can be supplemented with what has become known as an 'exceptions' policy. *PPG3* explains this as follows:

> In many rural areas there are particular difficulties in securing an adequate supply of affordable housing for local needs. A supply of such housing may be needed to secure the viability of the community. The existence of arrangements to ensure that new housing would be made available for local needs is a material consideration which the authority should take into account in deciding whether to grant planning permission. Such considerations should particularly be taken into account by authorities in considering the release of small sites which development plans would not otherwise allocate within or adjoining existing villages, and on which housing would not normally be permitted.
>
> (DoE, 1992a: Annex A)

Affordability and planning

Local authorities have thus been given some power, via the planning system, to address local needs housing in rural and urban areas. There are no reliable statistics available yet to determine whether this policy approach has made any significant inroads into the affordability problem on a national basis. The Joseph Rowntree Foundation held an *Inquiry into Planning for Housing* (1994) which, among other things, evaluated the various means by which the planning system was being used to achieve affordable housing in both rural and urban contexts. The inquiry panel consulted widely with the relevant organisations and concluded that while the various methods used to bring about affordable housing had some merit, there were also some disadvantages in terms of implementation and uniformity across the country. The panel also concluded that it was not the responsibility of the planning system to try to substitute for a non-existent council house-building programme:

> Such advantages as they possess are not so overwhelming as to divert us from the view expressed earlier [in the inquiry document] that the planning system should not be expected to play a leading role in the provision of social housing.
>
> (1994: 10)

There have often been criticisms, particularly from the House Builders' Federation, that planning makes housing more expensive (and therefore less affordable) by artificially restricting the supply of available housing land. The HBF's argument is that a general relaxation of planning controls to allow house-building on greenfield sites will dramatically increase the volume of the stock and reduce unit price. The Joseph Rowntree Inquiry considered this matter, as did a more extensive research project conducted by Bramley, Bartlett and Lambert (1995). The findings of the latter research were that if land allocations for housing in development plans were significantly increased across the country there would be only slight effects on price. Indeed the housing benefits of breaking into the green belt were calculated by the team to be a 2 per cent increase in overall output and a 1.2 per cent reduction in house prices (1995: 165–7). The research team also felt that a relaxation policy would have little effect on affordability in the short term. This is because new housing comprises a tiny percentage of the overall housing stock, so its price tends to be controlled in the short term by the price of existing homes. The Joseph Rowntree Inquiry had concurred with these findings, adding that:

> We recognise that there will always be a degree of constraint on land supply through the planning system and that this has a general impact by increasing prices. The recent boom and bust have shown, however, that prices are affected first by market changes rather than by planning controls. Levels of employment, changes in real income and the availability and cost of mortgages all affect the market.
>
> (1994: 7)

The development control system

Having considered the development plans system, we now need to consider the other half of the planning system, which is development control. The development control system essentially requires local authorities to scrutinise and grant or refuse planning consent for anything that is defined as 'development' under Section 55 of the 1990 Act. The all-embracing statutory definition of development is as follows:

> the carrying out of building, engineering, mining or other operations in, on, over or under land, or the making of any material change of use of any buildings or other land.
>
> (1990 Act: Section 55)

Section 13 of the Planning and Compensation Act 1991 added to the definition of development the demolition of dwellings, although the demolition of most other types of buildings (provided they are not listed or in conservation areas) does not come within the definition of development and therefore does not require planning consent.

The reference to 'change of use' in Section 55 above does require some explanation and reference to secondary legislation which is known as the Use Classes Order. This statutory instrument is kept under review by government and was last amended in 1995. The order defines eleven use classes within which most activities easily sit. For example Class A1 includes shops of all types, including hairdressers, undertakers and travel agents. This use class is separate and distinct from Classes A2 (financial and professional services) and Class A3 (food and drink). Residential use is contained in Class C3, which contains dwellings, small businesses at home and communal housing for the elderly and handicapped. Class C1 contains hotels and boarding houses, while Class C2 contains residential schools, colleges and hospitals. Uses which do not conveniently fall within a use class, such as theatres, are termed *sui generis* (in a class of their own).

The point of the use classes order is that planning consent is required to change the use of a building, or piece of land, from one use class to another (although there are specified exemptions). This allows local authorities to scrutinise proposed changes in their areas and if necessary to refuse consent for a proposed change of use which would be harmful to amenity in some way. For example, a local authority might refuse consent for a proposed change of use from a dwelling (Class C3) to an industrial unit (Class B2) on the grounds that the proposed industrial unit would harm existing residential amenity and result in the loss of much-needed housing in an area where housing was in short supply. However, approximately 80 per cent of the 500,000 planning applications received by local authorities each year are permitted by local authorities. This would indicate that the majority of proposed changes of use and physical development are not deemed to be harmful. As mentioned above, there are also exemptions from the need to obtain planning consent for some changes of use and minor developments (discussed further below); this is referred to as 'permitted development'.

The legislation thus defines what development is and, subject to exemptions, what therefore comes before local authorities for decision-making. It is government policy that provides the subtlety in how the legislation is implemented. The key policy advice for local authorities, as far as the operation of the system is concerned, is contained in *PPG1: General Policies and Principles* (DoE, 1997). A key extract from *PPG1* states that:

> If the development plan contains material policies or proposals and there are no other material considerations, the application or appeal should be determined in accordance with the development plan.
>
> (DoE, 1997)

The general advice in *PPG1* on how the system should work is reinforced by ministerial statements and keynote speeches at conferences from time to time and through other PPGs, Circulars and policy statements. Appeal decisions made by the inspectorate and supported by the Secretary of State also contain clues as to how government expects the system to operate, although individual development control decisions do not necessarily set precedents in the way that legal judgements may do. In development control each application is judged on its individual merits, since no two pieces of land or development proposals can be exactly the same. This explains the statement in *PPG1* requiring that applications be judged against the provisions of the development plan, although some account needs to be taken of other 'material considerations'. The scope of material considerations would include, for example, objections raised by members of the public, demonstrable need for the particular project, access, and design. However, the profit which would arise from the development for the owner is not normally considered to be material and nor is the extent to which a development might create commercial competition between users and investors in land.

The system purports to operate in the 'public interest'; that is to say, planning decisions should take into account the widest possible range of views that have a bearing on the particular decision and should not be influenced unduly by power, wealth or a narrow sectional interest. Sometimes it is not entirely clear where the public interest begins or ends, as objectors to planning applications may feel that they are the legitimate representatives of the public interest, while in fact they may be driven by quite selfish motives. It is for planning officers and planning committees to try to distinguish between what are legitimate objections 'in the public interest' and what are over-sensitive complaints against a perfectly acceptable development proposal.

There is also the rather delicate and enigmatic concept of 'amenity', which is often cited in planning decisions. A broad definition of 'amenity' is that it is the pleasantness or positive attributes of a particular place. Thus amenity is a variable concept, depending on location. Virtually everywhere, though, will have some degree of amenity by virtue of attractive views, peace and quiet, mature trees, open space or fine architecture. In some locations, such as a conservation area, the proposal to build a warehouse may be deemed to be detrimental to the amenity of the area. In another context, such as an industrial estate, the same development may not detract from amenity and will therefore be acceptable.

The development control system is thus complex, and there are those (including town planners) who feel that there is scope for rationalising some of the procedural aspects of the system. For those readers who are new to this system, a discussion on how a minor development proposal (an extension to an existing house) is processed may serve at this point to clarify matters.

Householder planning applications

Home-owners in established residential areas often seek to extend their houses to meet the demands of a growing household for increased space. Very often these needs derive from growing family size or the desire to add a 'granny annexe' so that families can live together in a satisfactory manner, without needing to move house. The desire to improve, particularly older dwellings, is also a legitimate reason for seeking to extend the available living space on a housing plot. The Town and Country Planning (General Permitted Development) Order 1995 provides thresholds of permitted development, which exempt from planning control house extensions up to a certain size. For example the GDO (as the order is more commonly known) allows an extension of 50 cubic metres to a terraced house and 70 cubic metres to a detached house (there are other detailed permitted development rights in the GDO). Of course the owner of a terraced house may wish to extend his or her house to a greater extent than 50 cubic metres and would in that eventuality need to seek planning permission from the local authority by completing the relevant forms, submitting a set of plans and an application fee. The local authority is required to issue a decision on the application within eight weeks, after which time the applicant may appeal directly to the Secretary of State on the grounds of 'non-determination' by the local authority. The majority of householder applications are decided within the statutory eight weeks, although, where there is an over-run, the agreement of the applicant is sought by the authority and the application is determined in the normal way.

The normal processing of householder planning applications involves some minimum publicity as required under the 1991 Act. In the case of a minor application such as an extension, there would need to be either a site notice posted outside the house for a minimum of twenty-one days or direct neighbour notification by the local authority. The purpose of the publicity is to provide members of the public (usually neighbours) with an opportunity to visit the local authority offices and inspect the proposals before the local authority has made a decision on the application. Members of the public may then make representations to the local authority in support of or opposition to the proposals. The planning case officer must then evaluate the responses to consultation, besides conducting a site visit and evaluating the submitted application and any other material considerations. The latter may include any relevant Local Plan policies on extensions or design guides issued by the local authority. The planning officer would then produce a short committee report summarising the relevant facts of the case, concluding with a recommendation for approval (with conditions) or refusal, depending on the merits of the planning application. For example, a modest extension which met the policy criteria and did not affect the neighbouring property would probably receive a recommendation for approval. On the other hand a large intrusive extension proposal which overlooked and overshadowed neighbouring properties and resulted in the loss of trees in the garden of the property might well receive a recommendation for refusal.

The final decisions on the majority of housing-related planning applications are taken by a planning committee made up of elected councillors. The councillors who make the decision are not necessarily bound by the recommendation of their planning officer, although councillors must demonstrate clear planning reasons as to why they have not followed their officer's advice. In the majority of cases and after a discussion during which the applicant and any objectors may be offered the opportunity to speak, the councillors generally follow the advice in their officer's report. The reasons for the decision have to be made explicit in the decision letter issued to the applicant by the authority following the planning committee meeting.

In the case of a refusal the applicant does have a right of appeal to the DETR and he or she may appeal by one of a number of different methods. If the applicant is to be successful with the appeal, he or she must successfully counter the reasons for refusal cited in the local authority's decision letter. In the majority of cases (over two-thirds) local authorities 'win' planning appeals; that is to say, the inspector dismisses the applicant's appeal.

Listed buildings and conservation areas

In conservation areas, or where there is a listed building involved, the stakes are effectively raised for the home-owner seeking an extension to a dwelling. There are now close to 500,000 listed buildings and an estimated 7,500 conservation areas in England and Wales. The vast majority of listed buildings are dwellings of one type or another and conservation areas cover vast tracts of attractive residential areas. For example, the 'urban village' of Blackheath in south-east London is almost entirely covered by conservation area status and a high proportion of the buildings within the village are listed.

The listing of a building records its 'architectural or historic interest', to quote from Section 1 of the Planning (Listed Buildings and Conservation Areas Act) 1990. The majority of the buildings on the list are Grade II, while buildings which are deemed to exhibit greater architectural or historic importance are listed as Grade II* or in exceptional cases Grade I. It is not unusual, for example, for a Georgian terrace designed by an unknown architect to attract listing at Grade II because of its group value, while a seventeenth-century church designed by Sir Christopher Wren might attract Grade I listing status. The latter has greater architectural and historic importance than the Georgian terrace and therefore attracts a higher listing status. There is, incidentally, no right of appeal against listing, although the Department of Culture, Media and Sport has introduced a consultation process for owners of buildings and members of the public to give their views when there is a proposal to add a building to the list. The presumption with regard to listed buildings is that they are preserved, and demolition consent will not normally be granted unless a number of tests have been gone through by the applicant.

Home-owners who occupy listed buildings usually respond to the enhanced prestige that listing implies and research suggests that listed houses hold their value better than unlisted buildings during turbulent market conditions. Estate agents also make great play of the fact that a building is listed, the implication being that the heritage value associated with listing makes the property in some way superior to an unlisted house.

The owner of a listed house needs to consider a number of issues if he or she is proposing to extend the building. For a start, the permitted development rights which are available to the owner of an unlisted house are restricted or removed for the owner of a listed property. Virtually any development affecting a listed building must go before a planning committee, where the application will be discussed, taking into account the views of local amenity societies that may have objected to the proposals. In some cases the planning committee might defer its decision by asking the applicant to resubmit revised drawings which overcome, for example, a perceived weakness in design. There is not a complete embargo on altering listed buildings or allowing new development in a conservation area, as many people think. There are numerous examples across the country where sympathetically designed extensions to listed buildings and modern development have been allowed to go ahead in conservation areas. Failure to obtain consent before carrying out works to listed buildings can, however, result in the owner having to reinstate the building to its original condition and there might in addition be fines of up to £20,000 for the owners.

To help home-owners and indeed local authorities implement the provisions of the Act, the Departments of Environment and Heritage have jointly published *PPG15* which is entitled *Planning and the Historic Environment* (DoE/DNH, 1994). *PPG15* contains clear advice on the procedures to follow and ensures that the 'rules' are implemented evenly across the country with regard to listed buildings and conservation areas. Readers who wish to find out more about this specialised area of planning should consulting an up-to-date planning law book or examine an excellent historical review and explanation of the workings of the listing procedure contained in Ross (1996). Ratcliffe and Stubbs (1996: 121–30) provide examples of conservation areas, a summary of the history of the concept and emerging issues associated with conservation area designations.

Planning applications for major developments

The ground rules for determining a major planning application, for example for a new settlement containing 3,000 dwellings and employment uses on a greenfield site, are similar in principle to the householder application discussed above. The differences are mainly related to scale. For example, the major house-builder who is proposing this hypothetical scheme would not want to undertake expensive design work and secure an option to purchase the site if there was no real prospect of obtaining planning permission to construct the development. The developer would normally undertake a careful analysis of the site and relevant planning policy contained in the Local Plan, as well as relevant national, regional and county planning policy. Once the house-builder was satisfied that there was potential for obtaining planning permission on the site, the company would begin a dialogue with the local authority to establish any specific planning requirements that the local authority might have. The negotiation process as Claydon (1996: 110–20) describes it between potential developer and local authority is an accepted and important part of the process. It is at this stage, when the scheme is in a state of flux, that the local authority can seek improvement, as it sees it, to a scheme.

The local authority can, for example, seek a legal agreement with the developer which is contingent on the grant of planning permission. The agreement may require the developer to upgrade access roads into the site in order to make the scheme more acceptable to the authority in planning terms. Such agreements are seen by local authorities as a legitimate means for implementing their development plan policies, often in the field of housing provision, although, as Allinson and Askew point out (1996: 62–73), agreements are difficult to negotiate where development pressure is weak. Major development proposals often end up being determined by way of a public inquiry. Ministers have powers under the 1990 Act to 'call in' major or controversial schemes. In general, developers have greater resources to bring to bear at such inquiries than third party organisations and local authorities, which cannot afford to employ the top planning barristers and expert witnesses to argue their case. However, it is on the planning merits of the case, and not the seniority of the legal representation, that an inquiry inspector will make his or her decision. The stakes are high in relation to unreasonable decision-making and appeals in the case of major development proposals. An unreasonable appeal might result in a developer meeting all the costs of a public inquiry and similarly an unreasonable refusal of consent on the part of a local authority which led to an appeal would see costs awarded against the authority.

ENVIRONMENTAL ASSESSMENT

In the case of a major housing development proposal, it is likely that there would be an additional requirement for an environmental assessment. This requirement is established by the Assessment of Environmental Effects Regulations (1987), which were themselves a direct result of a European Community Directive in 1985 (CEC 1985). The process of environmental assessment (EA) seeks to screen out potential sources of environmental pollution from a development project while it is still at the drawing board stage; that is, to prevent pollution at source. At the initial stages of an EA, the developer and the local authority agree the 'scope' of the assessment. In the case of a major housing proposal, the EA might involve expert appraisal of the following topics: traffic, groundwater and surface water, ecology and nature conservation, air quality, and visual impact. It is the developer's responsibility to assemble an expert team to conduct the EA and submit the ensuing environmental statement with the planning application for the proposed scheme. Failure to do so will invalidate the planning application where the regulations define a project as requiring an EA. The environmental statement should suggest mitigation measures which can form the basis of either a legal agreement or conditions to ensure that the project is implemented in an environmentally appropriate way. Of course if the EA process shows the development proposal to be environmentally damaging, the local authority is able to use the evidence submitted in the environmental statement as grounds for refusal of the project. In theory at least, the EA Regulations should lead to all major development projects being significantly more environmentally friendly and the screening out entirely of environmentally damaging projects. For a more detailed coverage of this specialist subject, readers should consult Glasson, Therivel and Chadwick (1994).

PLANNING AND HOUSING: THE FUTURE AGENDA

The discussion above on the development plans and development control systems highlights a number of critical points as far as the relationship between housing and planning is concerned. In the first place it is not planning that produces housing, but mainly the private sector supplemented by housing associations and local authorities. These are the customers of a planning system which has wider responsibilities than just the consideration of housing issues in isolation (important as those are). As the discussion above shows, planning does, however, engage with housing in a number of important senses and it is at these points of engagement that the debate will continue into the twenty-first centry.

Too many households, not enough land?

One of the critical points of engagement is in the matching of future households to future dwelling supply. The term 'household' is officially defined in the census as 'one person living alone or a group of people (who may or may not be related) living or staying temporarily at the same address with common house-keeping'. Households may seek dwellings through the market mechanism (i.e. demand) or they may need dwellings, but be unable to purchase or rent on the open market. The production of regional dwelling targets in Regional Planning Guidance seeks to provide dwellings for all future households; that is, those who can 'demand' a dwelling plus those households in 'need' of a dwelling. The regional dwelling targets, if added together, should in theory provide a national total for new dwellings so that the number of new dwellings provided should match the number of new households forming and looking for housing.

Unfortunately, the process which produces both sets of figures (households and dwellings) is not an exact science, because it involves the prediction of the future based on current trends. For example, before they are officially adopted the regional dwelling figures are subjected to debate and opinion by politicians and their expert advisers, and adjustments are made on the basis of assumptions about patterns of future migration into and out of a particular region. Thus a summation of all the regional figures for future dwelling requirements should be in the right 'ball park' but may not be entirely what is actually needed on a national basis to match new households coming forward. The government has resisted publishing future national housing targets and instead prefers to work through this regional target system.

The dwelling figures are revisited periodically, as government statisticians estimate future population based on available census data and then calculate the number of households that this will give rise to, on the basis of prevailing household formation rates. However, one of the key variables and assumptions that have to be made in the calculation of future households is the future rate of net in-migration into the country. The latter will always be subject to domestic and international political change as well as international patterns of movement. The variables cannot be predicted in advance with absolute confidence.

The most up-to-date household projections were published recently by the DoE (1995) and assume a net in-migration to the UK of 50,000 people per year. The latter assumption is much higher than that assumed in the previous figures published in 1989 and upon which existing dwelling forecasts were made. The 1995 household projections indicate that between 1991 and 2011 the country needs to provide for an additional 737,000 households on top of the 1989-based figures for the same period

(Breheney and Hall, 1996: 3–6). Put another way, the figures suggest that for each year over the twenty-year period from 1991, an additional 36,850 households will form that have not been planned for. This may seem like an alarming figure, but in reality there will not be 36,850 families added to the homelessness totals every year. What is most likely to happen is that the considerable land allocations for new housing made in plans across the country will be used up faster than was originally envisaged.

The new household forecasts have generated a lively debate in the various professions and the development industry. Davies (1996) on behalf of the RTPI, for example, advocates a review of green belt boundaries allied to sequential tests which would prevent the development of greenfield sites as the first and easiest option. The matter is felt to be of such importance that the Town and Country Planning Association conducted an inquiry into the issue, by consulting experts and practitioners in the field across the country. The results of this inquiry have been published (Breheney and Hall, 1996) in a document entitled *The People: Where Will They Go?* – a direct reference to the subtitle of Ebenezer Howard's 'three magnets' diagram of almost a century ago (Howard, 1985). The question thus arises again in the 1990s, Where will the additional households that have been predicted be housed? The TCPA has made a series of wide-ranging recommendations aimed at government and local authorities which include recommendations on producing, verifying and clarifying delivery mechanisms for housing targets in future. With regard to the location of future housing, the TCPA supports the recycling of urban sites advocated by the government and currently accounting for approximately 50 per cent of all housing land. However, this should not include the loss of urban green sites, which at present is accounting for another 12 per cent of all housing sites. The TCPA believes that the government should set targets for the use of rural land for housing and that greater attention should be given to housing need. The TCPA also believes that the role of the planning system in promoting and delivering social housing should also be strengthened. With regard to the location of new housing development the TCPA advocates that:

> The scale of required housing development is such that single solutions – infill, urban extensions and new settlements – will not be sufficient in most areas. Some combination of solutions will be required. These must be tailored to regional and local circumstance. Sustainability principles must underlie the chosen portfolio.
>
> (Holimans 1996: 73)

Proposals for private sector new settlements in the late 1980s were resisted by government (as discussed earlier, this chapter), since when the TCPA has constantly campaigned for a more relaxed approach to this type of housing solution. The current policy in *PPG3* gives support to new settlements only in very exceptional circumstances and when six strict criteria are met (DoE, 1992a: paras 32–37). No doubt the revised household figures will see a resurgence of interest on the part of government and the private sector in this modern-day manifestation of Ebenezer Howard's garden cities.

CONCLUSIONS

Town planning legislation in the United Kingdom has since its inception been closely related to housing development – housing being, in general, the largest single urban land use and the principal form of encroachment upon the countryside. Until the Town and Country Planning Act of 1968, *blueprint planning*, evolving from the technical skills of the architect, engineer or surveyor and a succession of earlier Acts, ensured that the relationship between planning and housing development was viewed largely in its physical context, and solutions to related problems were similarly physical – involving zoning, density controls, building regulations, planning standards and design. But since the 1968 Act, it has been increasingly recognised that the complex problems relating to planning and housing development cannot be examined, and solutions cannot be found, in purely physical terms. Continual reference to economic and social considerations is necessary. Whereas blueprint planning provided the framework in which the housing market operated, the market very largely provides the framework in which the successor to blueprint planning, *process planning*, is undertaken. A substantial increase in the number of households and the associated increase in the demand for housing, for example, thus calls for a response from process planning probably necessitating a revision of plans.

Reference to Chapters 1, 2 and 3 will enable the reader to understand how market forces, housing policy and finance relate to both blueprint and process planning discussed in this chapter, while the consideration of the problems of housing design and development in Chapter 6 can be enhanced by an understanding of past and present planning policies.

QUESTIONS FOR DISCUSSION

1. **To what extent do housing and town planning share a common history and philosophy?**
2. **Do the pioneering model settlements from the last century provide any lessons for those involved in housing management and development today?**
3. **Is the contemporary planning system a suitable vehicle for delivering social housing?**
4. **Should planning controls in green belts and other designated areas be relaxed to enable the private house-builder to meet housing demands in pressurised areas like the South East?**
5. **Given that demographic indicators predict a significant rise in the number of households seeking homes, should the new towns programme be restarted?**

RECOMMENDED READING

Adams, D. (1994) *Urban Planning and the Development Process*, London: UCL Press.
Cullingworth, J.B. and Nadin, V. (1997) *Town and Country Planning in the UK*, London: Routledge.
Greed, C. (ed.) (1996) *Implementing Town Planning*, Harlow: Longman.
Ratcliffe, J. and Stubbs, M. (1996) *Urban Planning and Real Estate Development*, London: UCL Press.
Ross, M. (1996) *Planning and the Heritage*, London: Spon.
Rydin, Y. (1993) *The British Planning System*, London: Macmillan Press.

REFERENCES

Adams, D. (1994) *Urban Planning and the Development Process*, London: UCL Press.
Allinson, J. and Askew, J. (1996) 'Planning Gain', in C. Greed (ed.) *Implementing Town Planning*, Harlow: Longman.
Arnstein, S.R. (1969) 'A Ladder of Citizen Participation', *Journal of the American Institute of Planners*, 35: 4.
Booth, P. (1996) *Controlling Development*, London: UCL Press.
Bramley, G., Bartlett, W. and Lambert, C. (1995) *Planning, the Market and Private Housebuilding*, London: UCL Press.
Breheney, M. and Hall, P. (1996) *The People: Where Will They Go?*, London: Town and Country Planning Association.
Brownill, S. (1990) *Developing London's Docklands*, London: Paul Chapman.
CEC (1985) Directive EC 85/337, Brussels, CEC.
Claydon, J. (1996) 'Negotiations in Planning', in C. Greed (ed.) *Implementing Town Planning*, Harlow: Longman.
Coleman, A. (1990) *Utopia on Trial*, London: Hilary Shipman.
Collins, G.B. and Collins, C.B. (1965) *Camillo Sitte: The Birth of Modern City Planning*. London: Phaidon Press.
Collins, M. (1994) 'Land-Use Planning since 1947', in J. Simmie (ed.) *Planning London*, London: UCL Press.
Cullen, G. (1971) *The Concise Townscape*, London: Architectural Press.
Cullingworth, J.B. (1996) 'A Vision Lost', *Town & Country Planning*, 65, 6: 172–175.
Cullingworth, J.B. and Nadin, V. (1997) *Town and Country Planning in the UK*, London: Routledge.
Darley, G. (1991) 'Tomorrow's New Communities: the Lessons from Yesterday', in G. Darley, P. Hall and D. Lock, *Tomorrow's New Communities*, York: Joseph Rowntree Foundation.
Davies, N. (1996) 'Institute Blocks Easy Option', report in *Planning* no. 1181.
Department of the Environment (1973) *Circular 24/73: Development for Agricultural Purposes*, HMSO.
—— (1991) *Planning Policy Guidance Note 17: Sport and Recreation*, HMSO.
—— (1992a) *Planning Policy Guidance Note 3: Housing*, HMSO.
—— (1992b) *Development Plans: A Good Practice Guide*, HMSO.
—— (1994) *Regional Planning Guidance Note 9: Regional Planning Guidance for the South East*, HMSO.

—— (1995) *Planning Policy Guidance Note 2: Green Belts*, HMSO.

—— (1996a) *Circular 13/96: Planning and Affordable Housing*. HMSO.

—— (1996b) *Planning Policy Guidance Note 6: Town Centres and Retail Developments*, HMSO.

—— (1996c) *Regional Planning Guidance Note 3: Strategic Planning Guidance for London Planning Authorities*, HMSO.

—— (1997) *Planning Policy Guidance Note 1: General Policies and Principles*, HMSO.

Department of the Environment and Department of National Heritage (1994) *Planning Policy Guidance Note 15: Planning and the Historic Environment*, HMSO.

Dunleavy, P. (1981) *The Politics of Mass Housing in Britain 1945–1975*, Oxford: Clarendon Press.

Dyos, H.J. and Wolff, M. (eds) (1973) *The Victorian City*, vol. II, London: Routledge & Kegan Paul.

Essex County Council (1973) *A Design Guide for Residential Areas*, Chelmsford: Essex County Council.

Fishman, R. (1992) 'The American Garden City: Still Relevant?', in S.V. Ward (ed.) *The Garden City*, London: Spon.

Glasson, J., Therivel, R. and Chadwick, A. (1994) *Introduction to Environmental Impact Assessment*, London: UCL Press.

Greed, C. (1993) *Introducing Town Planning*, London: Longman, Harlow.

Grimley J.R. Eve, Thames Polytechnic and Alsop Wilkinson (1992) *The Use of Planning Agreements*, DoE, HMSO.

Guiton, G. (ed.) (1981) *The Ideas of Le Corbusier on Architecture and Planning*, New York: George Braziller.

Hall, P., Gradey, H., Drewitt, R. and Thomas, R. (1973) *The Containment of Urban England*, vol. 2, *The Planning System*, London: Allen & Unwin.

Hebbert, M. (1992) 'The British Garden City: Metamorphosis', in S.V. Ward (ed.) *The Garden City*, London: Spon.

Holmans, A. (1996) 'Housing Demand and Need in England to 2011: The National Picture', in M. Breheney and P. Hall (eds) *The People: Where Will They Go?*, London: Town and Country Planning Association.

Howard, E. (1985) *Garden Cities of Tomorrow*, London: Attic Books (reprint of 1902 edition).

Jackson, A.A. (1991) *Semi-Detached London*, Oxford: Wild Swan Publications.

Jacobs, J. (1961) *The Death and Life of Great American Cities*, Harmondsworth: Penguin.

Joseph Rowntree Foundation (1994) *Inquiry into Planning for Housing*, York: Joseph Rowntree Foundation.

Kelly, R. (1995) 'High Hopes for Hated High Rise', in *The Times*, 8 November.

London Borough of Greenwich (1994) *Unitary Development Plan*, London Borough of Greenwich.

Newman, O. (1972) *Defensible Space: People and Design in the Violent City*, London: Architectural Press.

Ratcliffe, J. and Stubbs, M. (1996) *Urban Planning and Real Estate Development*, London: UCL Press.

Ross, M. (1996) *Planning and the Heritage*, London: Spon.

Rydin, Y. (1993) *The British Planning System*, London: Macmillan Press.

Sharp, T. (1968) *Town and Townscape*, London: John Murray.

Shucksmith, M. (1990) *Housebuilding in Britain's Countryside*, London: Routledge.

Skeffington Report (1969) *Report of the Committee on Public Participation in Planning*, HMSO.

Thomas, H. (1996) 'Public Participation in Planning', in M. Tewdwr-Jones (ed.) *British Planning Policy in Transition*, London: UCL Press.

Thornley, A. (1991) *Urban Planning under Thatcherism*, London: Routledge.

—— (1996) 'Planning Policy and the Market', in M. Tewdwr-Jones (ed.) *British Planning Policy in Transition*, London: UCL Press.

Urban Villages Forum (1992) *Urban Villages: A Concept for Creating Mixed-use Urban Development on a Sustainable Scale*, London: Urban Villages Forum.

Ward, S.V. (1992) 'The Garden City Introduced', in S.V. Ward (ed.) *The Garden City*, London: Spon.

Whitehand, J.W.R. (1990) 'Townscape Management: Ideal and Reality', in T.R. Slater (ed.) *The Built Form of Western Cities*, Leicester: Leicester University Press.

Whitehand, J.W.R., Larkham, P.J. and Jones, A.N. (1992) 'The Changing Suburban Landscape in Post-War England', in J.W.R. Whitehand and P.J. Larkham (eds) *Urban Landscapes*, London: Routledge.

Worskett, R. (1969) *The Character of Towns*, London: Architectural Press.

6

HOUSING DESIGN AND DEVELOPMENT

Jane Weldon

The purpose of this chapter is to provide an overview of the design and development of residential construction projects. The level of information provided is very broad in order to provide the reader with a basic understanding of the main processes involved. In relation to a range of residential properties, the chapter aims to examine:

- The groups involved in design and development.
- The design process.
- The construction process.
- Maintenance policy and practice.
- Building rehabilitation.
- Building failures.

Practical examples and case studies are used to help explain the above concepts.

INTRODUCTION

The UK construction industry is a large, highly fragmented sector of the economy which, like any other industry, produces a 'product'. The 'product area of the industry' includes the erection and repair of buildings of all types; construction and repair of roads and bridges; erection of steel and reinforced concrete structures; other civil engineering work such as laying sewers, gas or water mains and electricity cables, and erecting overhead power lines, line supports and aerial masts. Thus, a wide range of activities is involved in the construction industry, which is consequently very complexly organised. Additionally, each of the component parts identified above is required to be highly integrated.

Any construction project will have a range of common features; a similar range of people will be involved, standard planning and design considerations will be addressed and like activities undertaken, regardless of the type of project. Even so, each project may be considered as bespoke or unique.

A client has to initiate the decision to embark upon a project, then funds may be sought for the development, a site found and a design team appointed. The design team will usually be led by the principal designer, normally an architect or building surveyor, but may include a quantity surveyor for financial advice, services engineer for the planning of the services or structural engineer for the design of supporting elements of the structures. The scheme may be developed with the client's involvement, to standards required by the users of the building, in accordance with external constraints which may be imposed due to legislation, environmental factors and site influences. Documents will be produced by the team, including a specification, bill of quantities and drawings, all detailing the proposed project. A contractor is then appointed to undertake the construction works. Often subcontractors are employed to do specialist works. Ground works, laying the foundations and underground drainage will normally be carried out first. The building will then emerge from the ground, with external walls and roof providing a protective, weatherproof envelope, within which remaining internal works can be carried out. Once the construction works are complete the project will be handed over to the client. The client will then use the building, undertaking a detailed maintenance programme in order to keep it in a good standard of repair. At some point in the future the building may no longer be suitable for the purpose for which it was originally intended, in which case refurbishment works may be undertaken. Alternatively the building may be demolished, and the site redeveloped for a new use.

THE GROUPS INVOLVED IN DESIGN AND DEVELOPMENT

To facilitate the development and design of any construction project it is necessary that various groups of people are brought together to form a coherent team. Some groups will be required to be involved throughout the project, for example a residents' association, while others, for example contractors, may only be required to be involved at specific stages. The range of people concerned will depend on various factors, including the complexity of the project, sources of finance, the type of site selected for development and the range of legislation applicable to the development.

The principal designer of a residential construction project may be required to seek the views of groups about the end product; this may be either a relatively small group or a wide range of different groups. For example, the design and conversion of houses for private sale by a commercial contractor requires the views of few users' groups to be sought, and so is usually less complex in terms of the number of groups involved than the conversion of houses for a group of disabled people. The latter will usually have formed a residents' association, and special consideration will be required at the design stage to accommodate the needs of the proposed residents.

The actual scope of proposed construction works affects the range of groups required to be involved in a project and hence the complexity of this coalition. For example, construction projects undertaken on brownfield sites require special consideration of the use of the land prior to the proposed development. Contaminants may be present in the ground which require remedial treatment or removal. Specialist advice may be required from consultants specialising in environmental issues and specialist contractors may have to be employed to attend to the contaminant.

Thus, the broad groups of people who will be involved in a particular project may include:

1 the client;
2 residents;
3 designers and other construction professionals;
4 statutory undertakers, such as planners and building control officers;
5 building contractors;
6 manufacturers of construction projects; and
7 researchers.

The client

The 'client', who may be the prospective building user (or who may not in the case of speculative development) often has responsibility for various roles; for example, defining the needs of the proposed building users, establishing and providing the necessary finances, agreeing design and construction phases, and undertaking the management and maintenance of the completed project. The type of client varies considerably but can range from a family seeking a residential home to a carefully constituted team from a housing association. In some projects the client will involve different people within an organisation, each of whom has a specific role to play. Project managers may be appointed by a client who has limited knowledge of the construction industry. A project manager will assume the role of the client, and will usually be a construction professional. Clients often have developed their own standards, and may specialise in particular aspects of housing provision.

The residents

It is good practice to seek the view of the proposed residents when undertaking a construction project involving public or housing association-owned housing. This is particularly important when the housing is designed to accommodate people from a different cultural background. For example, a development of houses for families from South-east Asia could be designed with the ancient principles of *feng shui*, which is a system of arranging living and working environments so that people live more in harmony with their surroundings. The results could be successful: the residents would be aware that aspects of their culture had been accommodated within the design, as opposed to simply trying to adapt an existing western design to suit their needs.

Residents should also be consulted since they may have detailed local knowledge and understanding of the range of problems to be tackled on their estate. It is therefore important that residents' groups are involved from the very beginning, for example in local authority housing departments where residents could work with council officers to plan and implement a Single Regeneration Budget bid.

Residents should be involved in deciding what work the residents on the estates would prefer to have, what work is actually needed and how the housing organisation will consult with the residents as the project progresses. It is important that residents have access to independent advice and support right from the beginning of the project and that it is available to them throughout the project. This ensures that residents are able to challenge proposals with more credibility, they are able to obtain a second opinion on technical details and the advisers can negotiate on behalf of the residents where necessary.

Consultation with the residents about improvements and other major works on local authority properties is required by the Housing Act 1985. Consultation methods will range from regular meetings with all residents affected by the work on small estates to regular meetings with resident representatives. A residents' questionnaire and an analysis of day-to-day repair records can reveal what problems are found frequently and what changes the residents would prefer.

Housing organisations can use a variety of ways in order to keep residents informed of proposals, including the use of surgeries on the estate so that individual residents can go in and discuss their ideas. In addition, other means of informing residents of technical information can include the use of perspective drawings, cartoons or models. Certain items, such as new doors and windows to be fitted, can be displayed in the local housing office. A newsletter can be produced and distributed to let all the residents know what is being considered and decided. However, it is important that the amount of resident choice should be discussed and agreed at the design brief stage.

Other matters that may be considered at this stage are the provision of temporary or permanent rehousing with special arrangements for elderly or disabled people, and facilities for residents while the work is in progress, for example access to a community centre, or a spare flat to escape from the noise and dust. Compensation is an important issue and can include rent-free periods, home-loss and disturbance payments.

When the housing organisation decides to proceed with a building project, a large number of technical specialists become involved as mentioned above. One other important professional who many housing organisations insist that the contractors employ, or may be employed by the housing organisation, is a tenant liaison officer to deal with the residents over problems relating to the building works. The tenant liaison officer may also help with the programming of the building works by arranging appointments for the contractor. This person can also deal with issues such as emergency repairs and insurance claims.

The designers

The design of a building involves the consideration of a wide range of elements. Typically, a principal designer will be appointed by the client, usually either an architect or a building surveyor. Specialist designers will then be appointed; for example, landscape architects and services engineers. These will have responsibility for particular aspects of the design, such as the external features of the site, and the heating provision within a multi-storey development. Often, the working relationship between the client and the designers has developed over a long period of time during which both groups have been involved in various projects. This often enhances the design process, because the designer is aware of features and standards that the client requires or prefers the designs to meet.

Documents produced by the principal designers, such as the drawings and specification, as well as those from the specialist designers, for example the bills of quantities drawn up by the quantity surveyor, will form part of the contract documents.

Statutory undertakers

Housing developments in which negotiations are required with a range of groups involved in enforcing statutory legislation can be complex. Planners, building control officers, environmental health officers, conservation officers and engineers may all need to be involved and approval sought by the client in respect to different parts of the development.

The contractors

Construction of a development may be undertaken by a single contractor, or a group of contractors managed by a project manager appointed from the team of the main contractor. It is also commonplace for contractors to appoint sub-contractors for some of the activities to be undertaken. All contractors are required to undertake their work in accordance with a written contract. 'Standard forms' of building contracts are regularly used, which have been developed, published, monitored and amended by the Joint Contracts Tribunal (JCT). The number of versions and forms published by the JCT has grown considerably.

A client, or the client's appointed representative, will need to ask four primary questions prior to a decision being made on which contract is to be used (Speaight and Stone, 1996: 46–7).

First, who will design the works? A traditional approach may be used, in which an architect is engaged to design the building and then a contractor is subsequently employed to construct the proposals in accordance with these designs. Alternatively, the entire package may be handed to the contractor, who is employed both to design and to build on the basis of a statement of the client's requirements.

Second, will the works be completely designed before the start of the contract? In a traditional approach the works are fully designed and then a contractor is sought to build the project. This arrangement offers notable advantages, for example good cost control. In recent years different contractual arrangements have been adopted when speed of start and of completion have been important, particularly at times of high interest rates, where time taken to complete a full design adds interest rate charges to the overall construction budget.

Third, how firm a price is required? A traditional approach ensures certainty regarding the final cost of the development. Documents including sets of drawings, bills of quantities and a specification can be placed out to competitive tender. Prices are then returned from contractors, and normally the lowest will represent the cost of the construction of the development.

Finally, how large is the project? There are contracts published by JCT that suit projects of different sizes. In 1980, the Agreement for Minor Works was published for contracts up to the value of £70,000, although there are various circumstances where this will be unsuitable, for example when the client wishes to nominate specialist sub-contractors. The Intermediate Form of Contract, published in 1984, may be suitable where works are of a more straightforward nature and are not worth more than £280,000 and the contract period is not more than twelve months. Another standard form published in 1989, the JCT Measured Term Contract, is a suitable form of contract where an employer has a regular flow of minor works, such as regular maintenance jobs; for example, a single contractor may be appointed to undertake all small jobs which might arise in relation to a particular property over a specific period.

Some of the various trades involved in the erection of a building are listed in Table 6.1.

Due to the wide range of trades, it is necessary for the main contractor carefully to plan the activities to be undertaken. This may be done using a variety of methods, the most common being Gannt charts (which indicate the progress of a contract, month by month, showing where building works are taking place). Developments in information technology have also facilitated the modelling of projects on various software packages, for example Primavera. Clear planning of the construction process ensures that, for example, bricklayers commence work once the concrete foundations have been laid, or that decorators arrive once the plastering is completed and dried out. Organisational clarity also enables costs to be strictly controlled.

Table 6.1 Selected trades involved in construction projects

Trade	*Aspect of construction*
Asphalter	Roof, floor and wall finishes
Bricklayer	Laying brickwork
Carpenter	Structural and carcassing timber work
Concreter	Placing concrete
Drainlayer	Providing below-ground drainage
Electrician	Electrical installation
Excavator	Levelling site and excavation for drain and foundation trenches
Floor tiler	Internal floor finishes
Gas fitter	Gas installation
Glazier	Fixing glass
Joiner	Work with timber, e.g. fixing doors and windows
Metalworker	Sheet metal applications, e.g. roofing
Painter and decorator	Finishing of walls, ceilings, timber work
Paver	External paths, road finishes
Plasterer	Plastering walls/ceilings, screeding and rendering
Plumber	Plumbing installation, flashing, gas pipes
Scaffolder	Erecting scaffolding and working platforms
Steel erector	Erecting steel columns and beams
Steel fixer	Cutting, shaping and positioning steel reinforcement

The manufacturers

Manufacturers supply the materials, components and equipment which are used during the construction processes, and often recommend specific methods of installation, detailed in product literature. Traditionally, construction processes relied on the supply of readily available materials easily converted into forms and sizes that could be adaptable on site to suit a particular design.

Continual advancement of technology and increases in complexity and size of buildings have resulted in a more complex construction process. Subsequently manufacturers have extended their service from the supply of single components to the supply of much larger parts of a building and indeed whole buildings. Site operations are reduced to a minimum using mechanical plant, and methods of building become largely concerned with the organisation of the systematic supply and assembly of prefabricated items. Manufacturers play a vital role in the design of buildings and in some instances are included within the design team.

The researchers

Research to further understanding and development of current construction methods and materials is essential. There are various research organisations within the construction industry that undertake research. An example is the Building Research Establishment (BRE). The BRE, founded in 1921, has numerous fields of study, including research into building materials, structural engineering, geotechnics, mechanical engineering, fire and timber technology.

Other organisations undertaking research include the Construction Industry Research and Information Association (CIRIA), which often helps finance research into aspects of construction by allocating funds to universities and industrial organisations. The Building Cost Information Service (BCIS) provides a cost analysis service.

Research is also sponsored by organisations which promote the use and development of certain materials, for example the Brick Development Association (BDA) and the Timber Research and Development Association (TRADA).

The British Standards Institution prepares British standards on all aspects of the industry. The standards represent the recommendations made by specialist committees and interested parties in a particular subject and for the construction industry encompass various issues, such as performance standards, methods of assembly of components and minimum dimensions of structural sections.

The British Board of Agrement encourages the use of new materials, products, components and processes. The board offers an assessment service based on examinations and testing, and those innovations satisfying critical analysis relative to their proposed use are issued with an Agrement Certificate.

CASE STUDY 6.1

HOUSING DEVELOPMENT FOR UJIMA HOUSING ASSOCIATION

The range of activities and groups of people involved in a construction project is illustrated in the following case study, summarised in Table 6.2. The case study is based on a project undertaken by Ujima Housing Association, whose clients are young single homeless people, for a high-density block of flats in Hackney, east London (*Architect's Journal* 12 January 1994, 39–47).

Design

The following section describes the process by which the final design proposals emerge and also describes issues affecting the design, including the broad functions which domestic buildings are required to fulfil.

The design process

Design is a complex skill, requiring a thorough thought process and integration of many and varying considerations. To understand the design process, various maps have been produced by designers, to identify how designers think, and to provide formal exposition of the procedure, which is often wrongly viewed by the construction- or design-illiterate as ad hoc.

Lawson (1990: 23) describes one map of the process, which is used by the Royal Institute of British Architects (RIBA), designed for and by architects. The map (Figure 6.1) proposes that the design process may be divided into four phases:

- Phase 1: *Assimilation*, the accumulation and ordering of general information and information specifically related to the problem in hand.
- Phase 2: *General study*, the investigation of the nature of the problem and of possible solutions or means of solution.
- Phase 3: *Development*, the development and refinement of one or more of the tentative solutions isolated during second phase.
- Phase 4: *Communication*, the communication of one or more solutions to people inside or outside the design team.

The handbook points out that these four phases are not necessarily sequential, although it may seem logical that the overall development of a design will progress from phase 1 to phase 4. The map suggests that designers have to gather information about a problem, study it, devise a solution and draw it, though not necessarily in that order. The design process is iterative; information gathered during the

Table 6.2 Activities undertaken and range of groups involved during the design and construction phase of a housing development for Ujima Housing Association

Activity	*Holcroft Road, Hackney*
Initial decision to build	Ujima Housing Association was formed to provide good-quality affordable dwellings for young single homeless people and some homeless families nominated by local authorities.
Securing of financial resources	Ujima secured the finances from the Housing Corporation; budgets for the development were to be tightly controlled.
Selection of appropriate location	Land offered to Ujima by the local education authority. Some adjoining land also leased from the authority for use during the construction process.
Appointment and briefing of suitable members to be involved with the design and construction processes	Ujima appointed Walter Menteth Architects, as they had provided the client with imaginative concepts previously. A structural engineer, Stuart Richardson, and a quantity surveyor, Roderick Jackson, appointed. Contractors on Ujima's approved list invited to present tenders for the project. Afterheath appointed, after negotiations to arrive at an acceptable contract sum, using the JCT Intermediate Form of Contract 1984.
Definition of precise functional requirements	Ujima decided at an early stage to develop the site to a high density, providing three flats on three levels with an access stair on the street frontage. As a close working relationship was previously developed between Walter Menteth and Ujima, Walter Menteth was aware of the standards and requirements of the client.
Statutory legislation	Planning permission received, although restrictions included, ensuring that the eaves on the main facade aligned with those on the adjoining property and, due to the proximity of the playground, a shallow sloping main roof without an overhang to the playground was necessary.
Design process and decision on how to build	Scheme granted planning permission and the site purchased by Ujima in 1988. Tender documents and building regulation approval completed by 1989. Due to changes in Housing Corporation funding, the design process was complicated by site constraints including the adjoining building and restricted access to the site.
Planning and implementation of construction	Afterheath commenced work on site in July 1992. The stair tower is structurally independent of the main building and separated from it. This was to allow the tower to be built last so that the area at the front of the restricted site could be used during construction of the main building. Afterheath commenced work on site. Additional complications arose; for example, it was necessary to underpin the party wall.
Completion	Afterheath completed construction in July 1993.

general study may impact upon the problem in hand. Perhaps, during a meeting between designers, the solution to the problem will be further developed. Indeed, evidence suggests that there is a tendency for individual strategies to develop within practices or to be peculiar to the individual designer.

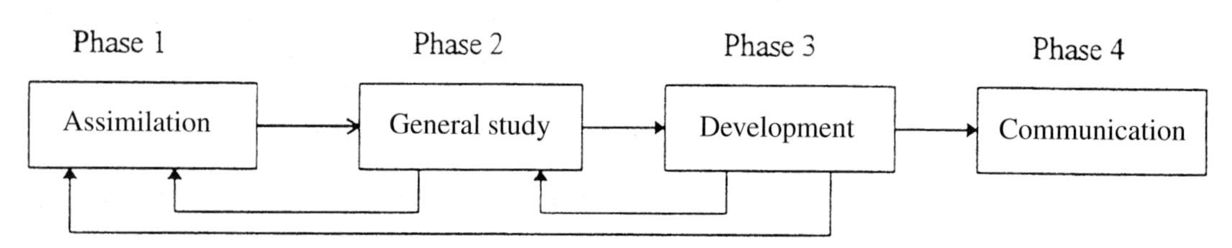

Figure 6.1 The RIBA map of the design process (after Lawson 1990: 24)

The RIBA also details a larger, more detailed map, called the Plan of Work, which is a description of the products of the processes (Lawson, 1990: 133–9; Osbourne, 1985). The Plan (Table 6.3) describes the purpose of each stage of work, the tasks to be done, and the people involved. The Plan of Work informs the client of what will be produced, the responsibilities of each member of the design team, and how they relate to the principal designer. Therefore, this analysis of the design process has enabled a structuring of the problem, exploration of the relationships and classification of objectives.

Determinants of the design

There are various influences that modify the proposed design of a development; for example, the functional requirements of the building, financial factors, environmental considerations, and legislation (Figure 6.2). The principal designer will also be required to consider the client's requirements, which are often standard to the client's organisations. These are some of the factors which determine the design of the residential development.

Functional requirements

All buildings and individual components of buildings are required to meet various functions, each of which must be considered in the design stage. The requirements are inherently linked together, with each factor clearly impacting upon others. Each of these functional requirements is briefly described below in relation to residential buildings.

The *appearance* will initially be determined by the activities to be accommodated as these strongly influence the scale and proportion of the overall volumetric composition. Aesthetic and technical criteria are affected by the composition, form, shape, colour and position of the materials employed.

Houses are required to be *durable*. Buildings are under the constant influence of climate (wind, snow, sunlight, sleet), attack from vandals, and damage by fire, explosions and structural movements. The design of residential buildings should incorporate elements which aim to ensure that the durability is not impaired. For example, to prevent fire spreading within a high-rise tower block of flats, walls and floors should be designed to contain the fire for a minimum period, normally one hour. This will allow people within the flats to escape as well as enable the firefighters to arrive and extinguish the fire before it can spread through adjoining flats.

The elements of the building, the walls, roof, windows, doors, and so on, should be *dimensionally suitable* for their intended use. Movement of materials during the lifetime of a building can cause variation in the way in which materials fit together. The materials that are used must be of an appropriate size and should be designed to suit manufacturing processes and methods of constructing the building. For example, calcium silicate bricks will be prone to thermal expansion and contraction over time. Within lengths of walls made of such bricks it is necessary that a movement joint is included, commonly every 6 metres, so that the wall can expand and contract without causing damage to materials in the vicinity.

Building products should also be designed to be easily assembled. *Dimensionally co-ordinated* products save time in both material (e.g. wastage) and labour costs. The clay brick is a standard size, designed to be easily manipulated in a single

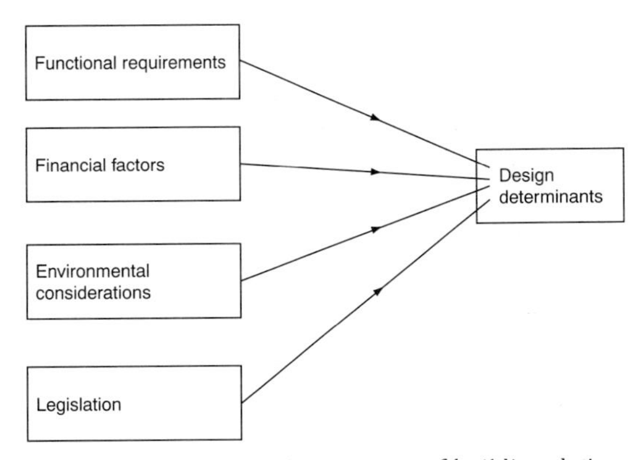

Figure 6.2 The primary determinants of building design

Table 6.3 The RIBA's Plan of Work

Stage	*Purpose of work*	*Tasks to be done*	*People directly involved*	*Common term*
A Inception	To prepare general outline of requirements and plan future action	Set up client organisation for briefing, consider requirements, appoint principal designer	All client interests, architect	Briefing
B Feasibility	To provide client with an appraisal and recommendation in order that they may determine the form in which the projects are feasible, functionally, technically and financially	Carry out studies of user requirements, site conditions, planning, design	Client's representatives, architects, engineers and quantity surveyor	
C Outline proposals	To determine general approach to layout, design and construction in order to obtain authoritative approval of the client on the outline proposals	Develop brief further. Carry out research on user requirements, technical problems as necessary to reach decisions	All client interests, architects, engineers and specialists as required	Sketch plans
D Scheme design	To complete the brief and decide on particular proposals including constructional method, outline specification and cost BRIEF SHOULD NOT BE MODIFIED AFTER THIS POINT	Final development of the brief. Submission of plans to statutory authorities for approval	All client interests, architects, engineers and all statutory and other approving bodies	
E Detail design	To obtain final decision on matters related to design, specification and cost ANY FURTHER CHANGES IN LOCATION, SIZE, SHAPE OR COST AFTER THIS TIME WILL RESULT IN ABORTIVE WORK	Full design of every part and component of the building by collaboration of all concerned	Architects, engineers and specialists, contractor (if appointed)	Working drawings
F Production information	To prepare production information and make final detailed decisions to carry out work	Preparation of final production information – drawings, specifications and sketches	Architects, engineers and specialists, contractor (if appointed)	

Table 6.3 continued

Stage	*Purpose of work*	*Tasks to be done*	*People directly involved*	*Common term*
G Bills of quantities	To prepare and complete all information and arrangements for arraign tender	Preparation of bills of quantities and tender documents	Architects, contractor (if appointed), quantity surveyor	
H Tender action	Action as recommended by the National Joint Consultative Council (NJCC)	Action as recommended by the NJCC	Architects, contractor, quantity surveyor, engineers, client	Site operations
I Project planning	To enable the contractor to programme the work in accordance with contract conditions; brief site inspection and arrangements made to start work on site		Contractor, sub-contractors	
J Operations on site	To follow plans through to practical completion of the building		Architects, engineers, contractors, sub-contractors, quantity surveyor, client	
K Completion	To hand over the building to the client for occupation, remedy any defects, settle final account, and complete all work in accordance with the contract		Architects, engineers, contractor, quantity surveyor, client	
L Feedback	To analyse the management, construction and performance of the project	Analysis of job records, inspections of completed building, studies of building in use	Architect, engineer, quantity surveyor, contractor, client	

Source: Osbourne (1985)

hand, leaving the other hand free to use a trowel. When used in large quantities the bricks with their joints and bonding method determine the scale and proportion of the overall shape and thickness of the walls, the sizes of the window and door openings. Windows and doors should be manufactured so that they can be fitted into an opening formed by the brick dimensions without adjustment.

Building products are required to *carry loads* and therefore must be sufficiently strong to carry the loads and resist collapse, distortion and movement. There are generally four loads required to be carried:

1 live or imposed loads, the weight of the people and the furniture within the building;
2 wind loads, the strengths of gusts of wind a building is required to stand;
3 snow loads, the weight of snow that a building may be subject to; and
4 dead loads, the weight of the building materials from which the building is formed.

There are various methods by, and materials from, which a building can be constructed to carry the required load. Traditionally constructed buildings are generally continuous structures, which rely on all the walls and floors to transfer the loads to the supporting soil. Alternatively, a frame may be built which will carry the weight of the building, separate from the elements in between the frame (the infill panels).

Exclusion of the elements of rain, snow, extremes of temperature and wind is essential to prevent damage to the fabric of the building and maintain a safe, comfortable internal environment. Water penetration should be prevented by adequate drainage of rainwater via gutters and downpipes to a storm-water drain. Penetration of damp from the surrounding ground within the lower walls of the building should be resisted by a damp-proof course.

The *elimination or reduction of unwanted sound* generated by sources within or outside the building is necessary, particularly within high-rise developments. Often, the construction of a building is sufficiently dense to resist the passage of sound within the building, although some elements require careful design. The design of a terrace of houses requires careful consideration of adjoining elements; for example, the dividing walls.

The *thermal comfort* of the occupants must also be considered. A balance of heat must be achieved: a balance between radiation from the sun, heating installations and heat from the occupants. In addition, provision should be made for ventilation of the building and the loss of heat through the elements of the building to the environment. Combined with considerations of thermal comfort and ventilation, it is necessary to consider lighting of the building. Maximum advantage should be taken of natural daylight and heating, to minimise the eventual cost of heating and lighting the building.

Any building is required to resist the effects of *fire*. The elements of the building should be designed to have a certain degree of fire resistance. Thus, materials that have limited combustibility should be used. The spread of fire within and between buildings should be prevented, and there should be adequate means of escape for the occupants of the building, as well as suitable access to the building for firefighters and their equipment.

Provision of satisfactory *sanitation* is required to ensure an adequate supply of drinking water, appropriate areas for food preparation and washing, acceptable methods of disposal of refuse and dirt, and drainage of both foul and storm-water.

Buildings are required to be *secure* and to provide a certain degree of privacy for their occupants.

Financial factors

The design of a building must be judged not only by its appearance and the way in which it performs, but also by how much it costs. Acquisition costs encompass the costs involved in the creation of a building, investment negotiations, professional fees, cost of land, building design and construction.

The cost of the land will vary widely according to location, quality, size and its value in terms of suitability for a particular development type. The consequences of these factors need to be carefully considered so that cost comparisons can be made with alternative sites. The cost of land is high and increasing, consequently it is essential that land is used with maximum efficiency.

The cost consequences of the construction methods used for the building must be examined to ensure that comparisons are made with alternatives. Traditional construction methods involve the shaping of materials on site by skilled and semi-skilled labour. Industrialised construction involves the erection of large, virtually finished factory-made components on site. Both forms of construction use the same basic resources, materials and labour, but the inter-related cost implication associated with each form affects their economic suitability.

The use of factory-made components for a building should result in considerable cost savings, due to the possibility of large production runs. Further, the quality of the factory-made components can be more easily assured, although the components will require transportation to site and protection against damage during erection. Traditional construction can be

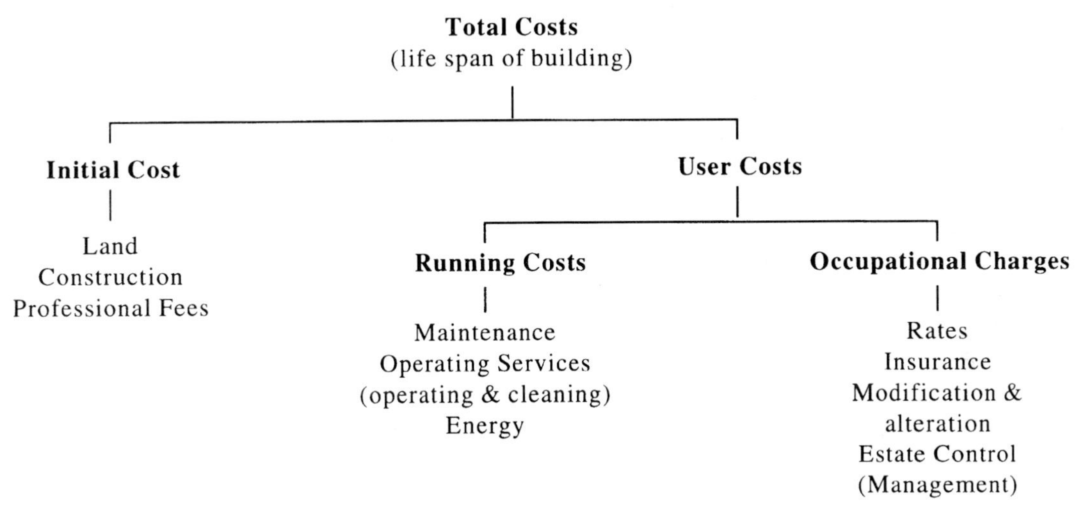

Figure 6.3 Total costs in the life-span of a building (after Seeley, 1985)

affected by weather conditions, and often working conditions can be awkward.

Running costs include expenses incurred for general maintenance, cleaning and servicing of a building and for renewing or repairing the fabric and fittings, in addition to payments for heating, lighting, ventilating and services. Allowance should also be made for insurance purposes.

The true cost evaluation of a building should involve a close examination of viable alternative acquisition, running and operational costs. The depth of investigation possible and therefore the accuracy in achieving the true value of a building will depend on the information available and on skill of interpretation during the design process.

Environmental considerations

Any site acquired for development requires analysis of a range of factors. The principal factors to be addressed are shown in Table 6.4.

Legislation

There are various statutes that affect construction projects.

The Town and Country Planning Act 1990 requires planning permission to be sought for developments except those alterations which are perceived to be permitted (see Figure 6.4). Development consists of carrying out of building, mining or engineering operations and/or the making of a material change of the land (including the buildings on the land). Reference should also be made to the General Permitted Development Order, 1995 and Planning Policy Guidance Notes (PPG) when considering submitting an application for planning permission.

There are two forms of application. Permission for outline development may be sought, which allows the developer to 'test the water' by submitting only a broad outline of the scheme proposals. If this is approved it will then be necessary to submit an application for full planning permission accompanied by detailed drawings and other relevant information. Generally the application will be considered within eight weeks. Full planning permission lasts five years, while outline planning permission expires after three years.

The Building Regulations 1991 require that all 'building work' must be carried out in accordance with the requirements stated in Schedule 1 of the

Table 6.4 Environmental factors to be considered prior to design

Factor classification	*Factor*	*Additional notes*
Natural factors	Geology	Bedrock and surficial
	Physiography	Geomorphology, relief, topography
	Hydrology	Surface and ground water
	Soils	Classification of types and uses
	Vegetation	Plant ecology
	Wildlife	Habitats
	Climate	Solar orientation, wind, precipitation, humidity
Cultural factors	Existing land use	Ownership of adjacent property and off-site nuisances
	Traffic and transit	Vehicular and pedestrian circulation on or adjacent to the site
	Socio-economic factors	
	Utilities	Storm and foul water systems, water, gas, electricity, telecommunications networks
	Existing buildings	
	Historic factors	Landmarks, archaeology
Aesthetic factors	Natural features	
	Spatial pattern	Views, spaces

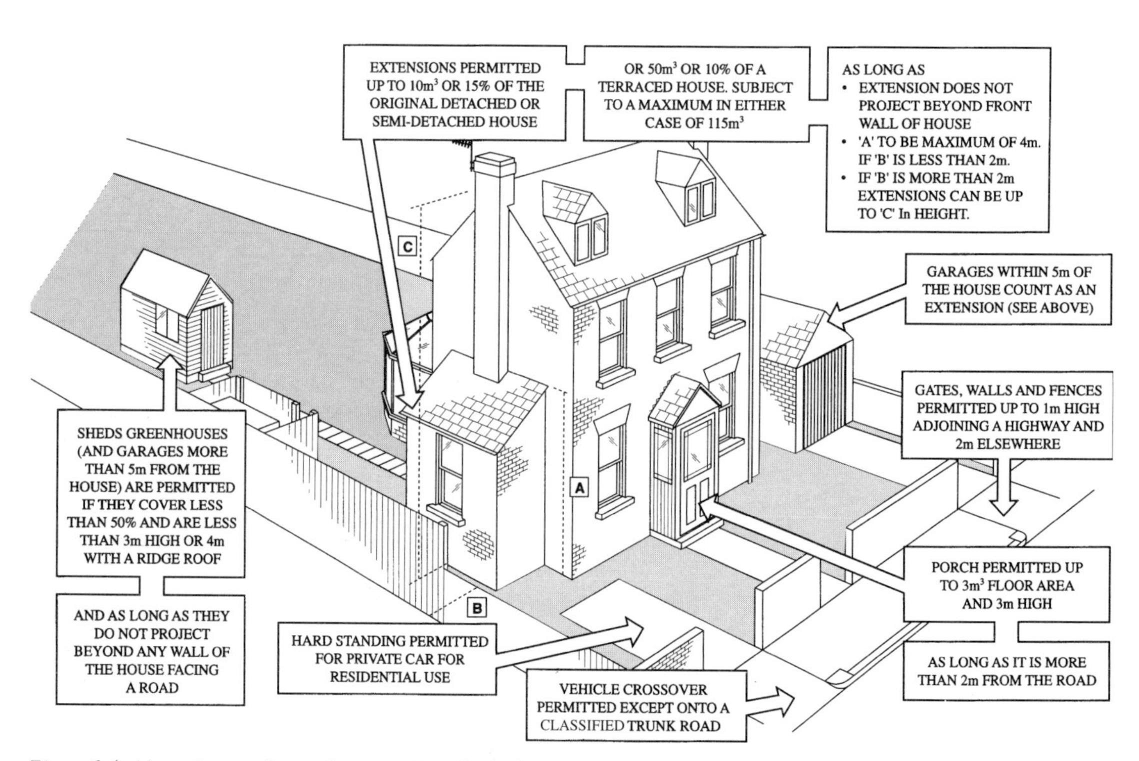

Figure 6.4 Alterations to domestic properties which do not require planning permission

Act. These requirements are brief and concise and merely state the minimum standards which must be attained (see Table 6.5).

The Building Regulations 1991 contain no technical detail so they are supported by a set of Approved Documents. Each Approved Document relates to a part of the Schedule 1 requirement. The Approved Document provides practical guidance on how designers may comply with the Schedule 1 requirements. The requirements within the documents do not have to be followed, provided that the technical design complies with the Schedule 1 requirement. Nevertheless, the Approved Documents are a useful design guide and source of reference when designing, particularly with respect to new build domestic properties, conversions, adaptations and refurbishment schemes for domestic purposes.

Subject to a number of exemptions, some of which are illustrated in Figure 6.5, most building work requires some form of building regulation approval.

Regulation 3 of the Building Regulations 1991 states that building work requires statutory approval. 'Building work' is defined as meaning the erection or extension of a building, the provision or extension of a controlled service or fitting, work required if a material change of use occurs, the insertion of insulating material into a cavity wall and work involving underpinning, hot water supply systems, sanitary conveniences, drainage and waste disposal, certain fixed heat-producing appliances and heating systems.

Approval for Building Regulations may be sought either from an approved body or from a local authority. An approved body is any organisation designated by the government as being suitable to

Table 6.5 Requirements of the Building Regulations 1991

	Title	*Main aspects covered*
A	Structure	Loading, ground movement
B	Fire Safety	Means of escape, internal and external fire spread, access and facilities for the fire service
C	Site Preparation and Resistance to Moisture	Preparation of site, dangerous and offensive substances, subsoil drainage, resistance to weather and ground moisture
D	Toxic Substances	Cavity wall insulation
E	Resistance to the Passage of Sound	Covers airborne and impact sound
F	Ventilation	Means of ventilation and prevention of condensation
G	Hygiene	Bathrooms, hot water storage and sanitary conveniences and washing facilities
H	Drainage and Waste Disposal	Deals with foul water drainage, cesspools, septic tanks and settlement tanks, rainwater drainage and solid waste storage
I	Heat Producing Appliances	Air supply, discharge of combustion products and protection of the building
J	Stairs, Ramps and Guards	Stairways and ramps, protections from falling and vehicle barriers
K	Conservation of Fuel and Power	Reasonable provision should be made for the conservation of fuel and power
L	Access and Facilities for Disabled People	Provision for facilities for the disabled, access and use of sanitary conveniences and audience and spectator seating
M	Glazing Materials and Protection	Reducing the risks associated with glazing at critical locations

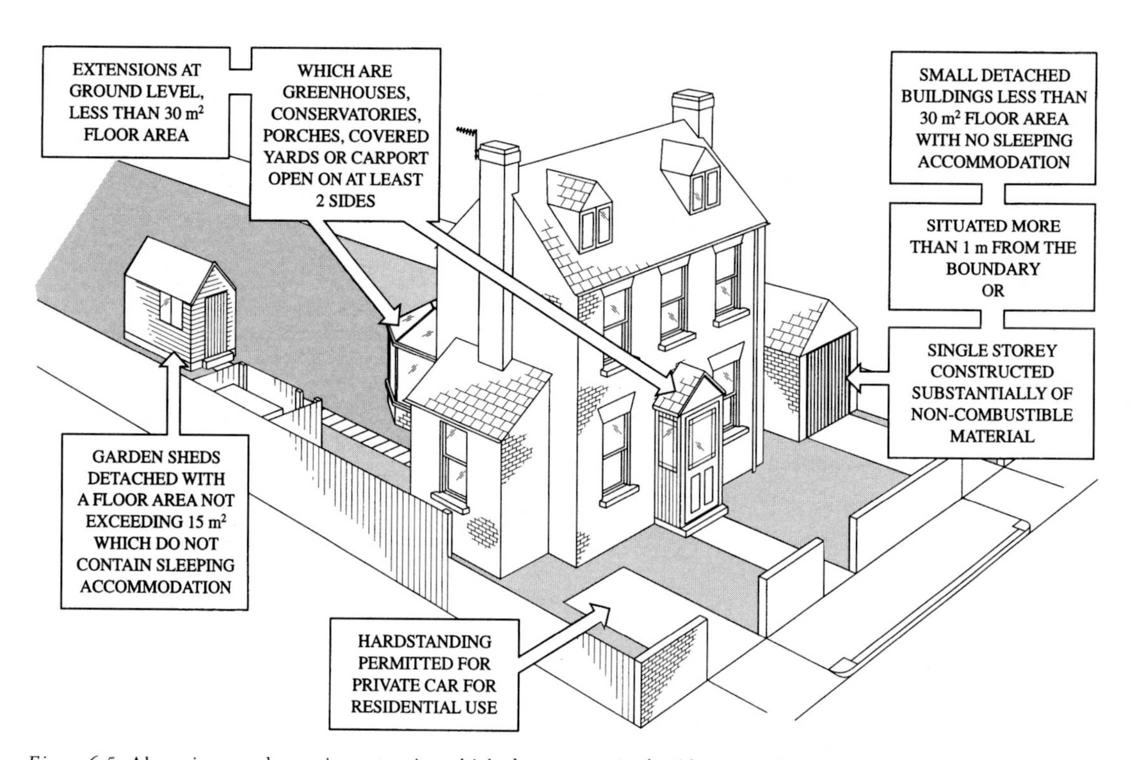

Figure 6.5 Alterations to domestic properties which do not require building regulation approval

inspect building works for compliance with the Building Regulations. In practice, most building work will be subject to control by the local authority because at present there is only one approved body, namely the National House Building Control Services Ltd (NHBC).

Approved bodies are only permitted to approve work to low-rise domestic dwellings. It is normal for contractors who specialise in large-scale domestic developments to apply for approval via the NHBC, where occasionally a more competitive price can be obtained from the NHBC rather than a local authority.

Approval for work on site will be required to be sought at specific points during construction (see Table 6.6). It is usual for a good working relationship to develop between the building control officer and the contractors; indeed building control officers are a useful source of advice with regard to local ground conditions, the existence of land drains and other matters.

The London Building Acts

Until 1986 the design, construction and use of buildings in Inner London were regulated by the London Building Acts and Bylaws, which formed a code of control different from that which operated

Table 6.6 List of inspections usually undertaken by building control officers on new build developments

Commencement of work – stripping of oversite
Concrete within foundations
Up to damp-proof course level
First-floor joists
Up to wall plate level
Roof prior to cover
Prior to plastering
Drains prior to backfilling
Completion

Table 6.7 Examples of powers provided for by the London Building Acts

Provision	*Detail*
Powers of entry	Wide powers of entry, inspection and examination to enable the district surveyor and others to carry out their role
Dwelling-houses on low-lying Land	Prohibits the erection or rebuilding of dwelling-houses on low-lying land without the consent of the appropriate Inner London borough council
Party structures	Special codes exist governing party structures and rights of adjoining owners

elsewhere. The Building (Inner London) Regulations 1985 came into effect on 6 January 1986 and repealed the London bylaws. This brought most of the national regulations into force in Inner London, and made some amendments to the London Building Acts. The remaining Building Regulations were applied to Inner London by the Building (Inner London) Regulations 1987.

None the less, Inner London is still subject to many additional controls under the remaining provisions of the London Building Acts 1930 to 1978 and so building control in the area is subject to many special features.

Other statutes that should be considered include the Construction (Design and Management) Regulations 1994, the Fire Precautions Act 1971 and the Highways Act 1980.

Construction

The following section reviews the processes involved in the construction of low-rise traditional dwellings and high-rise non-traditional dwellings. For both traditional and non-traditional construction the main differences occur in the way in which the external envelope of the building is constructed; that is, the walls, floors and roofs.

Initial site works

Prior to the site being cleared, consideration should be given to the way in which the entire construction work is to be carried out on the site and to the plant required. Plant should normally be used where it can be fully utilised, where there is room to manoeuvre and where the cost of using it can be favourably compared to the cost of doing the work manually.

The site should be cleared of debris, unwanted shrubs and bushes. Huts for the construction workers and storage facilities for the materials can be positioned usually following consideration of site requirements by the contractor. The areas of the site are then set out, by carrying out a survey of the proposed building and driving timber pegs into the ground, indicating the line of foundations and walls above.

Foundations

Functional requirements

The foundations of a building is that part which is in direct contact with the ground transmitting the live, wind, snow and dead loads of the building to the ground. Loads within a building should be transported to the ground in such a way that:

1 excessive settlement will not occur, for example, so the building will not sink into the ground;
2 differential settlement of various sections of the building will not occur, so that a part of the building does not collapse due to the limited load-bearing capacity of the soil; and
3 that the soil will not fail under its load, thus causing collapse or partial collapse of the building.

Construction of foundations

The topmost layer of soil at ground level (topsoil) is an unsuitable material on which to build because it has been weathered, is relatively loose and usually contains deleterious vegetable matter. The topsoil (or vegetable soil) varies in thickness usually from about 150 mm to 300 mm.

Soil beneath the topsoil, known as subsoil, varies widely in nature and characteristics. Classification of the soil type is necessary, to diagnose how the subsoil is likely to behave when it is required to support the loads of a building. There are four broad classes of soil types: gravels, clays, sands and silts. These may be seen in Table 6.8.

Foundations are usually constructed using concrete, either mass or reinforced. Reinforced concrete refers to steel mesh or bars which have been bedded within the concrete to increase the load-bearing capacity of the material.

The types of foundations used in building work may be divided broadly into shallow foundations, less than 1 metre deep, and deep foundations, in excess of 1 metre deep. Shallow foundations are usually less expensive to construct than deep foundations. The choice of foundation depends on the design of the structure and the soil to which the loads can be satisfactorily transferred. Shallow foundations are normally used for low-rise domestic dwellings.

Arguably, the simplest form of foundation is the strip foundation, which is a strip of concrete, with the wall built centrally above. A strip of soil is excavated to a depth where the soil has satisfactory bearing capacity, usually 1 metre in normal soil conditions. The strip is then filled with concrete (see Figure 6.6).

Another common form of shallow foundation is the raft, which is a large slab foundation covering the whole of the building area, through which all the loads from the building are transmitted to the soil (see Figure 6.7). Solid concrete floor slabs are common in areas where the soil is weak, for example where there has been mining, as the integrity of the underlying ground is reduced. Reinforcement may be required in the form of steel; at the top of the slab a light mesh may be incorporated to prevent surface cracking; a heavier mesh, or steel bars, may be needed under the walls and columns. The raft should be extended about 300 mm beyond the

Table 6.8 Classification of soil types and field tests

	Type of subsoil	*Condition of subsoil*	*Field test applicable*
I	Rock	Sandstone, limestone or chalk	Requires at least a pneumatic or other mechanically operated pick for excavation
II	Gravel	Compact	Requires pick for excavation
	Sand	Compact	
III	Clay	Stiff	Cannot be moulded with the fingers and requires a pick or pneumatic or other mechanically operated spade for its removal
	Sandy clay	Stiff	
IV	Clay	Firm	Moulded with substantial pressure with the fingers and can be excavated with the spade
	Sandy clay	Firm	
V	Sand	Loose	Excavated with a spade
	Silty sand	Loose	
	Clayey sand	Loose	
VI	Silt	Soft	Easily moulded in the fingers and readily excavated
	Clay	Soft	
	Sandy clay	Soft	
	Silty clay	Soft	
VII	Silt	Very soft	Natural sample in winter conditions exudes between fingers when squeezed in fist
	Clay	Very soft	
	Sandy clay	Very soft	
	Silty clay	Very soft	

Source: Department of the Environment (1991) *Supplement to the Building Regulations Act 1990*, HMSO, Approved Document A, table 12, p. 32 (adapted)

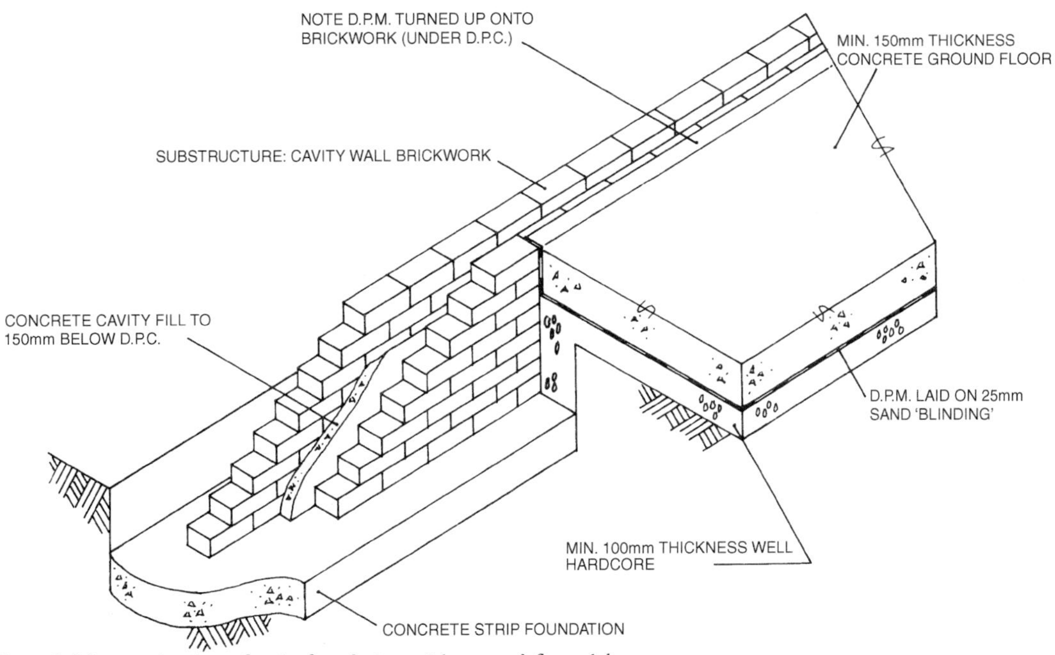

Figure 6.6 Isometric view of strip foundation with ground-floor slab

perimeter walls to help spread the load to the surrounding ground.

Deeper foundations are required for high-rise structures to transmit the loads to a firm stratum able to withstand the high loads without movement. Piles are columns, usually of reinforced concrete, which may be driven by a mechanical drop hammer or cast in place.

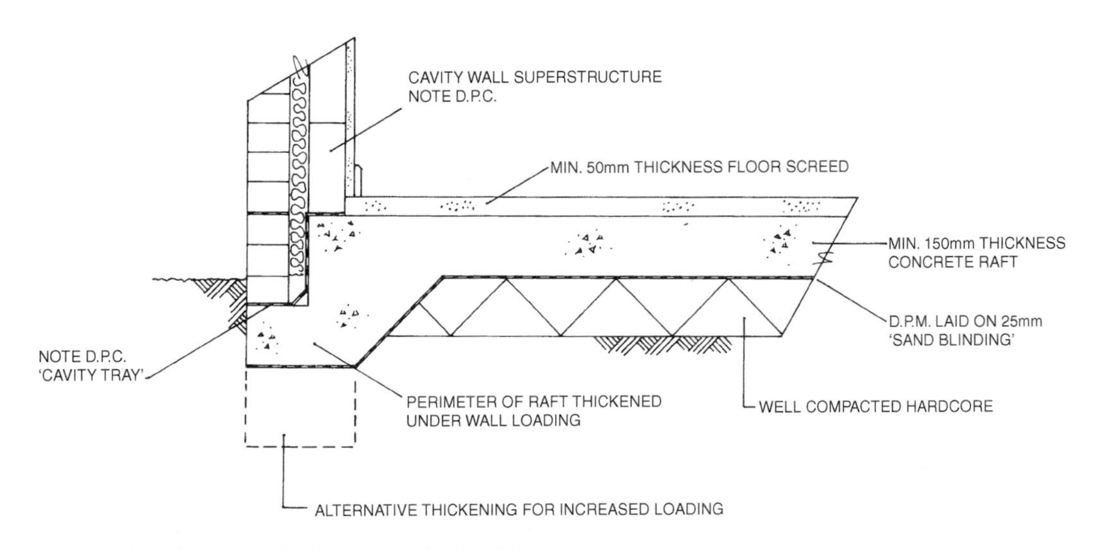

Figure 6.7 Typical section of perimeter of raft foundation

Walls

Walls may be divided into three broad categories:

1. load-bearing, supporting loads from floors and roofs in addition to their own weight and resisting side pressure from wind and sometimes stored material or objects within the building;
2. non-load-bearing, carrying no floor or roof loads, but simply serving to divide large areas of space; and
3. retaining walls, supporting and resisting the thrust of the soil and perhaps subsoil water below ground level.

Functional requirements of walls

Walls are necessary to *enclose or divide space*; walls may also be required to provide support; be impervious to the effects of weather; and afford a degree of fire resistance, sound resistance and thermal insulation.

The *strength* of a wall is measured in terms of its resistance to the stresses introduced by its own weight, by superimposed loads and by lateral pressure. A wall will fail if it is over-loaded and this will be evident as the wall will overturn or buckle. Thus provision of sufficient thickness is essential, and lateral support, by vertical piers, may be necessary.

Whatever their form, the external walls of a building are required to provide adequate *resistance to water and wind penetration*. The actual degree of resistance required in any particular wall will depend upon its height, locality and exposure. Wind penetration rarely presents a problem in masonry wall construction, as the materials used in domestic construction are sufficiently dense to be unaffected by winds. However, rain penetration can cause problems, which may be resisted in three ways:

1. ensuring limited penetration into the wall thickness;
2. preventing any penetration whatsoever, by, say, a waterproof rendering; and
3. interrupting the capillary paths through the wall, for example by the provision of a cavity.

Protection must also be provided against lateral penetration and vertical migration of water at the base of a wall, which can enter and rise by capillary attraction from the surrounding ground. This penetration is counteracted by the damp-proof course (DPC) and damp-proof membrane (DPM), components which resist the penetration of water. Dense polythene sheet is commonly incorporated as a damp-proof membrane. Asphalt is often used for damp-proof courses.

A degree of *fire resistance* is necessary, which is dependent on the location of the wall, use of the building and the activities accommodated within the building. Walls, like upper floors, are required to compartmentalise a building so that a fire is confined to a given area, to separate specific fire risks within a building, to form safe escape routes for the occupants and to prevent spread of fire between the buildings.

The external walls of a building, together with the roof and floors, must provide a barrier *to the passage of heat* to the external air in order to maintain satisfactory internal thermal comfort conditions without wasteful use of the heating system. Thermal insulation may be fixed to walls, to reduce the amount of heat that is lost through them. The degree of thermal transmittance through a wall is measured by u-values, and the Building Regulations 1991 state the acceptable u-values for domestic dwellings, not only for walls, but also roofs and floors. Thermal insulation may be applied on either the internal or the external face of the wall, or within the cavity. A cavity fill, for example urea formaldehyde foam, polyurethane foam, mineral fibre or granules of polystyrene or perlite, may be blown or injected into the cavity of the wall after construction. Where possible the cavity should be filled during construction with mineral or glass fibre slabs or quits or expanded polystyrene slabs. Where insulation is applied direct to the internal face of the wall this is known as dry lining. Insulation that is applied to the external face of the walls may be in the form of stiff boards or slabs, and may be protected by a waterproof covering of rendering.

Exceptional circumstances require the sound insulation qualities of an external wall to be a significant factor in its design, since the other function requirements which must be fulfilled, for example fire resistance, usually necessitate a wall which excludes noise significantly well in most circumstances. However, sound insulation is a significant factor in the design and construction of internal walls, particularly where the building is under shared occupancy, for example flats. Protection from two types of sound is essential: airborne (for example, a person shouting) and impact (for example, in the sound of footsteps on stairs). Internal non-load-bearing walls or partitions should rely on discontinuous construction to avoid sound transferral, for example cavity construction. Solid masonry walls rely on excessive thickness and weight to resist the passage of sound.

Construction of walls

To comply with these functional requirements, walls may be constructed from individual units, such as bricks or larger blocks; a frame with infill panels, for example a timber frame with rendered panels; membrane materials, for example profiled metal sheeting; or monolithic slabs of reinforced concrete. Commonly, walls of domestic low-rise dwellings are constructed from bricks and blocks; less frequently, timber-framed buildings are erected. It is common in high-rise dwellings to find a load-bearing structure of either a steel or a reinforced concrete frame with infill panels acting as walls.

Bricks are made from burnt clay or shale, sand or fling and lime (calcium silicate) or of concrete. There are three varieties of clay brick, known as common, facing and engineering, the difference between common and facing being appearance: common bricks are often used for walls which are to be plastered, whereas facing bricks are used for walls where the brickwork is to remain exposed. Engineering bricks are capable of withstanding additional loads and are therefore used where a high degree of strength is required; for example, walls below ground. The pattern (bond) in which the bricks are laid may be varied (Figure 6.8).

Blocks are produced in a wide range of sizes and thicknesses, although they are made to course with brickwork. They are made from dense and lightweight aggregate concretes and aerated concrete and may be solid, hollow or cellular in form. Precautions against cracking blockwork must be taken because of the considerable moisture and shrinkage movement which occur in concrete.

Bricks and blocks are joined together using mortar. The face edges of the joints of mortars in brickwork may be finished in various ways in order to compress and smooth the exposed surface and to give choice in the appearance of the final wall (Figure 6.9).

Walls may be constructed from a single skin of brick or block work, or alternatively, and more commonly, with a cavity. There are various functional advantages of a wall built in two leaves or skins with a cavity, including better resistance to rain penetration. The two leaves are connected by means of cavity wall ties. These are designed so that water cannot pass from the outer to the inner leaf, and so that mortar droppings cannot easily lodge on the ties during the building of the wall and bridge the cavity, because each tie is formed with a drip at the centre which prevents water passing across.

The outer leaf is usually a half-brick thick in stretcher bond and the inner leaf is the same or, more commonly, 100 mm thick lightweight concrete blocks. A cavity is incorporated of at least 50 mm but usually 100 mm width, with 50 mm of insulation held by small discs attached to the cavity wall ties to the internal face. Building in block work is quicker and cheaper than in brickwork. The base of the cavity is filled with concrete, the top of which must be kept at least 150 mm below the level of the lower damp-proof course. This provides a space as a precaution against moisture rising above the damp-proof course. Every third vertical joint in the outer leaf at the base of the cavity is left open as a means of discharge for any water which might collect at this point. These holes are called weep holes.

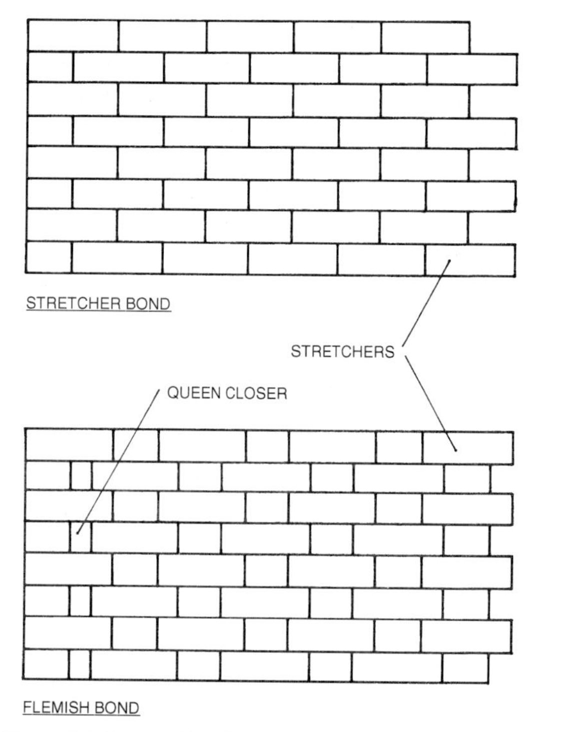

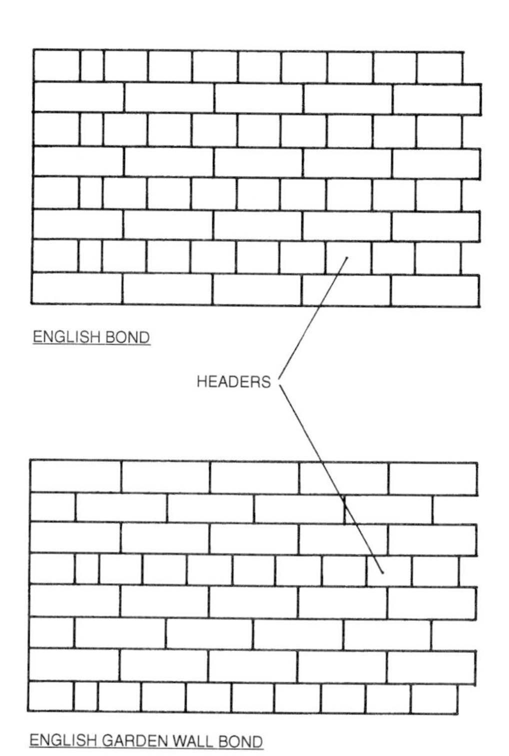

Figure 6.8 Types of brickwork bond

In framed buildings external walls may be constructed from solid or cavity walling of stone, block or brick facings applied to solid or cavity background walls or cladding panels of precast concrete. Standard panel sizes can be cast in a factory and transported to the site; erection of the panels can be relatively quick and the finish of the panel, rugged or otherwise, can be varied. It is common to find facing materials of brick or stone on the panels. Care is required when designing joints between the frame and panels to allow for movement. Selection of a system for fixing the panels also requires careful consideration.

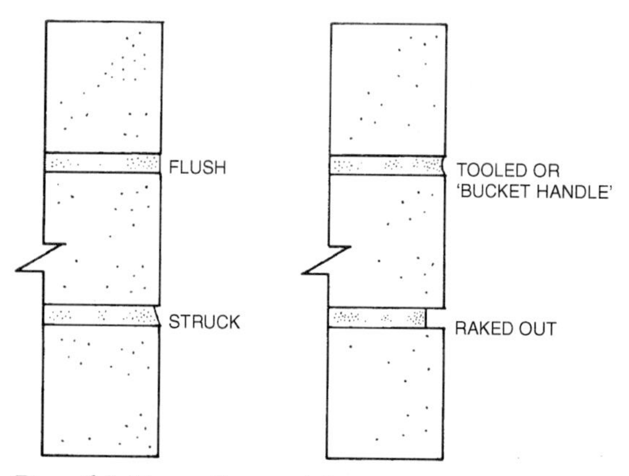

Figure 6.9 Types of mortar joints

Horizontal damp-proof courses are used at the head and base of a wall to prevent vertical movement of moisture. Suitable materials for this purpose are sheet metal, of which the most commonly used are lead and cooper, bituminised felt, polythene, asphalt, slates and engineering bricks. Stepped damp-proof courses are required when a building is set into a sloping site or steps down. The stepping must maintain the damp-proof course at least 150 mm above ground level at all points.

Openings in walls are required for windows and doors. A number of considerations determine the dimensions of openings and their position in a brick wall. Structural considerations are concerned with the relative widths of openings and adjacent walling. The greater the width of openings the greater the weight of the wall above that is transferred to the walling on each side, which must be strong enough to carry it. This is particularly important where a series of closely spaced openings leaves relatively narrow piers of brickwork between them.

Economic considerations are concerned with the width and position of openings relative to the normal bonding of the wall in which they occur. To avoid irregular bond above an opening, its width and height should be a multiple of the brick size. Thus, the correct face appearance is maintained throughout.

Support for the wall above an opening is provided by a horizontal beam called a lintel. The most common materials used for lintels are reinforced concrete and steel. Arches are also used to span an opening with components smaller in size than the width of the opening. Arches consist of wedge-shaped blocks, which by virtue of their shape mutually support each other in spanning the opening between the supports.

Careful detailing round openings in cavity walls is essential (Figure 6.10). It is mainly at these points that the cavity is bridged and damp may penetrate to the inner leaf. Jambs of openings may have square or rebated jambs as in solid walls. In either case the cavity must be closed to preserve its integrity as a thermal insulator and to form a finish at the jamb. The closure must be carried out in such a way that moisture cannot penetrate to the inside of the wall.

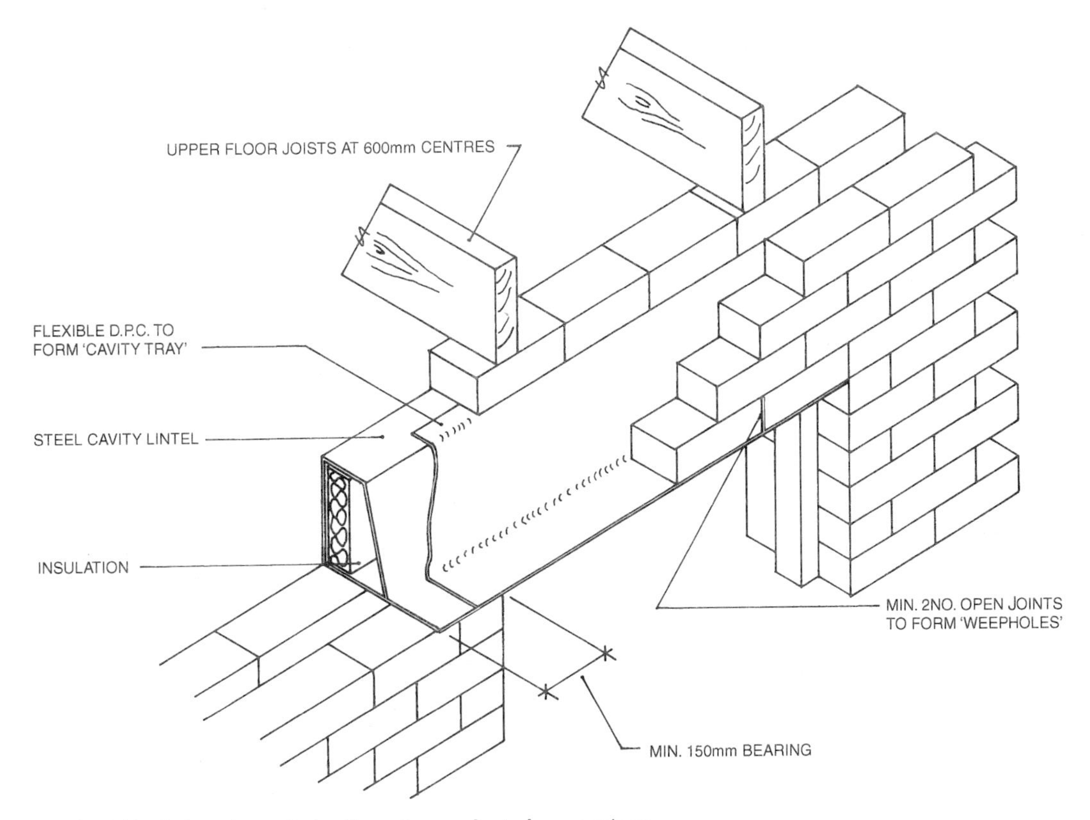

Figure 6.10 Typical cavity wall details at door and window openings

Sills are applied to the external opening, providing protective covering to the wall. The wall below a window opening is particularly vulnerable to water penetration as the rain falling on the window flows easily from the glass. A sloping sill should be designed so that it also prevents driving rain penetrating the joint at the seating of the window frame. Suitable materials for the construction of sills are stone, concrete, brick, or roofing or quarry tiles laid in cement mortar.

Thresholds below a door opening are to form a firm and durable base to the doorway and to exclude water; materials used include stone, concrete, brick, quarry tiles and timber.

Floor structures

At ground and basement levels full support from the ground is generally available at all points and a slab of concrete resting directly on the ground may be used. This is known as 'solid floor' construction. At upper levels the floor structure must span between relatively widely spaced supports in order to leave the floor area below unobstructed. This is known as suspended floor construction.

Functional requirements of floors

The primary function of a floor is to provide support for the occupants, furniture and equipment of a building. To perform this function and others the floor must satisfy a number of design requirements. It must have strength and be stable; provide fire and damp resistance; and provide a degree of sound and thermal insulation.

Problems of *strength and stability* are usually minor ones at ground or basement level where the floor can use the support of the soil below for support. A suspended upper floor is required to be strong and stiff enough to bear its own weight and the dead weight of any floor and ceiling finishes, together with the superimposed live loads which it is required to carry.

Fire resistance is important with respect to upper floors, which are often required to act as highly resistant fire barriers between the different levels of a building.

The *sound insulation* need not be considered in ground or basement floors unless there are sources of excessive sound vibration in close proximity to the building. Generally, the degree of insulation required will vary with the type of building and the noise source likely to create a nuisance. The form of sound insulating construction will vary with the type of floor that is used; for example, whether it is constructed from concrete or timber.

Thermal insulation is not normally required in upper floors except in those over external air or ventilated space. Thermal insulation is usually incorporated within the construction of ground floors, as this can considerably reduce the amount of heat loss through the floor; insulation is also necessary to comply with the u-values required by the Building Regulations 1991, which must be adhered to for all new and extensions to domestic dwellings. U-values measure the thermal conductivity of a particular element, for example a cavity wall.

Solid ground-floor construction

In the design of ground floors consideration of the level at which the floor is to be placed relative to the surrounding ground is important. A number of factors govern the floor level, including the nature of the site and the form of floor construction. The damp-proof course is usually placed about 150 mm above ground level. Thus, the level of the floor is fixed slightly higher than the damp-proof course in the walls and at a level so as to permit the damp-proof membrane in the floor to connect with the damp-proof course in the walls. This is common to all ground floors.

A bed of well-consolidated hardcore is laid, to a minimum of 100 mm, to reduce the capillary rise of ground moisture, to act as a filling to provide horizontal surface at the appropriate level for the concrete slab, and to form a firm, dry working surface. The hardcore is usually covered or 'blinded' with sand, to prevent the hardcore from piercing the polythene. Polythene sheeting in roll form may be laid directly on the sand blinding, and this acts as a damp-proof membrane (DPM), to prevent moisture from penetrating the floor slab; it should be connected with the damp-proof course, if necessary by means of a vertical damp-proof course. Thermal insulation may be laid either below or above the concrete slab. The concrete slab is not less than 100 mm thick. The edges of the floor slab should not be built into the surrounding walls nor should they rest directly on foundation slabs.

Suspended timber ground floors

A suspended floor (Figure 6.11) is generally of limited span; the floor structure is supported on low walls built off the ground and thus out of direct contact with the ground moisture. After the topsoil is removed, the ground must be covered with a 100 mm layer of concrete on a hardcore base to exclude ground moisture and to prevent vegetable growth. Dwarf half-brick sleeper walls are built off the surface concrete to support the floor structure. The floor structure consists of timber bearers called common or bridging joists bearing on the dwarf walls; the size of the timber members will depend upon their span, spacing, loading and the grade of timber used. The joists bear on timber wall plates, bedded in mortar on top of the sleeper walls. A damp-proof course is placed in the sleeper walls immediately below the wall plate, preventing rising damp from penetrating the floor timbers. The height of the underfloor space above the surface of the concrete must be at least 75 mm to the underside of any timber wall plate and at least 125 mm to the underside of the suspended timbers. The space must be adequately ventilated to prevent the air becoming humid and giving rise to conditions favourable to the growth of fungi. Thus, airbricks must be provided in all external walls if possible and in at least two, to allow a cross-flow of air, and all dwarf walls should be in honeycomb construction to permit free flow of air. Ventilating holes must be formed in all partition walls through the underfloor space. The airbricks should be placed well free of the ground and free of any obstructions when they are situated in cavity walls.

Intermediate floors

An upper floor in timber differs in some respects from a timber ground floor. In the latter there is no restriction on the number of supports in the form of sleeper walls, so that the span and size of the joists may be kept small. Relatively large unobstructed areas are, however, required under upper floors, resulting in the need for a wider floor span and larger joists.

Single floors in timber

These consist of common or bridging joists spanning between walls or partitions and bearing usually on wall plates or other members to distribute the load. Material and minimum thickness of joists are the same as for ground-floor joists. The depth of the joist will depend on loading, span and spacing and it may be calculated taking into account these factors. Stiffening is required when joists are deep in order to avoid winding or buckling at the top or compression zone. Herring-bone or solid strutting is required. Timber is easily cut and drilled for pipes and conduits but must be done with care, so as not to affect the structural integrity of the timber.

When a timber or other lightweight non-load-bearing partition bears on a timber floor the joists of which can run parallel with the partition, the joists are commonly doubled up under the partition, a pair being spiked together.

The floor around openings, such as stairs or hearths, must be so constructed as to be self-supporting at these points. This is accomplished by cutting short some of the bridging joists to form the opening.

In a building with a reinforced concrete or steel frame, systems of precast reinforced concrete slabs, beams and infill concrete block floors are used. Precast, hollow, reinforced concrete floor units are relatively lightweight, as they are hollow, and mechanical lifting equipment can lift them into

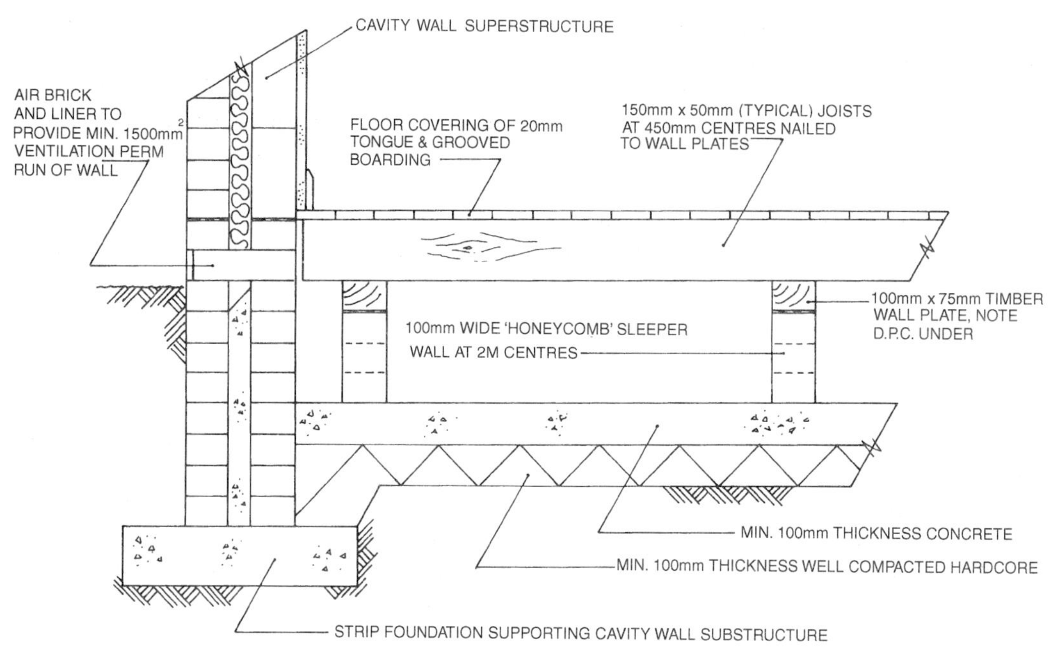

Figure 6.11 Typical suspended timber ground floor

place. However, there are a wide range of systems, for example precast, prestressed concrete units and hollow clay blocks with in situ cast reinforced concrete between the blocks.

Roofs

Functional requirements

The main function of a roof is to *enclose space* and to protect from the elements the space it covers. A roof is required to be stable and to carry various loads, be resistant to the effects of weather, provide some thermal and sound insulation, and be fire resistant.

Strength and stability are provided by the roof structure and a major consideration in the design and choice of the structure is that of span. It is necessary to keep the dead weight to a minimum so that the loads can be carried with the greatest economy of material. The roof should be able to carry the dead weight of the materials from which it is constructed, plus the superimposed loads of snow and foot traffic for, say, maintenance purposes, and must resist the effects of wind.

Adequate weather resistance should be considered in the design stage, to provide the correct specification of shape, construction and materials of the roof. It is necessary to ensure that excessive water is prevented from penetrating through the covering, and that it is allowed to drain freely from the covering, via gutters and downpipes.

The provision of thermal insulation in the roof is essential, as with floors and walls, to reduce heat loss to the external air. Insulation is also required by the Building Regulations 1991. However, increased thermal insulation results in an increased risk of condensation, hence adequate ventilation of the roof space is necessary to prevent condensation.

The degree of fire resistance which a roof should provide depends upon the proximity of other buildings and the nature of the building which the roof covers. Generally for domestic structures it is unnecessary to consider fire resistance, as the materials which are used are sufficiently fire resistant.

Most forms of roof construction provide for the majority of residential buildings an adequate degree of insulation against external noise sources.

Construction of roofs

Roofs may be either pitched or flat. A roof is called a flat roof when the outer surface is horizontal or is inclined at an angle not exceeding 10 degrees and called a pitched roof when the outer surface is sloping in one or more directions at an inclination greater than this.

Climate and covering materials affect the choice between a flat or a pitched roof. In hot, dry areas flat roofs are common because they are not exposed to heavy rainfall and the roof forms a useful out-of-doors living area. In areas of heavy rainfall a steeply pitched roof quickly throws off rain, while in areas of heavy snowfall a less steeply pitched roof with a slope of, say, not more than 35 to 40 degrees preserves a useful 'insulating blanket' of snow during the cold season but permits thaw water to run off freely.

Covering for roofs may consist of individual unit materials such as tiles or slates laid close to one another. Alternatively, a membrane or sheet material is used, with specially formed joints, for example asphalt, bituminous felt or metal sheeting with sealed or specially formed watertight joints. Sheet materials must be laid to a slight fall to allow adequate drainage of water from the area.

Flat roof construction

A flat roof structure consists of joists spanning between supports and is basically the same as for a suspended timber floor structure. The thickness of joists and factors affecting their size are also the same as for suspended timber floor construction. It should be noted that a greater imposed loading must be assumed for a roof with access not limited to repair.

The necessary gentle slope or fall in the roof may be obtained by laying the joists to fall in the required direction. If a level ceiling is required, battens of timber tapering in depth, called firrings, are laid on top of the horizontal joists to support a sloping top surface.

A flat roof may be insulated in two ways: by placing the insulating material on the ceiling lining between the roof joists (warm deck construction) or by placing it above the roof decking (cold deck construction).

In cold deck construction (Figure 6.12), the insulating material is placed immediately above the ceiling. The roof deck and roof space are at or near the external air temperature and for much of the year substantially below that inside the building. There is considerable danger of warm humid air from inside the building penetrating and condensing within the cold roof space. For this reason, a vapour barrier or check should be incorporated on the underside, the warm side, of the insulation to prevent the passage of water vapour into the roof. This involves the provision of a layer of material such as polythene sheeting. As some leakage of water vapour is inevitable, the roof space must be adequately ventilated in order to disperse such vapour and thus prevent it condensing on the underside of the cold deck. Cross ventilation is necessary by means of permanent vents on two opposite sides of the roof. The insulating material may be in the form of mat, quilt or loose fill and may need to be secured against displacement by the flow of air through the roof spaces.

In warm deck construction (Figure 6.13), the insulating material is placed between the waterproof covering and the roof deck with a vapour barrier on the warm side of the insulation. The material must be capable of withstanding normal roof loads, and resin-bonded mineral slabs and expanded polystyrene slabs meet this requirement.

An alternative to warm deck construction is an inverted warm deck, in which the insulation is placed above the waterproof covering, which avoids some potential problems associated with the warm deck form. The insulating material must be capable of withstanding normal roof loads; for example, expanded polystyrene.

When the supporting walls carry up as parapets, the roof will be enclosed by the walls. If the roof is carried over the top of the walls its edge will be

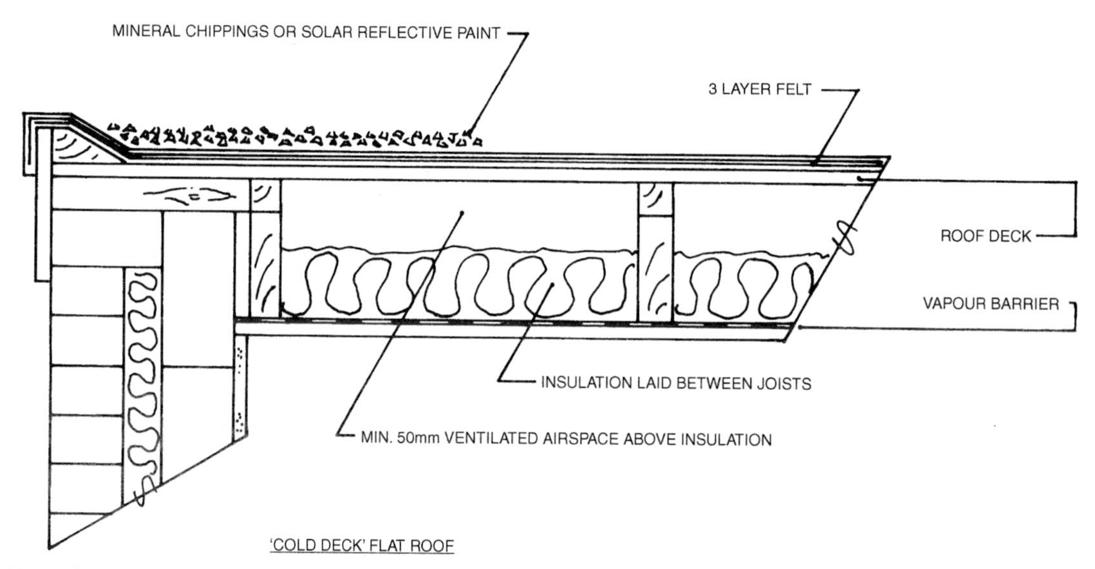

Figure 6.12 Typical section through a cold deck flat roof

exposed as eaves which can be finished either close to the outer wall face to form flush eaves or beyond the outer wall face to form projecting eaves.

Where the ends of the joists terminate at the outer wall face they may be covered or finished with fascia board. In projecting eaves, where the ends of the joists project beyond the outer wall face they may also be finished with a fascia board, as in flush eaves. Closed eaves are formed by fixing a soffit to the underside of the joists.

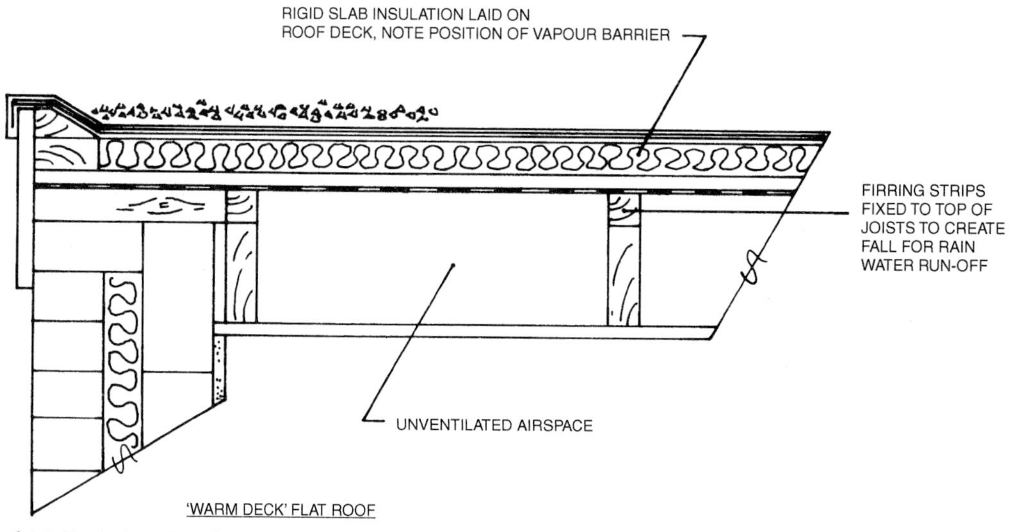

Figure 6.13 Typical section through a warm deck flat roof

Pitched roofs

Pitched roofs may be constructed by a traditional method, using individual timber members, or by using factory-made timber trusses. The production of trussed rafters under factory conditions employs stress-graded timbers joined with precision by galvanised mild steel plates. The trusses are economical and lightweight, and their delivery to the site can be timed to fit to the overall programme of works.

Roofs which are traditionally constructed may consist of a number of different elements including a ridge board, ceiling joists, rafters, purlins, collars and wall plates, depending on the roof shape and loadings. Each of the elements must be of a size and a grade of timber suitable for the load which they are to carry.

Pitched roofs are commonly covered with tiles or slates. Felt is laid below the covering, to provide additional protection to the structure. Horizontal rows are laid starting from the eaves, and each tile or slate overlaps. On completion, the rainwater gutters and drainpipes are installed. Work can now commence on the interior of the building, as it is now watertight and a dry interior is ensured.

Doors and windows

Doors and windows may be constructed from various materials including timber, aluminium and uPVC. There are alternative types of glass that can be used, for example laminated and toughened. When one is designing windows and doors, their location, use and maintenance should be considered carefully.

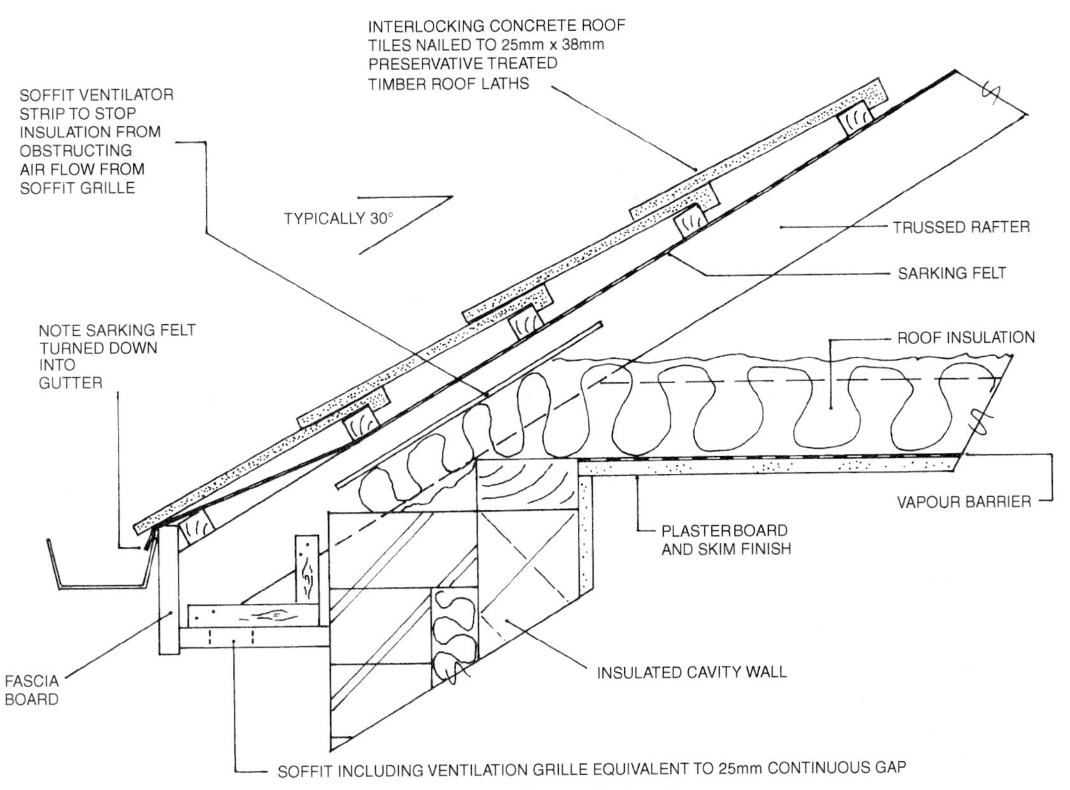

Figure 6.14 Typical section showing eaves detail in a timber pitched roof

On lower floors, glazing is particularly vulnerable to impact, so laminated or toughened glass should be used. Upper-storey windows may be difficult to clean externally, so windows that pivot 180 degrees, enabling access to the outside face, could be installed.

The frames for doors and windows may be positioned prior to the construction of the wall around them. Weatherproof joints must be made between the wall and the frame, and damp-proof courses are installed around the opening. As stated previously, lintels are required to support the wall above the opening.

Stairs

Stairs are generally constructed from timber in low-rise dwellings and reinforced concrete in high-rise buildings. Staircases may be of various designs, including spiral, dog-leg and open treads. The primary consideration in the design of any staircase is that it should be safe. The width and depth of each step should be the same, as should the height of the step. The overall pitch of the staircase should not be too steep, and there should be adequate headroom above the stairs. These should be handrails on at least one side of the treads and preferably both. There should not be an excessive number of treads in each flight of stairs, so there should be a change in direction, and a second flight, if there are a large number of steps leading in the same direction. The treads should be covered with a non-slip surface.

Services

Drainage is required for both foul and storm-water. Storm-water is water from snow or rain, collected from roofs, flat or pitched, and hard areas, such as garage forecourts or patios. Foul water has been used for washing or cooking, and includes water from the w.c., hand basins, baths, washing machines and dish washers.

Storm-water may either drain into the drainage system, which in some areas may be combined with the foul water, or, in a garden, to a soakaway, a hole in the ground filled with clean rubble to which the drain runs, the water filtering through the rubble to the garden. Alternatively, the water may be collected in rainwater butts for use in the garden.

Foul water may drain into the drainage system, to a cesspool or a septic tank. Cesspools and septic tanks are common in rural areas.

The principles of drainage are similar for the design and construction of both foul and storm-water systems. In essence, the system must be adequate to carry the waste, and must be laid with a sufficient fall to prevent both backwash up the system and settlement of solids in the system. Access should be provided at every change in direction, to clear blockages should they arise; this may be a rodding eye or an inspection chamber. Foul water systems also require ventilation of the system, which is undertaken by extending the stack upwards, ensuring the outlet is clear of openable windows, to prevent foul smells entering the building. Gutters and downpipes should be kept clear of debris and vegetable matter, particularly during the autumn. Common materials used for drainage systems are uPVC and clay. uPVC is easy to handle and quick to lay. Clay is a more traditional material, and is not often used today, although it can still often be found in older properties.

The drains of a building are often laid at the same time as the foundations, as the machinery is present for excavation of the drainage run. The trench should be backfilled with pea shingle and the drain-run laid, ensuring that there is adequate fall to the run. All drainage work requires inspection, particularly where the drain connects into the public drainage system.

The drainage above ground is installed usually after the basic structural work has been completed. Often the sanitary appliances are fixed towards completion, to prevent the tradesmen having to work behind awkward spaces.

The water supply is taken from the local water board, with a stop valve located at the edge of the site, or within individual flats. The gas pipe and electric service cable are brought independently to the site to a point of termination by a meter. Once the gas and electrical service have been installed an inspection should be undertaken by the relevant Gas and Electricity Boards, to ensure compliance with the CORGI (Council of Registered Gas Installers) and IEE (Institution of Electrical Engineers) regulations. Telecommunications are installed by the relevant company, to current regulations.

Finishes

Surface qualities have a critical influence on the aesthetic qualities of a building. In the selection and use of a finished material the colour, its texture and pattern, and the way it meets and joins other materials should be carefully considered.

The primary function of a plaster is to cover up any unevenness in the surface of an internal wall or partition. The surface provides a hard-wearing surface which is then suited to application of the desired decorative finish, whether paint, wallpaper or tiles. Plaster also enhances the thermal, acoustic and fire-resistant properties of a wall. Ceilings and some walls are usually covered with plasterboard, which is easily handled and can contribute to fire resistance. A wet plaster finish provides a smooth surface and covers any irregularities in the wall surface. Gypsum plasters are widely used. The type of plaster and number of coats required will depend on the nature of the background and the proposed use. Wet plastered walls require time to dry out prior to decoration. Paint, wallpaper, tiles or other surface finishes may then applied. The decoration selected should be suitable to the overall use of the room. For example, hallways will be subject to considerable wear and tear. The finish in bathrooms will be required to be resistant to high concentrations of condensation.

CASE STUDY 6.2

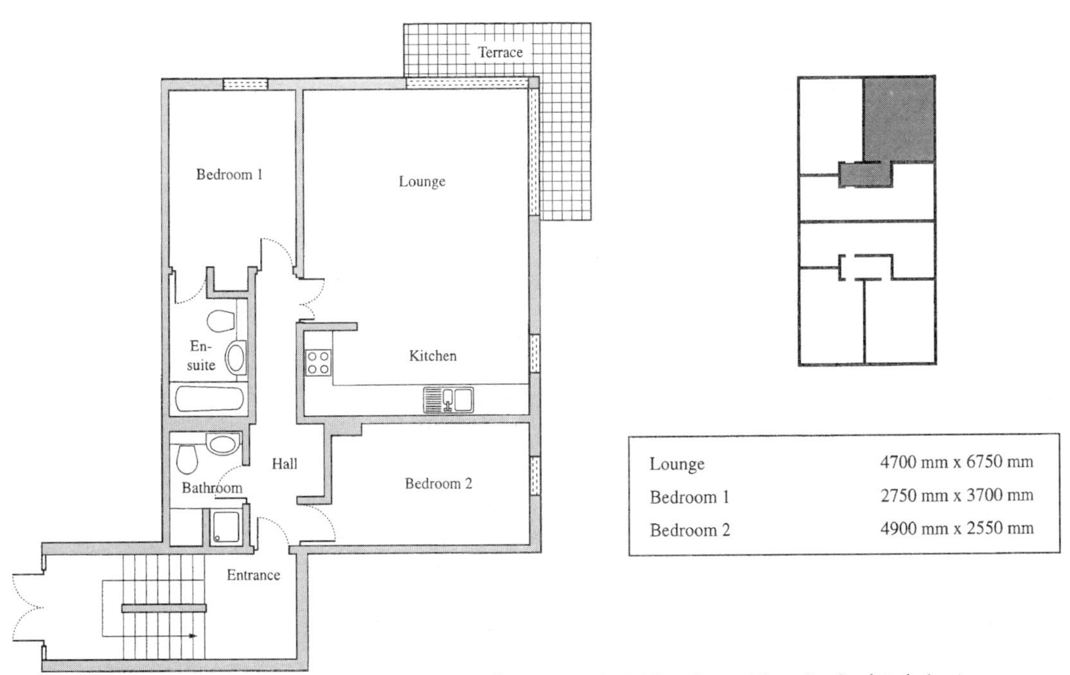

Figure 6.15 Plan of typical layout of a two-bedroom flat in King's Bridge Court (Case Study 6.2, below)

KING'S BRIDGE COURT, ISLE OF DOGS, LONDON

The pictorial case study in Plates 6.1–6.4 is based upon a £3.5 million development of forty one- and two-bedroomed flats (Figure 6.15). The flats have been, in the main, purchased by private investors and are leased to private tenants, although they are managed by letting agents. The development is typical of those constructed in recent years in London's Docklands.

MAINTENANCE POLICY AND PRACTICE

It is unlikely that a building can be designed and constructed that is maintenance-free; that is, one which requires no servicing or cleaning, rectification of defects or replacement of worn-out elements. Buildings comprise a number of elements and the constituent materials and components have a range of life expectancies which, in most cases, will be shorter than the life of the building as a whole. Therefore maintenance is inevitable and it is thus

Plate 6.1 King's Bridge Court, Isle of Dogs, London, rear elevation just prior to completion of the development

Plate 6.2 King's Bridge Court, front elevation during construction
The protruding steel beams will support balconies similar to those that can be seen in the background. The blockwork is to be rendered, as on the rear elevation

essential that property-owners and managers allocate sufficient time and resources to the management of maintenance of the building fabric. In addition, poorly maintained buildings have an adverse effect on the overall value of the property, and on the health and morale of the occupants.

The following section provides an overview of building maintenance: the essential nature of maintenance; varying types of maintenance, the manner in which maintenance is undertaken; and the resources required to manage an effective policy.

The necessity of building maintenance

Various wide-ranging surveys have been undertaken of the housing stock in the United Kingdom. They indicate that generally there is a considerable backlog of maintenance work to be undertaken, particularly on local authority housing, and an inefficient allocation of resources, indicated by the large amount of responsive and emergency maintenance undertaken. These and other factors which have emerged from this recent research indicate that many householders live in poorly maintained buildings, which

Plate 6.3 King's Bridge Court, rear elevation during construction
The pediment element is made of concrete and is purely decorative. Polystyrene board froms the cavity wall insulation, and can be seen behind the brickwork. Cavity wall ties are visible, with small plastic discs which keep the insulation securely fixed to the internal face of the brickwork

have a wide range of defects such as condensation and dampness penetration (Seeley, 1987: 6). There is evidence to suggest that properly maintained public sector housing can promote the well-being of the occupants. For example, where the external landscaping is maintained in good condition, with the grass regularly cut, rubbish removed, benches cleaned and maintained, and so on, the people living on the estate may be encouraged to make further use of the grounds. Where the access corridors and staircases are clean and tidy, where rubbish is cleared, graffiti painted over or removed, the people using the buildings may feel safer, and to a certain extent people may dispose of rubbish properly or not deface walls.

Building maintenance is therefore essential, primarily for three main reasons:

1 To retain the value of the investment and to ensure longevity of the components of the building. For example, a property that has missing roof tiles, a boiler that requires replacing, or windows that need to be repainted cannot retain its value as much as one where regular maintenance has been undertaken. Lack of maintenance can also lead to further problems. For example, missing

Plate 6.4 King's Bridge Court, pitched roof during construction
Timber trussed rafters pre-formed in a factory have been used. Toothed plate connectors join the ceiling joists and rafters together; these may only be used in factory production, as they must be driven with a hydraulic press or roller. Such factory-made trussed rafters can be constructed simply and quickly on site, and are therefore easier to use than traditionally constructed roofs

tiles can cause water to penetrate to the roof structure, possibly causing rot within the timbers; eventually the roof structure may need replacing, before the time predicted during the design of the roof.

2 To maintain the building in a condition in which it continues to fulfil its function. For example, where condensation has been persistent and extensive, and no rectifying measures have been adopted, the health of the occupants may be severely affected; eventually the area of the building affected becomes a health hazard, thus no longer providing a clean, safe environment to inhabit.
3 To present a good appearance, by regular decoration of walls, doors, windows, and so on, which indicates that a building is probably well maintained, as well as being a more pleasant to live.

Definitions of building maintenance

There are three separate constituent parts to the maintenance of a property: servicing, essentially cleaning; rectification work, needed as a result of errors in the design of a component or the incorrect use of a material in a particular location; and replacement, of the components no longer fulfilling their functional requirements. Replacement of various components can be predicted by assessing the life expectancy of similar components used in the same circumstances.

Maintenance is defined by BS 3811 as 'the combination of all technical and associated administrative actions intended to retain an item in, or restore it to, a state in which it can perform its required function'. There are various types of maintenance, which are classified in BS 3811 as indicated in Figure 6.16.

The BS 3811 categories are defined as follows:

1 Planned maintenance: 'The maintenance organised and carried out with forethought, control and the use of records to a predetermined plan.'
2 Unplanned maintenance: 'The maintenance carried out to no predetermined plan.'
3 Preventative maintenance: 'The maintenance carried out at predetermined intervals or corresponding to prescribed criteria and intended to reduce the probability of failure or the performance degradation of an item.'
4 Corrective maintenance: 'The maintenance carried out after a failure has occurred and intended to restore an item to a state in which it can perform its required function.'
5 Emergency maintenance: 'The maintenance which it is necessary to put in hand immediately to avoid serious consequences.'
6 Condition-based maintenance: 'The preventative maintenance initiated as a result of knowledge of the condition of an item from routine or continuous monitoring.'
7 Scheduled maintenance: 'The preventative maintenance carried out to a predetermined interval of time, number of operations, mileage, etc.'

This is a very detailed way to divide maintenance works, but, in essence, maintenance should be *planned*, and may be carried out according to a schedule developed from a survey of the building fabric. Of course, not all maintenance can be planned, as occasionally emergency maintenance may be necessary; for example, if the lift within a high-rise block of flats breaks down. Some mainte-

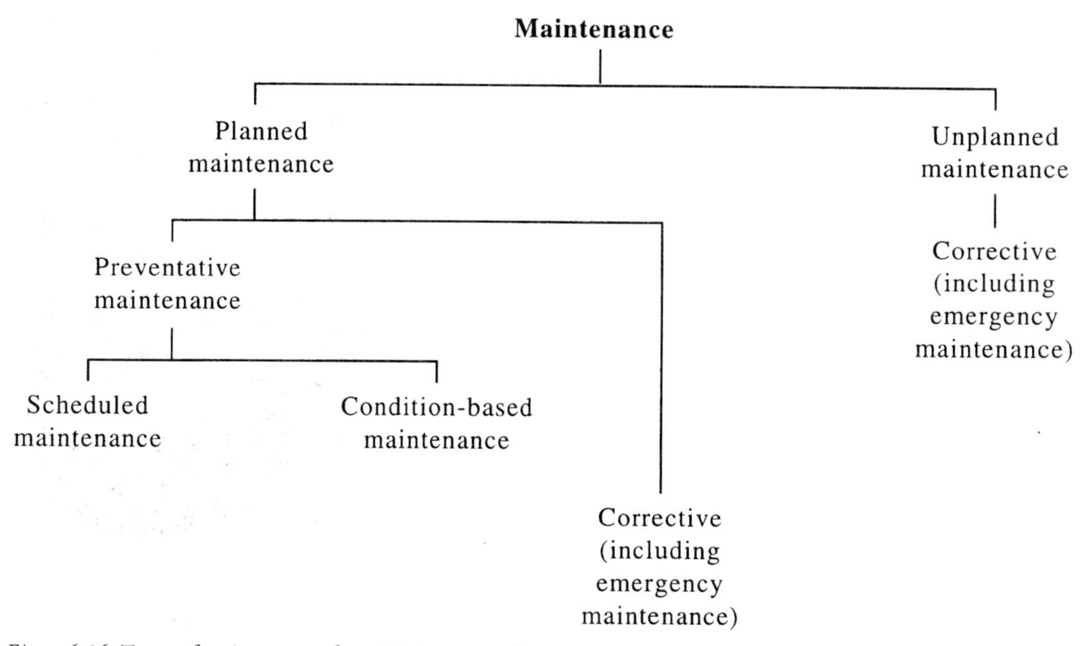

Figure 6.16 Types of maintenance (from BS 3811: 1984)

nance will be corrective; for example, the lift may be regularly serviced to prevent it from breaking down.

Maintenance management

Maintenance management requires a systematic approach, to ensure high standards, value for money and management control. Standards of building maintenance have been defined by government, local authorities, housing associations and other owners of property. If the maintenance standard is to be met then it is necessary to provide sufficient resources in finance, manpower and technology.

When there is little or no maintenance, the fabric of the building will initially be unattractive, then be unacceptable to the occupants and eventually dangerous and consequently uninhabitable. An optimum level has to be decided upon to preserve the fabric of the building to an acceptable standard. Factors influencing this optimum level may include:

1 the use and condition of the building, for example the number of people using the building, the way in which the building is used, the age of the building, the quality of the workmanship within the building, etc.;
2 the comparative cost and the effectiveness of different types of repair; while it may be expensive to undertake a regular inspection of a particular component, for example a lift in a multi-storey building, the real cost of not carrying out regular inspections can be high: when the lift breaks down, there is considerable hardship endured by the occupants, particularly for the elderly and for families with small children, in having to climb the stairs.
3 the expected future life of the building; for example, if the local authority is expecting in the near future to replace, say, a development of flats built during the 1960s, it may only be financially viable to spend money on urgent, emergency repairs, recognising that the flats will shortly be vacant; and
4 the acceptable standards of maintenance, which should be clearly defined and justifiable, and be to a similar standard to those of similar organisations, although this is dependent on the policy of the individual organisation, providing that statutory standards are met.

In defining a maintenance policy for an organisation there are various factors that need to be considered. They include the aims of the organisation, related to the nature of the end product, which in terms of housing provision will vary primarily according to the type of lease arrangements to be made between the occupiers and the housing association or local authority. The policy will be influenced by the standards required, although these may vary between different buildings; by the need for compliance with statutory legislation, particularly that relating to health and safety standards; and the way in which maintenance is to be undertaken, whether it is to be executed by employees of the organisation (direct labour) in a Direct Labour Organisation (DLO) or by outside contractors (see below, p. 190). The cost and method of financing the work will also affect maintenance policy, which may be funded by, for example, government grants, council tax, and rents paid by the tenants, and may need to be justified through cost–benefit analysis, which allows a comparison to be made where there are alternative solution, so the optimum solutions can be selected.

In addition to formulating the overall policy of the organisation, it is also necessary to form a more detailed policy for specific properties. For this, Seeley (1985: 339) recommends an analysis of the present condition of buildings, their nature and use, carried out by a desk survey, and then a detailed site survey, where a schedule of the condition of the property may be prepared. Various information can be recorded, usually on standard inspection forms, for example the locality and identity of elements; the type and extent of work needed; estimated total cost; and estimated year of treatment. Various categories can be used to detail the urgency of the work. The survey may show a need to deal with a backlog of general disrepair; to plan major restoration works some years ahead, to deal with year-to-year painting and associated repairs; and to operate a system of regular inspection and minor repairs.

The schedule of condition prepared, which will take into account all the works to be done, can then be analysed and costs can be attributed to each of the elements requiring maintenance. Unfortunately it may not be possible to calculate complete budgets for the maintenance work, although the calculations can be altered to take into account inflation in future years, which can help in preparing budgets in the medium and long term.

Following on from the schedule an outline programme of work can be developed, consisting of work necessary to put and keep the building in satisfactory condition. Once this work has been carried out to place the building back into a satisfactory condition, a programme of planned maintenance should be implemented, to maintain the building to the standard required.

The planned maintenance programme for the property will then consist of three maintenance elements: preventative running maintenance, which can be done while the building is still in use; corrective shut-down maintenance, carried out when the building is empty, or partially occupied; and corrective breakdown maintenance, carried out after a failure, for which advance provision should have been made in the form of spares, materials, labour and equipment being available. Both preventative running maintenance and corrective shutdown maintenance should be predicted and accommodated, although corrective breakdown maintenance cannot be predicted and thus, in the case of housing, is carried out when the occupants of the building report the defect to the maintenance organisation. The categories included within the programme may include the external fabric, prepainting maintenance, external decoration, external works, plumbing, internal fabric, internal finishes and fittings, heating installations and electrical installations.

Thus, maintenance programming should ideally be preventative as far as is practically possible, based on regular inspections at intervals designed to prevent trouble from developing or accumulating.

Maintenance work may be undertaken by contractors, direct labour organisations or a combination of both systems, and the decisions will be based on a number of criteria. The structures of maintenance organisations are examined together with programming and operational activities. The training of maintenance staff and the operation of incentive schemes are considered.

The cost of directly employed labour is made up of wages and materials; consumable stores; administrative overheads such as labour on costs and associated clerical, travelling and supervisory and depot costs. The cost of employing a contractor consists of the contractor's charges plus administrative overheads, such as inviting and comparing tenders, drawing up contracts, work supervision and checking invoices. The main advantages of employing a direct labour force are that it:

1 allows full control of activities of operatives, permitting reasonable flexibility, a more rapid response and direct quality control; and
2 should ensure a good standard of workmanship by craft operatives who enjoy continuity of employment and are suitably trained, although recruitment may be a problem.

Maintenance feedback should be an essential part of any maintenance programme; it should be mainly injected into the planning system directly into the design team, particularly information on design faults, faulty workmanship and materials failures; there should be general discussion with the maintenance team.

The frequency of follow-up inspections and the feedback from the people undertaking the maintenance are the mainstays of an up-to-date programme, although this will be very dependent on the allocation of resources and, as previously stated, resources – time, labour and financial – within many housing associations and local authorities can be limited.

Design issues

It is essential during the design stage of a project for a building designer to consider the implications of selecting various components and materials, in terms of the overall maintenance strategy of the client's organisation.

The amount of maintenance work can be considerably reduced through the careful selection of materials and components. Consequently it is beneficial if the designer has a detailed knowledge of the maintenance implications of the selection of various materials and components. Unfortunately designers often have a limited involvement with the building in the long term, and consequently their responsibility towards maintenance issues is reduced. In addition, poor communication, particularly in terms of feedback about the success or failure of components, can lead to repeated errors of judgement on the designer's part. Designers should also take full advantage of the manufacturer's advice when specifying a particular project. Consideration of maintenance issues should be integral to the overall design process.

Maintenance manuals

A convenient form of communicaticn is maintenance manuals and there is growing awareness of the need for them to be prepared for new buildings by the design team. Many local authorities provide their tenants with manuals and commonly the information within the manual will be arranged to include the following (Seeley, 1987):

1. contract and legal particulars including the design team, contractor and sub-contractors, details of easements (from statutuory regulations), statutory consents;
2. housekeeping details of surface finishes and decoration both internally and externally;
3. the means of operating mechanical, electrical or solid fuel plant or fittings, with details of requisite periodical routine maintenance and servicing and location of meters and stopcock valves;
4. full details of materials, components and constructional processes: all hidden features should be described and special items noted such as jointing, replacement techniques, and the methods of dismantling and re-erecting demountable components. The names and addresses of subcontractors of components should be given together with catalogue numbers, colours and other relevant information which may ease the task of ordering replacements or spare parts;
5. a record of maintenance executed, and provision for a maintenance log to permit constant updating;
6. emergency information: names, addresses, telephone numbers of contacts in the event of fire, theft or burglary; and
7. manufacturers' leaflets to give an after-sales service with technical information on cleaning, operating and maintenance.

BUILDING REHABILITATION

During any building's lifetime it may be necessary to alter the building fabric in some way to improve the way in which the building functions. These improvements may be necessary for various reasons, perhaps to extend the economic life of the building, to improve the living conditions so that they comply with current standards, or perhaps to change the use of the building or some part of it.

The following section reviews the way in which buildings may be rehabilitated, the purposes of various different ways of rehabilitating buildings, and the techniques used. Case studies illustrate examples where buildings have been successfully adapted, improved, extended or refurbished.

Refurbishment techniques

Refurbishment is possibly most important where a building no longer complies with statutory standards or even current expectations. Domestic dwellings should have access to hot and cold water, have an internal w.c., a hand basin and bath or shower, and indeed many householders would expect this standard of service provision. Thus many terraced dwellings that were previously without hot and cold water and an internal bathroom have had these installed. Grants are available to carry out these types of improvements. It may be necessary to accommodate the new bathroom within an extension, and often this may be on the ground floor of the property, extending into the rear garden. Alternatively, it may be possible to provide a two-storey extension, accommodating a new kitchen on the ground floor, with a new bathroom on the first floor.

Improvements may also be essential where the accommodation can no longer be used by the occupants, perhaps due to an occupant's disability. Where a disability severely affects the way in which a person can live, as they may have to use a wheelchair, it is extremely desirable to adapt the building specifically for the use of that individual. Disabilities are of such a personal nature that it is essential that the user is involved in the design of the improvements to be undertaken. Doorways may need to be widened; electrical sockets and light switches placed at a reasonable height to permit access even when sitting in a wheelchair; baths, showers and sanitary facilities may need to be altered, and hoists and leaning bars installed.

Extensions are an easy way of making additional space available, which may be easier than trying to find new suitable accommodation. Low-rise traditional dwellings may be extended according to the amount of space available around the site. It may only be necessary to provide a porch, providing extra storage space for coats and shoes. A single garage could be constructed to the side of a dwelling, perhaps with a bedroom above. Alternatively the foundations may be dug sufficiently deep to support another storey in the future. A utility room could be added to the rear of a property, with space for a washing machine, dryer, storage space and perhaps a ground-floor w.c. with hand basin. Conservatories are an attractive way of providing additional living space, although it is essential to ensure the glazing is suitable, and should be either strong enough to resist breaking on impact, or designed to shatter safely into small pieces with no sharp edges. In older properties, where there is a cellar, this may be converted to a workroom or storage space, although often dampness can be a problem.

Living and storage space can be made within the roof space. Attic conversions can be attractive, with sloping ceilings and dormer or velux windows. Access can be a problem, as often there is not enough room within the first floor to provide a full-width staircase, although the building regulations now permit space-saver stairs to be used, thus providing a viable alternative. In converting an attic to habitable space, there must be satisfactory provision for escape from the building in the case of fire. Installing additional smoke alarms can increase the chances of people being aware should fire break out. Where roof conversions have more than two habitable rooms, it may be necessary to ensure that there is a second way of exiting from the building, just in case one of the staircases is found to be unusable. The local authority building control officers provide advice on these types of issues.

Within non-traditional high-rise dwellings, it is not often possible to extend, although sometimes alterations to balconies, by enclosing the space and making it weatherproof, can provide some additional living space.

The services within a property can be overhauled or replaced, which may improve the quality of life of the occupants. For example, a central heating system may be installed. Gas wall-mounted instantaneous boilers are common, providing hot water without the need for preheating and storage in a hot water tank, and the cost of space heating is reasonable. A property may be rewired, providing the opportunity to install additional electrical sockets. Over recent years it has become necessary to install a large number of sockets, due to the increase in the use of electrical items.

Various improvements can be made to reduce the cost of heating the dwelling. Insulation may be laid within the roof space, although care should be taken not to lay the insulation within the eaves, in order to ensure that there is adequate ventilation of the void, to prevent condensation. If the walls are of cavity construction and were not built with insulation, insulation may be added, by cavity wall fill. Double-glazed uPVC window units now commonly replace old timber windows. uPVC is a material that requires little maintenance and double glazing considerably reduces heat loss from a building, as well as increasing the amount of sound insulation.

Old properties that are in a very dilapidated state, or where it is simply not economical to carry out minor repairs, may require complete refurbishment, to convert, or return, into comfortable living accommodation. Barn conversions are common in the countryside, and are often used to provide holiday accommodation, providing a second income for the farmer. In urban areas churches that have been successfully converted into living accommodation can often be seen.

With properties requiring extensive refurbishment work, there is often a difficult decision to be made by the client and the designer, as to whether it is more economic to demolish the property and construct a new building, although sometimes planning legislation will place restrictions on the old property, and consequently demolition is not a viable alternative.

CASE STUDY 6.3

EXTENSION TO TERRACED VICTORIAN HOUSING

Extensions to terraced Victorian properties are common, to provide additional accommodation, usually a bathroom and kitchen. Figure 6.17 indicates a typical arrangement where there was an external w.c. and internal bathroom, which will have been added approximately thirty to forty years ago.

In Figure 6.18 a new extension has been constructed in the old back yard, with a bathroom, lobby and kitchen. The rear ground-floor room has been enlarged by the removal of the old bathroom, and the adjoining wall between the living and rear room has been removed, thus making a larger, lighter living area.

CASE STUDY 6.4

MAJOR REFURBISHMENT OF TWO BLOCKS OF FLATS AND MAISONETTES

Complete modernisation is often required of older local authority housing. The following case study reviews a major refurbishment project. Figure 6.19 shows the proposed external elevations to a block of flats and maisonette. The work involved the complete modernisation of forty flats and maisonettes and internal and external refurbishment of high-rise buildings. The contract value was £1.5 million, and the work was undertaken in the late 1980s. Works to the individual flats included:

- complete modernisation including the provision of a fully fitted kitchen and replacement of all bedroom fittings;
- provision of independent plumbing, full central heating and hot water systems to each unit; and
- complete rewiring of electrical installations and provision of additional power sockets.

Works to the two blocks included:

- enclosing the front of the communal staircase with uPVC window units and improvement of the existing door entry system to improve security;
- provision of a new pitched roof to both blocks, with additional insulation to improve the thermal insulation and overcome condensation;

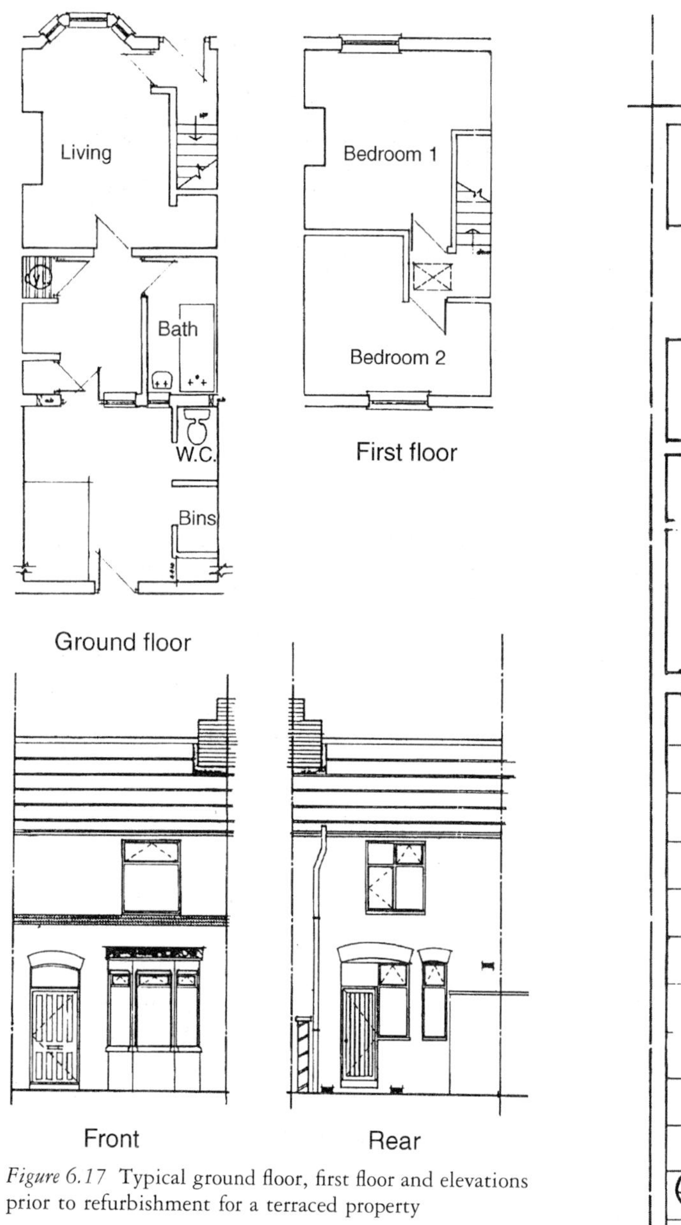

Figure 6.17 Typical ground floor, first floor and elevations prior to refurbishment for a terraced property

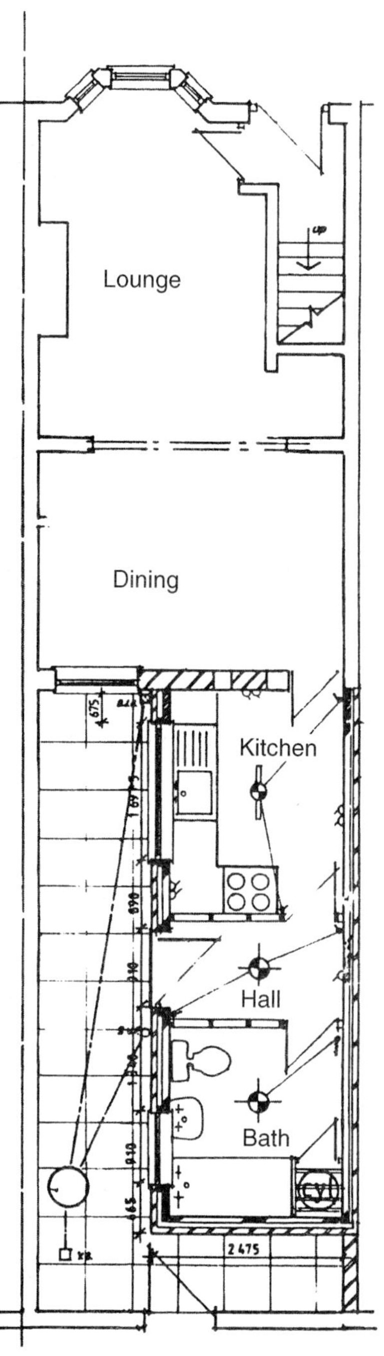

Figure 6.18 Plan of typical ground-floor extension providing a new kitchen and bathroom

Figure 6.19 Elevations of maisonettes as proposed

- extensive concrete repairs, together with brickwork and general repairs to the external fabric;
- replacement of all existing windows and access doors to private balconies with double-glazed uPVC units; and
- external areas to both blocks were landscaped and individual gardens provided to the ground-floor flats.

The work was to be undertaken while the building was fully occupied and consequently the contractors who undertook the work were chosen specifically bacause they had experience of working within occupied premises. The programme of works was formed by the contractors in conjunction with the residents' association, and stated that, subject to minor items, each living unit could be modernised in twelve working days, a factor which particularly appealed to the client.

While the work was being undertaken, the contractors had to be particularly mindful that the tenants were in occupation. Hence care had to be taken with a range of health and safety issues – for example, unattended drills, trailing electrical wires, wet paint, storage of excess building materials – as well as the tenants' belongings – for example, dust sheets had to be used to cover all carpets and furniture. Tenants were, however, decanted to other flats within the same blocks while the major work was being carried out within the twelve-day period. Once the major works were complete, the tenants were able to return to their homes.

CASE STUDY 6.5

Plate 6.5 Hall and Braithwaite Tower, Paddington Green Estate, City of Westminster, external elevations with full scaffolding
The scaffolding to the top three floors is covered with protective netting, to limit the fall of materials or tools that might drop

HALL AND BRAITHWAITE TOWER, PADDINGTON GREEN ESTATE, CITY OF WESTMINSTER

This was a £2.5 million refurbishment programme including external decorations and repairs; new uPVC double-glazed window units; a new flat roof; and removal of asbestos from the old communal heating system. On the roof of the tower, the asphalt had become defective; there was ponding on the roof during and after periods of heavy rainfall and the waterproof rendering to the raised elements was inadequate (Plate 6.5). The cables needed to be re-routed so that they did not trail across the flat roof. The concrete had spalled, exposing the reinforcement. Patch repairs needed to be undertaken, to prevent further deterioration of the concrete. The defective lightning conductor also needed to be replaced, and a lightning mast needed to be installed. Damp-proof courses were incorporated to the raised elements.

CASE STUDY 6.6

CONVERSION OF REDUNDANT BARNS WITHIN THE PEAK DISTRICT NATIONAL PARK

Within the Peak District National Park there are several hundred barns, most of which are neglected and considered to be redundant. Recently there has been a rapid increase in the number of barns being converted into additional living accommodation and holiday accommodation. With regard to statutory authorities, both the Peak Park Joint Planning Board and relevant local authority, for example, Staffordshire Moorland District Council, require formal notification of any barn conversions in the form of planning permission and Building Regulations approval, respectively. When planning permission is granted for the conversion of a barn, it is often subject to a range of conditions (see Table 6.9). The aims of these restrictions are to minimise the impact of the developments and to safeguard the landscape character of the area.

Many barn conversions are successful, providing unusual accommodation. Plates 6.6 and 6.7 show two interesting conversion schemes.

Table 6.9 Examples of restrictions imposed on barn conversions undertaken in the Peak District National Park

All new stonework to be reclaimed natural stone matching the existing in terms of stone size, colour, coursing and pointing.
No window or door frame to be recessed less than 100 mm from the external face of the wall.
All window openings to be provided with natural gritstone lintels and sills.
All soil and vent pipes to be completely internal and to discharge either via a ridge vent or by means of an anti-siphon valve in the roof space.
The gutters to be fixed directly to the stonework with brackets and without the use of fascia boards.
Roof lights to be fitted flush with the roof space.

Source: Peak District National Park Authority

Plate 6.6 Peak District National Park, barn prior to conversion
The barn, originally an eightennth-century dwelling, had been used since the 1940s to store farm materials and to shelter animals. One of the two lean-tos had to be demolished in order to comply with Park Park planning requirements, as they were not part of the orignal structute but were added in the 1950s. The building is now converted into a dwelling and includes a third storey

Plate 6.7 Peak District National Park, complete barn conversion
Three barns were converted at the farm to form holiday accommodation. One of the units was specially designed to provide facilities for disabled people. The barns were enclosed by walled courtyard, which provided privacy both to the adjacent farmhouse and to the holiday accommodation

CASE STUDY 6.7

ALTERATION AND ADAPTATION OF A GROUND-FLOOR FLAT FOR A DISABLED PERSON

People who are severely disabled may, on account of their own problems or the effects that their disability has on those who are caring for them, have special housing needs. In addressing the problems of disabled people, local housing authorities have in recent years been encouraged by government policy and associated legislation and regulations to work in conjunction with social services authorities for in-house adaptations. Statham, Korczak and Monaghan (1988) have shown that management of a major house adaptation scheme will be better exercised where there is a project team comprising all those who have a direct interest. For the preliminary work this will usually mean the client, the occupational therapist, the designer and the local authority grants officer, where the scheme is grant aided, or the housing administrator, in the case of a public sector property.

The aim of the refurbishment of the disabled person's ground-floor flat (Figure 6.20 and 6.21) was to provide facilities tailored to his specific needs and to maximise independence and safety in daily living, giving opportunities for development of work, leisure and social interests.

The property was a pair of flats built in the 1890s as part of a terrace of traditional construction. The conversion was to make the flat more suitable for one-hand wheelchair use. Internal reorganisation of the rooms was required: the position of the doorways caused very difficult turns into rooms from the hallway; the original opening to the bathroom precluded wheelchair access; neither of the bedrooms nor the kitchen was large enough for reasonable wheelchair accommodation. In the alterations, internal walls dividing the bathroom, the bedrooms and hall were removed and realigned to give a larger hallway; the wall dividing the kitchen and dining room were removed and external walls dry lined; the decaying timber floor was replaced in concrete, 200 mm below the previous level; and bonded, heavy-duty waterproof nylon carpeting was laid to all floor surfaces, except the tiled kitchen and shower area; the rear boundary wall was realigned where the external w.c. was removed.

Good working relationships were formed between all those involved in the project, and the comprehensive specification was closely adhered to by the contractor. The adaptation was successful: the occupier has been able to undertake a more independent lifestyle.

BUILDING FAILURES

The following section (Table 6.10 to 6.27) examines some principal defects and the symptoms and the primary causes. The information is presented in tabular format, and is merely a brief description of some of the defects that may arise in both traditional and non-traditional low-rise construction. Building defects or failures may arise due to a variety of factors, including poor design, failure of the material, poor construction, and lack of maintenance. Correct identification of the cause of the defect is essential to ensure that the cause is properly addressed, to prevent further damage to the fabric of the building.

Table 6.10 Possible defects in a pitched tiled roof

Defects	*Symptoms*	*Causes*
Slipping and deteriorating of interlocking roof tiles.	Some areas of surface eroded, looking different to others, tiles falling off. Patches of damp shown on ceiling.	*Either* surface deterioration caused by frost action *or* too shallow a pitch to the roof *or* slipping tiles because of inadequate field fixing *or* rotten battens *or* broken tiles.
Deformation and sagging of pitched roof.	Tiles out of line or slipping. Ridge line sagging.	*Either* overloading due to retiling the roof with heavier tiles *or* weakening of timbers due to movement of supporting walls.
Deteriorating and slipping of roof slates.	Dampness visible in the roof space or on the ceiling.	*Either* attack of the slates by air pollution *or* slipping occurs due to rust nails, broken nail holds *or* fungal woodworm attack on battens.

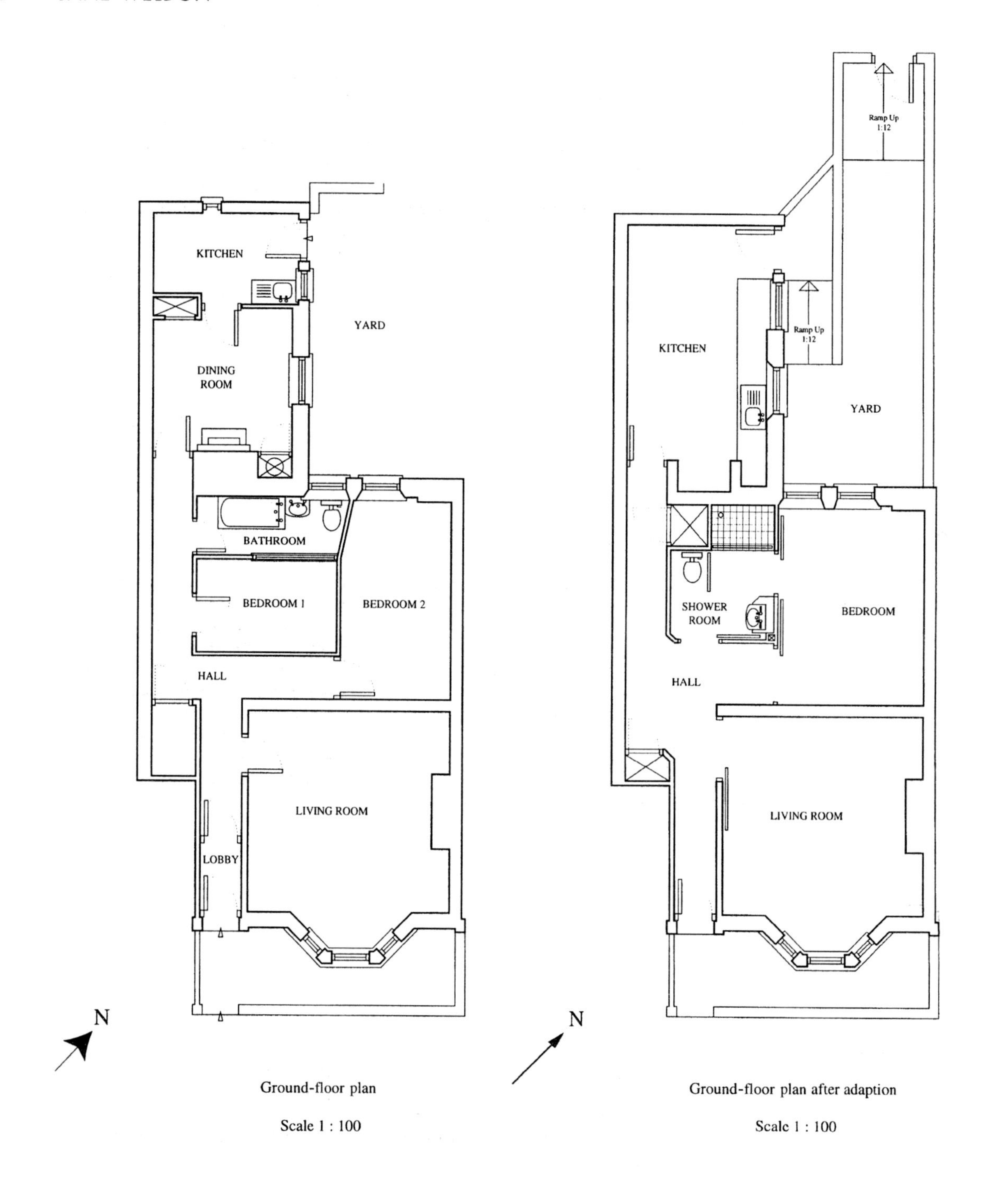

Figure 6.20 Ground-floor plan prior to conversion

Figure 6.21 Ground-floor plan after conversion

Table 6.11 Sagging and deformation of pitched roof

Defect	*Symptoms*	*Causes*
Sloping surface shows sagging or rising along slope.	Roof tiles sag, gaps visible between tiles or slates.	*Either* undersized rafters or insufficient purlins *or* roof re-covered with heavier slates or tiles.
Ridge tiles sag.	Roof sags along ridge or roof sags along hips or valleys.	*Either* wall movement causing eaves timbers to move *or* undersized timbers or roof re-covered with heavier slates *or* fungal or insect attack has weakened timber.

Table 6.12 Flat roofs: general defects

Defect	*Symptoms*	*Causes*
Blisters	Blisters/bulges of various sizes over specific or random areas.	*Either* moisture trapped during construction *or* moisture vapour (condensation) trying to escape from within building.
Ponding	Water lying in pools on roof after rain.	*Either* slope of roof too shallow *or* deflection of roof in isolated areas.
Random splits, cracks and tears	Splits, cracks and tears at random.	Loss of elasticity due to *either* excessive heat from sun or other source *or* loss of elasticity *or* mechanical damage.

Table 6.13 Defects in flat roof coverings

Defect	*Symptoms*	*Causes*
Splitting	Random or over whole area.	Movement of base or no isolating membrane.
	Occurs at edges or parapets.	Differential movement.
Brittle surface	'Dry' brittle surface, cracked or crazed.	Exposed to ultra-violet light or ageing.

Table 6.14 Cracking of rendering and brickwork of external walls

Defect	*Symptoms*	*Causes*
Cracked brickwork	Horizontal or diagonal cracks across external wall are visible.	*Either* due to sub-soil movement *or* foundation failure *or* expansion of brickwork *or* failure of wall ties *or* spread of roof structure.
Cracking, splitting, bending of chimney stacks	Cracking of mortar joints, vertical splitting, bending or distortion, cracking of rendering. Damped or stained areas. Displaced material falling within stack.	*Either* due to condensation of water vapour in the flue gases above the roof level *or* changes in temperature causing cracking allowing water penetration.
Cracking of rendering on brickwork	Cracks form without any formal pattern. If tapped the rendering around the crack may sound hollow and in some instances it might fall off.	*Either* as a direct result from cracks in brickwork *or* shrinkage in the rendering *or* chemical action.

Table 6.15 Loose rendering on external brickwork

Defect	*Symptoms*	*Causes*
Top coat of render is loose	Some areas of render fallen from wall, render sounds hollow when tapped. Cracking and bulging visible.	*Either* moisture penetration behind impervious top coat *or* frost action *or* base coat porous *or* incomplete top coat *or* differential movement between coats.
Full thickness of render is loose	In some areas full thickness of render has come away exposing brickwork.	*Either* moisture penetration through cracks in impervious render *or* frost action *or* movement of structural background causing loss of adhesion *or* render incompatible with background.

Table 6.16 Leaking gutters and rainwater downpipes

Defect	*Symptoms*	*Causes*
Leaking gutters	Water leaking or overflowing from points along gutter, possible dampness on external wall or at eaves or internal wall.	*Either* the gutter or the downpipe is blocked *or* old gutter may be perforated *or* have ineffective joints *or* deformation of gutter due to heat if too close to a flue.
Blocked or leaking rain-water downpipes	Water fills the pipe and hopper or gutter overflows. Signs of discoloration near joints, growth of fungus visible.	*Either* pipe is blocked by leaves, bird nest, general rubbish *or* growth of plants inside the pipe *or* pipe may have become dislodged due to impact.
Missing support brackets	Guttering distorts and sags between supports.	Missing support.

Table 6.17 Dampness on internal walls

Defect	*Symptoms*	*Causes*
Dampness of walls Mould growth on walls	Damp appears after rain. Damp appears after occupation, particularly in cold weather.	Rain/moisture penetration through wall. *Either* condensation due to insufficient heating and ventilation *or* kitchen or bathroom doors left open so steam migrates through property. Calor gas or paraffin heating compounds the problem.
	Damp appears during cold weather especially behind furniture and at corners of room.	*Either* condensation due to lack of proper ventilation *or* cold walls with limited thermal insulation *or* lack of heating *or* building not occupied during daytime.

Table 6.18 Cracks on internal surfaces

Defect	*Symptoms*	*Causes*
Cracks in wall	Crack corresponds with cracks on other side of wall. Crack increasing in size.	Movement of foundation due to changes in moisture content of soil, i.e. due to drought.
Cracks in ceiling	Cracks may be long and fine, appear in new work after it dries out. Usually seen in junction of floors and walls/ceilings and sometimes at plasterboard joints.	Shrinkage in timber and cement-based products such as concrete block walls due to changes in moisture content.
Blister in applied decoration	Dampness visible. Changes in the atmospheric condition of the room. Discoloration near the ventilators, airbricks or old flue. Blisters are temporary and when broken may create sticky yellow runs.	*Either* presence of moisture *or* colour changes are caused by chemical reaction *or* mould growth *or* deformed flue can allow fumes to find their way into rooms causing discoloration and danger to occupants.

Table 6.19 Damp on ceiling

Defect	*Symptoms*	*Causes*
Internally below parapets	Damp visible only on external walls/ceiling junction after heavy rain.	*Either* absence of damp-proof course *or* splits in junctions and flashings *or* blocked gutter overflow.
On ceiling under pitched roof	A damp patch visible after rain or snow.	*Either* defective roof tiles/slates *or* defective battening *or* defective guttering *or* leaking water tank in loft.
Underside of flat roof	Damp patches visible after rain. Cold during occupation. If severe will drip from ceiling.	*Either* direct rain penetration through leaking roof *or* condensation *or* residual water trapped during construction.

Table 6.20 Loose plaster and loose tiles on internal walls and ceilings

Defect	*Symptoms*	*Causes*
Top coat of plaster is loose	Top coat of plaster comes away from the base coat; plaster sounds hollow if tapped. Cracking and bulging are apparent.	The base structure has shrunk after being plastered *or* a strong expanding top coat has been applied to a weak or shrinkable base coat.
Ceramic tiles are falling off	Surrounding tiles sound hollow when tapped. Tiles start falling off shortly after fixing.	*Either* poor workmanship *or* inappropriate adhesive *or* shrinkage in the background structure *or* use of non-adhesive and grout in wet areas, i.e. around baths, sinks.
Full thickness of plaster is loose	Full thickness of plaster has come away showing the sub-structure. Surrounding areas when tapped sound hollow.	*Either* the plaster has failed due to the use of an unsuitable undercoat *or* if the plaster is on timber lathe (in older properties) then the cause may be movement, vibration or ageing *or* skim coat has lost its adhesion to the plasterboard.

Table 6.21 Failure of paintwork against plaster background

Defect	*Symptoms*	*Causes*
Blistering, cracking, uneven colour, loss of gloss	Small blisters or irregular cracks over painted surface.	*Either* vapour trapped under paint surface *or* excessive heat *or* premature drying of oils in paint *or* use of wrong paint system.
Efflorescence	White powder on surface of paint.	Trapped moisture escaping and soluble salts deposited on wall.

Table 6.22 Deterioration of paintwork on timber

Defect	*Symptoms*	*Causes*
Running	Thick vertical ridges or runs of paint.	Too much paint applied in one coat.
Brush strokes	Visible small ridges showing brush marks.	*Either* too much paint applied in one coat *or* too much pressure on brush *or* brush strokes not in one direction.
Blistering	Small blisters over surface.	*Either* damp or unseasoned wood *or* knots incorrectly treated *or* surfaces not properly prepared or cleaned.
Peeling, poor adhesion, irregular cracks	Paint peels away from background.	*Either* damp or unseasoned wood *or* knots incorrectly treated *or* surfaces not properly prepared or cleaned.
Chalking, powdering, cracking	Surface of painted area appears to be covered in powder.	No primer used.
Wrinkling	Wrinkles visible on surface.	Paint film dried too quickly.

Table 6.23 Decay in timber

Defect	*Symptoms*	*Causes*
External doors and windows	Woodwork soft and breaks off easily. Joints open up. Noticeable deformation. Paintwork discoloured and flaking. Loose or missing putty.	*Either* fungus in timber, usually wet rot due to lack of treatment to timber, *or* penetration of water through joints or from adjoining wet walls *or* moisture trapped in between timber and putty or paint.
External joinery, tongued and grooved boarding, cladding, fascias, verges and barge boards	Flaking or peeling of paintwork.	Lack of maintenance or incorrect preparation of timber.
Internal joinery and doors	Outer ply of the door is wrinkled, distorted and does not fit in the frame. Cracks in the paintwork.	*Either* change in the moisture content of timber *or* poor joinery work *or* in old doors broken or rusty split hinges.

Table 6.24 Defective floor coverings

Defect	*Symptoms*	*Causes*
Lifting of wood block flooring	Localised areas raised away from the rest of the floor.	*Either* trapped moisture absorbed by wood causes expansion *or* loss of seal allows moisture to penetrate surface of flooring.
Lifting of clay floor tiles	Arching or ridging of tiles.	*Either* shrinkage of screed causes tiles to move *or* missing or defective expansion joint at perimeter.
Vinyl floor tiles loose	Curled edges to tiles, some tiles loose.	*Either* poor workmanship in fixing tiles or insufficient adhesive *or* water penetration from surface *or* water present in base or shrinkage of base.
Vinyl floor covering blistered	Raised bubbles or blisters locally or over large area.	Moisture vapour trying to escape from base or shrinkage of floor base.
Uneven or broken clay tiles, loose tiles	Localised cracked or broken tiles, some loss of adhesion.	Age of tiles – wear and tear.

Table 6.25 Defective timber floorboards

Defect	*Symptoms*	*Causes*
Gaps between floorboards	Draughts from space below ground floor. Sound transmitted easily between floors.	*Either* floorboards contain moisture when laid *or* drying out and shrinkage *or* poor workmanship when laying floor.
Warped or twisted floorboards	Floor noisy when walked on, localised ridges or depressions on floor.	Insufficient fixing when floorboards nailed *or* insufficient strutting to joists.
Floorboards not fixed properly after removal for service installation	Noisy or localised movement of loose boards.	Insufficient fixing when boards relaid.

Table 6.26 Defective solid concrete floors

Defect	*Symptoms*	*Causes*
Uneven floor surface	Sagging at middle of floor, with possible damage to floor finish.	*Either* movement or deflection of the earth or hardcore below concrete, due to poor compacting of hardcore, *or* leaking pipes causing soil to wash away.
	Sagging at edge of floor.	*Either* possible movement of wall has allowed earth to move *or* excessive hardcore *or* fill incompletely compacted
Cracking of floor	Floor cracking with some hollow areas with cracking around edges of slab.	*Either* screed separated from concrete floor, due to incorrect thickness or composition of screed, *or* failure of screed to bond to concrete floor *or* sulphate attack *or* differential movement caused by heating pipes in screed.

Table 6.27 Defective timber floor joists

Defect	*Symptoms*	*Causes*
Movement in timber floor in middle of room	Sagging of timber floor.	Floor joists undersized or overloaded.
Timber floor moves at edges of room	Sloping of timber towards supporting floor.	Support to floor failed.

CONCLUSIONS

The design and development of residential construction projects is undoubtedly a complex technical process. To understand this process, it is necessary to be aware of the respective roles and responsibilities of the various groups involved in design and development, to recognise the constraints placed on designers and design standards, to be familiar with the basic aspects of construction technology in different forms of housing, and to appreciate the attributes of effective maintenance and rehabilitation.

Although this chapter concentrates on the above aspects of design and devlopment, reference should also be made to Chapters 2 and 3 for an understanding of the economic and financial background to housebuilding, and to Chapter 5 for an appreciation of the planning context in which development takes place.

QUESTIONS FOR DISCUSSION

1 For a housing project of your choice, chart the development of the project from inception to completion, identifying the various groups involved and the range of documents produced, including drawings, specification and contracts.

2 Compare and contrast the design features you might expect to find in developments for:
(a) a sheltered housing scheme;
(b) housing association accommodation for young people; and
(c) private development of executive-style detached houses.

3 For the flat illustrated (Figure 6.21), prepare a schedule of finishes using a tabular format similar to the one shown below.

Element	*Description*
Ceiling	
Walls	
Floors	
Windows	
Doors	
Skirting-board	

4 For a semi-detached house of your choice, prepare sketches to illustrate how you could extend the building to include a self-contained annexe for an elderly relative.

5 Prepare a maintenance schedule for a detached three-bedroom property, identifying external and internal works to be undertaken over a five-year period.

RECOMMENDED READING

Osbourne, D. (1985) *Introduction to Building*, Mitchell's Building Series, London: Batsford.
Powell-Smith, V. and Billington, M.J. (1992) *The Building Regulations Explained and Illustrated*, 9th edition, Blackwell Scientific Publications, Oxford.
Seeley, I.H. (1987) *Building Surveys, Reports and Dilapidations*, Macmillan Press, London.
Speaight, A. and Stone, G. (1996) *Architect's Legal Handbook*, 6th edn, Butterworth Architecture, Oxford.
Stroud Foster, J. (1983) *Structure and Fabric Part 1*, Mitchell's Building Series, Mitchell Publishing, London.

REFERENCES

Addleson, L. (1992) *Building Failures: A Guide to Diagnosis, Remedy and Prevention*, 3rd edn, Butterworth-Heinemann, Oxford.
Architect's Journal (1994) 'Housing the Homeless in Hackney', 12 January, 39–47.
Gibson, E.J. (ed.) (1979) *Developments in Building Maintenance*, Applied Science, Barking.
Hawkesworth, R. (1994) *Housing Design in the Private Sector: An Architect's View towards a Design Philosophy*, Serious Graphics, University of Portsmouth Enterprise, Portsmouth.
Hereford and Worcester Local Authority, *Building Control Handbook*, HWLA, Worcester, 19.
Kernohan, D., Gray, J., Daish, J. and Joiner, D. (1992) *User Participation in Building Design and Management: A Generic Approach to Building Evaluation*, Butterworth Architecture, Oxford.
Lawson, B. (1990) *How Designers Think: The Design Process Demystified*, Butterworth Architecture, Oxford.
Moore, C., Allen, G. and Lyndon, D. (1974) *The Place of Houses*, Henry Holt, New York.
Osbourne, D. (1985) *Introduction to Building*, Mitchell's Building Series, London, Batsford.
Powell-Smith, V. and Billington, M.J. (1992) *The Building Regulations Explained and Illustrated*, 9th edn, Blackwell Scientific Publications, Oxford.
Poyner, B. and Webb, B. (1991) *Crime Free Housing*, Butterworth Architecture, Oxford.

Property Services Agency (1981) *Flat Roofs: Technical Guide*, HMSO, London.

Quiney, A. (1986) *House and Home*, BBC, London.

Regnier, V.A. (1994) *Assisted Living Housing for the Elderly: Design Innovations from the United States and Europe*, Van Nostrand Reinhold, New York.

Seeley, I.H. (1985) *Building Maintenance*, Macmillan Press, London.

—— (1987) *Building Surveys, Reports and Dilapidations*, Macmillan Press, London.

Speaight, A. and Stone, G. (1996) *Architect's Legal Handbook*, 6th edn, Butterworth Architecture, Oxford.

Statham, R., Korczak, J. and Monaghan, P. (1988) *House Adaptations for People with Physical Disabilities*, Department of the Environment, HMSO.

Stroud Foster, J. (1983) *Structure and Fabric Part 1*, Mitchell's Building Series, Mitchell Publishing, London.

Tutt, P. and Adler D. (1979) *New Metric Handbook: Planning and Design Data*, Butterworth Architecture, Oxford.

Taylor, L. (ed.) (1990) *Housing: Symbol, Structure, Site*, Smithsonian Institution, New York.

7

ENVIRONMENTAL HEALTH AND HOUSING

Pauline Forrester

Healthy housing is a primary objective of environmental health activities. In the nineteenth century, public health reformers recognised that poor living standards were an important agent in the causes of ill-health. This recognition initiated a series of legislative responses and activity designed to alleviate basic problems such as lack of sanitation, contaminated water supplies and gross overcrowding. Building on these foundations, current environmental health roles are based on the underpinning principles of health, risk assessment and sustainability. This chapter examines these aspects and also considers the main legal provisions enabling local authorities to control disrepair, unfitness and inadequate standards of management in private sector housing. Specifically, the chapter focuses on:

- Health and housing.
- The legal background to environmental health control in the private sector.
- Housing conditions.
- Enforcement powers.
- Unfit dwellings.
- Disrepair and nuisance.
- Houses in multiple occupation.

HEALTH AND HOUSING

The relationship between poor housing and ill-health has been recognised for centuries. Indeed, it is suggested by Bynum (in Lowry, 1991) that these links were recognised as far back as the Hippocratic treatises of ancient Greece. However, while both the direct and the indirect impacts of inadequate housing on health status are generally accepted, less is understood regarding the exact nature of the relationship. This is in part due to the number of confounding factors which influence an individual's health status. Factors such as poverty, socio-economic status, lifestyle and genetic predisposition will each have an impact on health, as well as the housing and environment in which the individuals live. In addition, housing conditions can also themselves have multiple health impacts. For example, inadequate cooking facilities can both potentially result in an increased level of home accidents and act as a barrier to positive lifestyle choices such as eating a healthy diet. Thus, our understanding of the exact role played by combined environmental and social factors in influencing variations in health is still developing. The complex nature of these relationships is such that the correlation of health outcomes with specific measures taken to improve housing can sometimes be difficult to establish (Hopton and Hunt, 1996).

For these reasons, key areas for action under the Health of the Nation strategy have tended to be to measure health outcomes from environmental health improvements in terms of health risks or potential causes of ill-health (DoH, 1996). The following section of this chapter considers some of the highest risk factors which have been associated with housing

(Raw and Hamilton, 1995; Cox and O'Sullivan, 1995). An illustrative listing, ranking the various health and safety hazards occurring in the domestic environment, is given in Table 7.1.

Accident prevention is one of key areas in government's national strategy for improving health (DoH, 1992). Accidents in the home account for more than 4,000 deaths each year (OPCS, 1989) and around 28 million non-fatal treated accidents (DTI, 1990). Clearly, the home can be a dangerous place. Poor housing design, overcrowding and inadequate play facilities for children can all lead to accidental injuries and deaths. A significant proportion of these accidents are likely to be preventable and both environmental health and housing professionals have an important role to play in prevention. However, while local authorities have discretionary powers under the Home Safety Act 1961, relatively

Table 7.1 Health and safety hazards in domestic dwellings, grouped by relative risk ranking

Health hazards	*Safety hazards*
Highest risk	
Hygrothermal conditions[1]	Slips, trips and falls
Radon[2]	Burns and scalds
House dust mites	Drowning
Environmental tobacco smoke	Fire[5]
Carbon monoxide	
Second level of risk	Electric shock/burn
Fungal growth[1]	Collision/entrapment: architectural glass and windows
Security and the effects of crime	Explosions
Noise	
Lead[1]	
Third level of risk	Collision /entrapment: doors
Sanitary accommodation[3]	Struck by objects: structural collapse
Sources of infection other than sanitary accommodation[3]	
Space	
Volatile organic compounds[4]	
Oxides of nitrogen[4]	
Particulates	
Fourth level of risk	
Sulphur dioxide and smoke	
Landfill gas	
Pesticides[3]	
No clear basis for risk assessment	
Lightning	
Electromagnetic fields	

Source: Raw and Hamilton (1995); Cox and O'Sullivan (1995)

Note:

[1] Mainly in older dwellings
[2] Localised in certain geographic regions
[3] Potential risk controlled by current standards
[4] Further research may increase risk ranking
[5] Forthcoming research will further define risk rating in this area

few authorities have made a strategic commitment to their home safety roles.

The risk of fire in the home is another key area of home safety. In this case, detailed risk rating data are under development. One of the main fire-related hazards, in domestic settings, relates to smoke inhalation and asphyxiation, as domestic fires tend to produce significant amounts of smoke and toxic fumes. In the United Kingdom, around 500 people die each year as a consequence of domestic fires, in addition to which there are between eight and ten thousand related non-fatal casualties each year, ranging from very serious effects through to mild smoke inhalation (Home Office, 1995). The danger of fire affects dwellings in all tenures and it is thought that available statistics may well underestimate the level of risk, as many domestic fires go unreported. In particular, houses in multiple occupation (HMOs) are recognised as presenting an increased threat to occupants, the risks of death from fire in this sector being eight times higher than in other dwelling-types. Because of this, local authorities have specific powers to tackle fire safety problems in HMO accommodation. Additional controls are provided under the Furniture and Furnishings (Fire Safety) Amendment Regulations 1993, which place a duty on landlords who supply rented furnished accommodation to ensure that any new or second-hand furniture satisfies standards for fire resistance.

In developing its strategy for health the government identified a range of 'environmental' issues which might provide significant health gains (DoH, 1996). In their consultation document the overall objective 'to bring a decent home within the reach of every family' was stated. Surprisingly, hygrothermal conditions were not identified as a specific target area. Nevertheless, the majority of the proposed environment key indicators relate to housing and covered aspects such as indoor air quality, radon, noise pollution and lead in drinking water.[1]

Indoor air quality is a growing area of concern. On average, people in the United Kingdom spend between 60–90 per cent of their time inside their homes, thus, the desirable limits of exposure for indoor pollutants in domestic settings are below those established for occupational environments. There are of course hundreds of different chemicals which can be identified in air inside any home. Fortunately, many of these chemicals do not pose any risk to health as the levels generally encountered in most homes are low (Harrison, Humfrey and Shuker, 1996). Nevertheless, indoor air quality was identified as a key area for action in the national Environmental Health Action Plan (DoE, 1996a). This identifies a range of indoor pollutants for potential action, including carbon monoxide, radon, nitrogen dioxide and volatile organic compounds, as well as allergens such as fungi, moulds and house dust mites. The following paragraphs consider some of the highest-risk pollutants encountered in housing.

Carbon monoxide is a colourless, odourless gas produced by incomplete combustion of fuels and is responsible for as many as seventy accidental deaths in the United Kingdom each year. The danger of carbon monoxide is that it is exceedingly toxic and causes asphyxiation or long-term effects at low concentrations. Most fatalities have been linked to faulty, or inadequately ventilated, gas fires and many of these have occurred in low-cost rented accommodation, such as houses in multiple occupation (HMOs). Recognising the threat to health posed by faulty gas appliances, the government has introduced specific legislation to control their installation and maintenance (Gas Safety Installation and Use Regulations 1994). Cut-off devices are required for gas appliances in rooms where people sleep and all landlords are required to have appliances checked on an annual basis and to inform their tenants of the results.

Another indoor pollutant of public health concern is radon, which is a naturally occurring radioactive substance. It is estimated that, annually, 2,500 deaths from lung cancer can be directly attributed to domestic radon exposure. Radon is a significant problem in certain areas of the country and these geographic variations depend on the type of ground soil and rocks and also the building materials used in the construction of dwellings. Concern about the health impacts of radon exposure has increased, as the extent of radon-affected areas is more widespread than was previously thought (Lomas, Green *et al.*, 1996). It is estimated that as many as 100,000 homes in the United Kingdom are affected by radon levels above the action level of 200 becquerels (Miles, Cliff *et al.*, 1992). Protection against excessive radon exposure typically relies on measures such as the provision of sumps combined with ventilation of the dwelling and sub-floor spaces. The government is keen to encourage home-owners to carry out remedial measures to protect their homes, making assistance available through the house renovation grant system.

The health risks of active smoking have been well publicised. More recently, however, environmental tobacco smoke has been highlighted as an indoor pollutant of public health concern. Research indicates that non-smokers have a 30 per cent increased risk of lung cancer where they are passively exposed to environmental tobacco smoke in the home (Raw and Hamilton, 1995). Young children appear to be particularly susceptible to the effects of exposure and hospital admissions for respiratory disease can be as much as 50 per cent higher where parents smoke.

Other indoor pollutants of current concern include nitrous oxides and volatile organic compounds. In the case of nitrogen dioxide, the main source of domestic exposure arises from inadequately ventilated gas-fuelled cooking and heating appliances. Exposure to nitrous oxides has been linked with respiratory effects, particularly among asthmatics and young children. However, the evidence at present appears a little uncertain. In contrast, the home environment presents a virtual cocktail of several hundred different volatile organic compounds (VOCs). These are products with low boiling-points and they produce vapours at room temperatures. Sources include furnishings and fabrics, building materials and paints, products used in damp-proofing, cavity wall insulation and timber treatments, as well as day-to-day household products such as cleaning agents, solvents and cosmetics. There are differing levels of sensitivity to the effects of exposure to these compounds, varying from discomfort from bad odours and sneezing through to more general irritation of the skin and membranes in the eyes, nose and throat. At higher levels of exposure headaches can also occur. While some VOCs have been linked with increased risks of cancers and genetic defects, these have occurred in occupational settings, at levels of exposure which would not normally be encountered in the home. Other indoor pollutants include allergens, such as house dust mites, moulds and fungi. These are now considered in association with hygrothermal conditions.

As a nation, a large proportion of our homes are relatively cold, a situation that is exacerbated as the problems of cold and dampness in homes often coincide. Cold, damp homes are recognised as representing the primary health risk associated directly with the condition of the housing stock (DoE, 1996a; Raw and Hamilton, 1995). The risks posed by hygrothermal conditions (factors which affect thermal comfort; that is, temperature, humidity and air movement) are most commonly associated with older buildings and inter-war dwellings which are often poorly insulated and are also more likely to have costly or inefficient forms of heating. In contrast, newer homes tend to be warmer, because of improved insulation standards achieved through Building Regulation control. However, for low-income occupiers, even those fortunate to live in newer dwellings, affordable warmth can be a significant problem especially during colder spells. Particular problems of fuel poverty[2] exist in the private rented sector, with 95 per cent of dwellings failing to meet minimum heating standards (that is, 18°C for main living rooms and 16°C in other habitable rooms) when the external temperature is less than 4°C, and more than 20 per cent of dwellings in this sector are damp (DoE, 1996a).

It is estimated that several thousand of the excess deaths (those over and above the normal mortality rate for the rest of the year) that occur each winter are associated with cold homes (Boardman, 1991).[3] The elderly are most at risk from cold housing, and the causes of excess winter mortality include respiratory and cardiovascular illnesses, and increased risk of trips and falls. The health risks posed by cold, damp homes are further increased as these conditions also encourage the growth of mould, fungi and house dust mites.

More than 20 per cent of all households experience problems with condensation and mould growth. The incidence and severity of this is greatest in the private rented sector (where approximately one in five homes is affected) and least in the owner-occupied sector (where one in sixteen homes is affected) (DoE, 1996a). The mould and fungal growth which is so often associated with condensation can cause allergic responses in susceptible individuals and sensitisation appears greatest among children. These allergic responses can also hasten the onset of symptoms such as asthma and hay fever. Apart from the direct health impacts of cold, damp housing and associated mould growth, these conditions cause both significant distress to occupiers and damage to furnishings and clothing. Damp conditions also encourage the proliferation of house dust mites, which can trigger severe allergic reactions, particularly asthma. In this case, the allergenic material is concentrated in the mite's faecal pellets, which are found in house dust. It is estimated that as many as 50 per cent of the adult human population may be affected (Raw and Hamilton, 1995).

From this brief overview of the key health risks posed by housing, it can be seen that the provision of a healthy housing stock by the year 2000[4] (WHO, 1986) provides a formidable task for environmental and housing professionals. Implicit in achieving this target is an understanding of the potential risks to health posed by poor housing conditions, as this is an essential element in the development of effective risk assessment strategies which will enable prioritisation and targeting of scarce resources.

Sustainability and health

The areas of health and sustainability are increasingly being inter-linked in policies affecting environmental health and housing, many of the approaches being based on the principles of sustainable development. The precise definition of sustainable development remains a contested area. However, one most widely accepted definition is 'development which meets the needs of the present without compromising the ability of future generations to meet their own needs (Brundtland in World Commission on Environment and Development, 1987). Environmental debates such as the Earth Summit in 1992 have focused attention on the consequences of global warming and depletion of finite resources. The 1992 summit provided the wellspring for a range of sustainability initiatives including processes such as Local Agenda 21, which is being used by many local authorities and environmental groups to promote partnerships for sustainable development.

In working towards sustainability it is important that initiatives embrace multi-sectoral approaches and that they are integrated across management, policy and regulatory processes. Many organisations and institutions are starting to consider the role of housing in achieving sustainable development. It could be argued that the values underpinning sustainable development (that is, concern about the well-being of future generations (futurity); recognition that the health and integrity of the environment is critical to future human well-being; and values relating to the quality of life and equity which extend beyond conventional measures of economic wealth (LGMB, 1995a)) could be readily applied to the context of housing. As part of the Local Agenda 21 process, many authorities have developed sustainability indicators to measure performance (LGMB, 1995b). These can include various housing-related measures, for example access to local facilities (with implications for public transport usage); homelessness statistics (which relate to access to safe and healthy shelter); as well as domestic carbon dioxide emissions (which contribute to global warming).

The planning system is a primary tool for promoting sustainability. Projections suggest that between 1991 and 2016 the number of households in the United Kingdom will increase by around 4.4 million. As pressures on land use increase, our planning systems will be increasingly important in limiting the environmental impact of new housing. This includes encouraging the use of urban and brownfield sites for new housing, in addition to controlling the treatment and future use of contaminated land sites. Construction and design processes are also important in enabling sustainable development. These aspects include the use of renewable resources in construction, such as timber, in preference to non-renewable resources, such as aggregates, as well as approaches to housing design which enable conservation of natural resources such as water usage and which meet the lifetime needs of occupiers.

There are of course strong links between energy efficiency and health and safety issues in housing, which have been highlighted earlier in this chapter. The improvement of hygrothermal conditions in our homes and in particular increasing the energy efficiency of our dwelling stock is a key aspect of sustainable development initiatives. The Home Energy Conservation Act 1995 recognises both the environmental and the welfare benefits of domestic energy efficiency. Emissions from the domestic sector account for more than 25 per cent of all emissions of the main greenhouse gas, carbon dioxide (DoE, 1996b).[5] The Act requires local authorities (energy conservation authorities) to prepare reports on their plans to achieve a 30 per cent improvement in the energy efficiency of the local housing stock over a ten-year period. Because much of the older housing stock is relatively inefficient in terms of energy conservation, the long-term target recognises that many initiatives still tend to result in increased thermal comfort levels, as opposed to achieving real reductions in the levels of domestic energy consumption.

Clearly, the containment of urban sprawl and the protection of rural areas will be one of the major issues as demand for housing increases. Likewise, the development of sustainable towns and cities where people choose to live provides a challenge for policy-makers. Sustainable development thus provides an opportunity for all housing and environmental health professionals to consider the impact of their own roles on the quality of life and the environment.

THE LEGAL BACKGROUND TO ENVIRONMENTAL HEALTH CONTROL IN THE PRIVATE SECTOR

The public health reform movement of Victorian times initiated the early legislative responses to the problems of overcrowding and insanitary housing which accompanied the Industrial Revolution and urbanisation. These early controls provide the historical basis of the current legislative framework, which is discussed in this section.

There is a strong strategic emphasis in contemporary policy on private sector renewal and repair (DoE, 1996c) and in policies relating to HMOs (DoE, 1997). This combines both proactive and responsive approaches and embraces the interests of both the people and the properties concerned. Typically, environmental health roles in housing renewal and repair adopt both enabling and direct enforcement approaches. Enabling activities include encouragement of repair through the education of owners and the home improvement grants system. Enforcement roles, on the other hand, rely on statutory powers ranging from individual repairs notices, through to regeneration initiatives such as the declaration of renewal areas. The current powers controlling conditions in private sector housing are contained in a complex patchwork of legislation, the extent of which is illustrated in Table 7.2.

While Table 7.2 illustrates the significant legislative underpinning for environmental health activ-

Table 7.2 Illustrative list of Acts controlling conditions in private sector housing

- Public Health Act 1961
- Local Government (Miscellaneous Provisions) Act 1976
- Building Act 1984
- Housing Act 1985
- Local Government and Housing Act 1989
- Environmental Protection Act 1990
- Housing Act 1996
- Housing (Grants Reconstruction and Regeneration) Act 1996

ities, it should be noted that, although the enforcement powers exercised by local authorities extend to cover housing associations, case-law effectively prevents a local authority from taking legal action against itself (*Regina v. Cardiff City Council ex parte Crosse 1983*). Thus, control of conditions in council-owned premises tends to rely on in-house procedures or, alternatively, independent actions under the Environmental Protection Act 1990 (Section 82) and civil proceedings such as those under the Landlord and Tenant Act 1985 (Section 11).

The following sections of this chapter summarise the main features of the enforcement provisions for securing the repair and renewal of private housing.

HOUSING CONDITIONS

Local authorities have a duty to consider housing conditions in their area on an annual basis (Section 605 Housing Act 1985). This informs both their enabling role and the use of powers for housing renewal and repair. While local authorities tend to use various sources of information in considering how best to target resources for private sector renewal, the majority have used local house condition surveys as the central means of collecting information on which to base their strategic policy-making (Barlow, Chambers and Forrester, 1995).

The evidence from both national and local house condition surveys indicates that disrepair and unfitness remains a persistent problem affecting our housing stock. There are approximately 1.2 million unfit private sector dwellings in England alone (DoE, 1993a). The main grounds for unfitness are serious disrepair, dampness and instability. It is generally accepted that by far the worst conditions are found in the private rented sector, where levels of disrepair are as much as 85 per cent greater than in owner-occupied dwellings. Nevertheless, significant problems exist for home-owners and it is estimated that there are around 2 million owner-occupied dwellings in England which either are unfit or require essential repairs costing more than £2,000 (1996 prices) (Davidson and O'Dell, 1996). Clearly, issues of housing maintenance have become a key factor in supporting the ideology of home ownership and a viable private rented sector.

Certain household types are disproportionately affected by poor-quality private sector housing. These include single-person and lone-parent households, the elderly, the unemployed (DoE, 1993a), ethnic minority groups (DoE, 1993a; Ratcliffe, 1996), and owner-occupiers who have lived in their homes for a long time (Davidson and O'Dell, 1996). From this household typology it can be seen that one of the common themes linking many of the people in these groups is low income, and these are key target households for authorities when developing renewal and repair strategies.

Statutory fitness standard

There are a range of standards applied to housing, which include the Building Regulations as well as target standards for new developments. However, it is the statutory standard of fitness which provides a benchmark for the assessment of housing conditions and is one of the main triggers for enforcement actions. The standard is based on the minimum habitable conditions which are consistent with health and safety and was introduced by the Local Government and Housing Act 1989, with detailed guidance on its application being provided in *Circular 17/96* (DoE, 1996c). Criteria are provided which must be satisfied if a dwelling is to be considered fit for human habitation. These cover the following aspects:

- structural stability;
- freedom from serious disrepair;
- freedom from dampness prejudicial to health;
- provision of adequate heating, lighting and ventilation;
- an adequate piped supply of wholesome water;
- satisfactory facilities for food preparation and cooking, including a sink with hot and cold running water;
- a suitably located WC;
- suitable washing facilities, including a bath or

shower and hand basin, both supplied with hot and cold water; and
- an effective drainage system.

Where a dwelling fails any of the criteria and is not considered *reasonably suitable for occupation*, the premises will be considered unfit for human habitation and the local authority will have a duty to consider the most satisfactory course of action to deal with the property. Flats which are unfit because of the state of the building envelope are also included. However, the government is currently undertaking a review of the statutory fitness standard and recommendations have been made to extend the criteria to include aspects such as fire safety, internal arrangement and radon, as well as security and home safety.

Where a private dwelling fails the statutory standard of fitness, the authority has a duty to take the *most satisfactory course of action* for dealing with the problem. This involves a choice between the following options:

- repair;
- deferred action;
- closure;
- demolition;
- clearance.

When deciding the best course of action to deal with unfit dwellings, authorities follow detailed guidance contained in DoE *Circular 17/96* (DoE, 1996c). The methods used are based on cost–benefit analysis techniques and take account of both the economic consequences and the socio-environmental outcomes (non-quantifiable costs and benefits) of the action, over a thirty-year period. As part of this process an authority will need to consider the views and aspirations of the people affected by the proposed action, as these are an important element in deciding the best course of action for unfit dwellings.

Renovation grants

Public funding through the grant system has shifted away from the broad-based eligibility which existed prior to the Local Government and Housing Act 1989. It is now targeted at low-income home-owners and towards areas of run-down housing. Nevertheless, the house renovation grant system remains an important enabling tool and varying grant strategies and approaches are adopted by local authorities to meet the particular private sector housing needs in their area. Around 98 per cent of the improvement and repair of privately owned homes is now carried out without public funding (approximately £28 billion at 1991 prices) (DoE, 1996d). Thus the strategies adopted by local authorities should also recognise the important role of home-owners in the maintenance process.

The current grants system was introduced by the Housing (Grants Reconstruction and Regeneration) Act 1996. Its predecessor, under the Local Government and Housing Act 1989, was largely mandatory and demand-led. However, this earlier scheme had suffered as a consequence of long-term underfunding which left the policy and legislative framework under-resourced. Despite persistent levels of unfitness, there has been a significant reduction in funding available for private sector rehabilitation over recent years. For example, between 1991 and 1993 there was a 25 per cent reduction in expenditure, and Figure 7.1 illustrates the longer-term pattern of decline in grant-aided renovations for private sector housing.

By breaking the mandatory link between unfitness and grant aid, it is intended that councils will be able to achieve more effective targeting of scarce resources. The new system comprises five main grants:

- renovation grants;
- common parts grants;
- HMO grants;
- disabled facilities grants;
- home repairs assistance.

In addition to these mainstream grants there is a block-based group repair grant and a relocation grant which has been designed to help occupiers displaced by clearance. The system is now largely discretionary, with the exception of disabled facil-

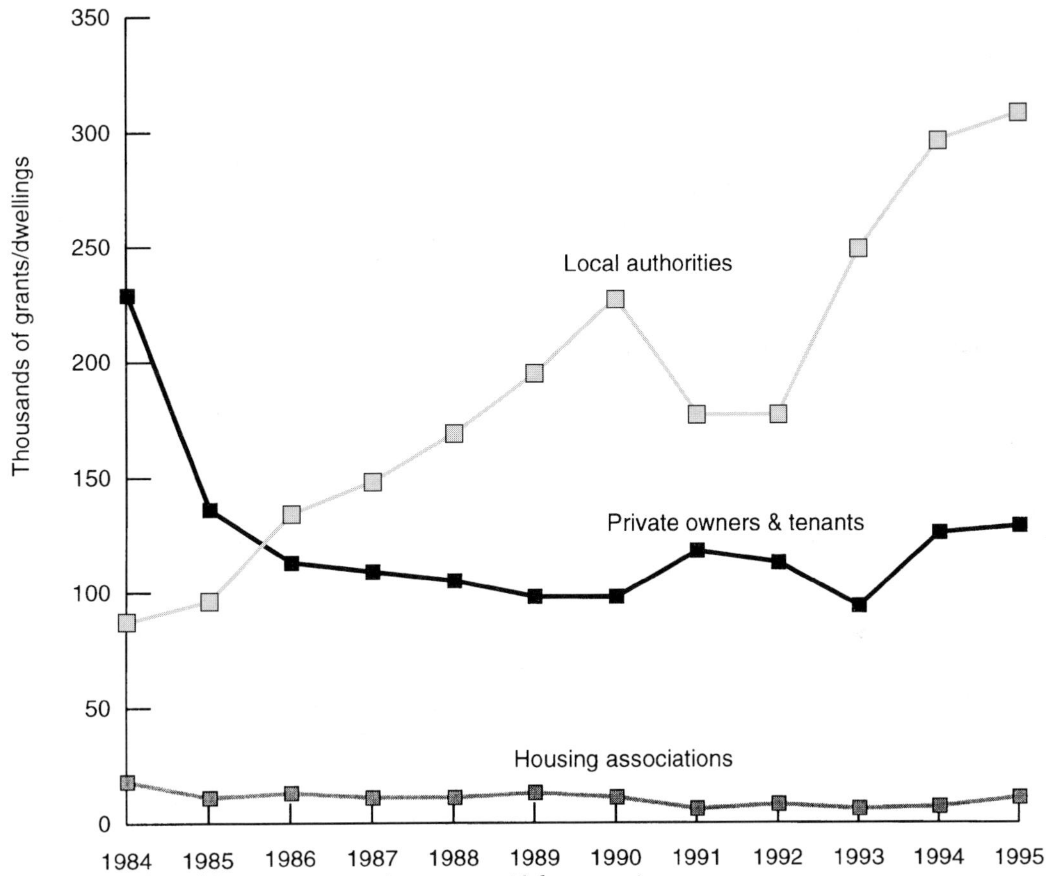

Figure 7.1 Number of dwellings renovated or grants paid for renovation
Source: Department of the Environment, *Housing and Construction Statistics*

ities grants, which have a mandatory element covering access to and the provision of basic amenities. In most cases, the amount of grant aid is based on the cost of eligible works and the financial means of the applicant. The applicants are means tested using a system which is very similar to that used for housing benefit purposes. The exception to this rule is home repairs assistance, where applicants are 'passported' by receipt of a means-tested welfare benefit.

The structure of the grant system under the 1996 Act is generally similar to its predecessor under the 1989 Act. There are certain standard conditions governing eligibility which apply. For example, applicants must be at least eighteen years of age and as a general rule grant aid is not available for dwellings which are less than ten years old (with the exception of disabled facilities grants). To prevent abuse of the system, there is a five-year grant condition period for most grants and applicants are required, generally, to certify the future occupation of the dwelling. This has the effect that, in most cases, if the property is sold within a five-year period the applicant will have to repay any grant which may have been paid. The following sections outline the key features of each of the mainstream grants in the current system.

Renovation grants are mainly used for bringing a dwelling up to the fitness standard or where a repairs notice has been served (Section 189 Housing Act 1985). These grants may also be awarded for improvements beyond the fitness standard (that is, putting the property into reasonable repair) and eligible works include energy efficiency measures, provision of space heating, improvements to internal arrangements, and radon remedial works. Renovation grants may also be paid to bring empty properties back into use and for the provision of dwellings by conversion. This grant is discretionary and there is a three-year residency requirement in most cases, although this does not apply in the case of landlords' applications or to owner-occupied dwellings inside a renewal area.

As their name suggests, common parts grants cover the common parts of a building containing flats. These may include stairways and the external building envelope. They are discretionary grants, and cover similar aspects to renovation grants, although there are differences in operational and administrative details such as preliminary conditions.

The HMO grant provides the main form of grant aid for multiply occupied dwellings and is discretionary. It is important to note that the definition of HMO for grant purposes is a modified version of that used for enforcement under Part XI of the Housing Act 1985 and excludes any part occupied separately by a single household. This grant is only available to HMO landlords and may also be awarded for the provision of an HMO by conversion. Otherwise the HMO grant covers similar aspects to renovation grants. On completion of the grant-aided works, the HMO must be both fit for human habitation (Section 604 Local Government and Housing Act 1989) and also fit for the number of occupants (Section 352 Housing Act 1985). This type of grant is only available for means of escape works required under the Housing Acts; fire safety requirements under other statutory provisions are not included.

Disabled facilities grants are widely available to owners, occupiers and tenants, including both council and housing association tenants. This grant is mandatory for provision of access around the home up to a maximum expense limit of £20,000. Discretionary disabled facilities grants may also be awarded for welfare or employment purposes and for works in excess of £20,000. Typical works which are eligible for mandatory assistance include access to, or the provision of, essential amenities such as:

- WCs;
- bathrooms;
- kitchens;
- adaptation of heating and lighting power controls;
- safety measures in rooms; and
- those enabling care of a dependent resident.

In the case of disabled facilities grants, councils must consult with the social services authority (occupational therapist) on the housing adaptation needs of the person with the disability and it is important that these services are well co-ordinated. For these types of grants, the conditions governing the minimum age of the property and other restrictions such as residency requirements do not apply. The means test for the disabled facilities grant is also different from that used for other forms of grant aid in that it focuses on the resources of the disabled occupant. Because this grant is mandatory, local authorities are allowed to delay payment of the grants for up to twelve months, in order to facilitate management of over-stretched budgets.

Home repair assistance is a discretionary grant which is intended to streamline works where full-scale renovation is considered inappropriate (it replaces the former minor works assistance which existed under the 1989 Act). This approach complements policy developments in other welfare service areas relating to care in the community and social service departments' responsibilities under the Chronically Sick and Disabled Persons Act. It is often administered through home improvement agencies such as care and repair organisations which are able to provide significant levels of support to the applicants. The assistance provided is discretionary and is aimed at elderly disabled or infirm people wishing to stay in their home. It is also available to people on certain welfare benefits such as income support. Home repair assistance also covers houseboats and mobile homes as well as more traditional dwellings and takes the form of a grant or materials for small-scale works and adaptations. Eligible works include thermal insulation, and patch repairs to properties affected by clearance proposals. Unlike other mainstream grants, prior residence requirements do not normally apply except in the case of non-standard dwellings such as houseboats and mobile homes, where there is a three-year prior residence requirement.

ENFORCEMENT POWERS

Local authorities have a range of powers which enable them to enforce repair and renewal of private housing. These powers generally contain provisions enabling the local authority to carry out works in default and to prosecute where the person responsible fails to comply with the terms of a notice. As part of the government's deregulation initiative, councils are required to serve a notice of intention informing the owner that they are minded to take formal action. This has come to be known as the 'minded to' notice procedure and applies in the case of actions covering unfitness and disrepair, and where works are required to make an HMO fit for the number of occupants. There are exceptions to this process; for example, where there is an imminent risk to health and safety, or where a landlord has a poor management record or has failed to respond to pre-formal action notices in the past. However, where an owner fails to respond to this type of notice and the council takes formal action, a charge of up to £300 can be made to cover administration costs.

Where a local authority is considering formal action, it will often serve a notice requiring information on the ownership and interests in the property (Section 16 Local Government (Miscellaneous Provisions) Act 1976), in order to ensure that any notices are correctly served. Notices will usually be served on the person in control of the premises (the owner or the person in receipt of the rent if the dwelling is tenanted). In the case of HMOs, notices may alternatively be served on the person managing the HMO. The majority of enforcement actions and grants are registered as local land charges, so that prospective owners are warned of any repairing or grant obligations which may be attached to the property. The overall enforcement processes relating to the improvement and repair of poor housing are illustrated in Figure 7.2.

UNFIT DWELLINGS

Repair notices (Section 189 Housing Act 1985) deal with unfit properties and require works to make the property fit within a specified time period. These provisions play a key role in managing unfitness levels in the private sector stock; in England more than 39,756 dwellings were reported as having been made fit during 1993/4 (DoE, 1995). The council may approve an application for a grant where a notice has been served, subject to the means test and the eligibility criteria described earlier.

Deferred action notices were introduced by the Housing (Grants Reconstruction and Regeneration) Act 1996 and allow local authorities to delay action on an unfit dwelling for up to two years. These powers have been designed for use in cases where, for example, home-owners could not cope with comprehensive works to make their dwelling fit and may only wish to undertake basic repairs. The powers will also assist local authorities in targeting their resources and in managing their grant budgets within the new funding arrangements for private sector renewal (DoE, 1996c). However, local authorities are expected to use these deferral procedures with care because of concerns about the potential blighting effect of these actions.

Closing orders apply to unfit properties where renovation is not the best course of action and where demolition is not feasible. This might apply in the case of a listed building or a mid-terrace property. The order prevents the house from being used for human habitation although the authority may

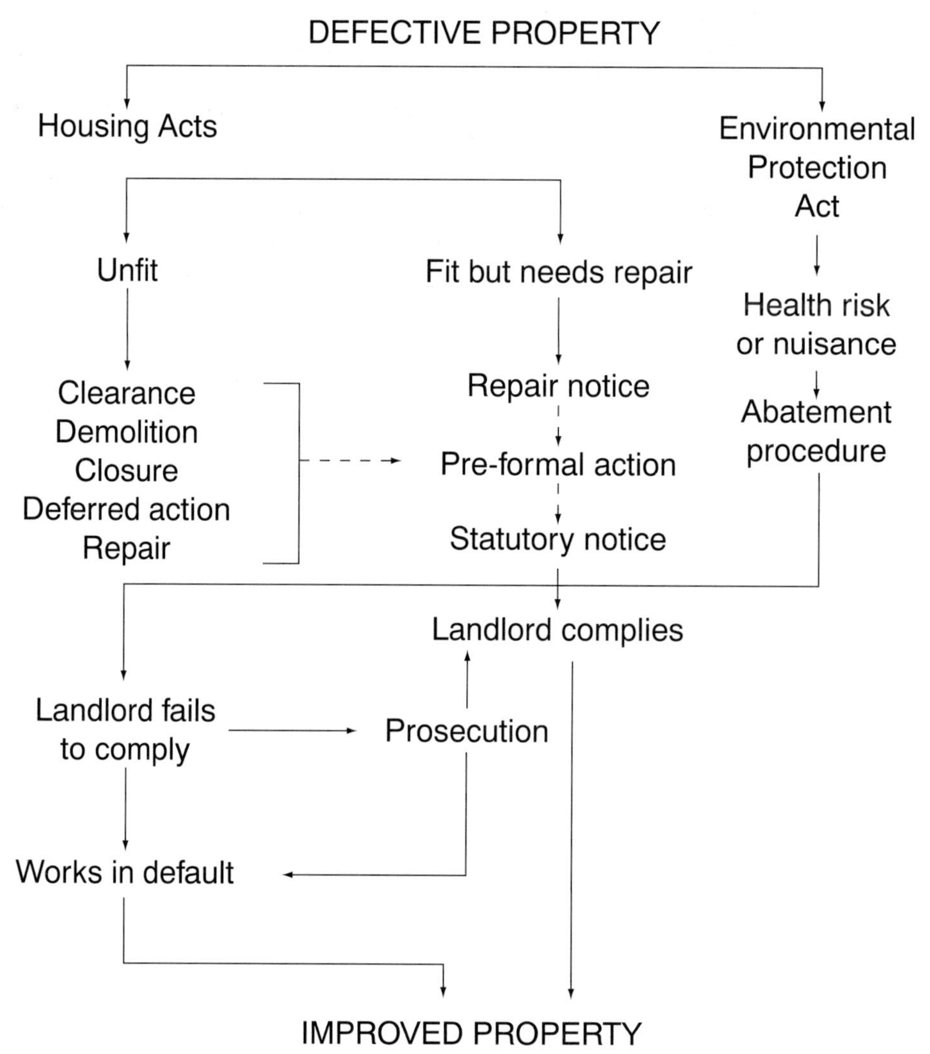

Figure 7.2 Enforcement processes relating to improvement and repair of poor housing

approve other uses such as storage. Displaced residents are offered compensation and alternative housing by the authority. Rehousing obligations apart, one of the key concerns about closure is the potential blighting effect on adjacent houses. However, dwellings affected are often subsequently improved for sale or rent. More than 2,400 dwellings were closed in England during 1993/4 (DoE, 1995). This was a significant increase on the previous two years, which had followed the introduction of new procedures for grants and fitness actions under the Local Government and Housing Act 1989. The increase may reflect some of the difficulties encountered by authorities in managing the formerly demand-led renovation grants system and it will be interesting to observe future trends in unfitness actions following the breaking of the link between unfitness and entitlement to grant aid.

Demolition orders enable authorities to require the vacation and demolition of an individual premises and, as in the case of closure, displaced residents are offered rehousing and compensation. During 1993/4 around 1,560 dwellings were demolished in England, most of these actions (93.2 per cent) occurring in, or adjoining, clearance areas (DoE, 1995).

While demolition and clearance actions are often closely linked, large-scale Housing Act clearance was generally abandoned in the early 1970s and slum clearance is generally at a low level, following a continuous pattern of decline which is illustrated in Figure 7.3. This pattern has evolved largely as a consequence of the difficulties posed by large-scale clearance in terms of both political and financial implications. More recently, however, policy has recognised the need to get small-scale selective clearance back on to the housing agenda and authorities have been advised to adopt an even-handed approach in balancing choices between renovation, demolition and clearance (DoE, 1996c).

It is often the case that unfit dwellings are situated in areas of houses in similar poor condition. In

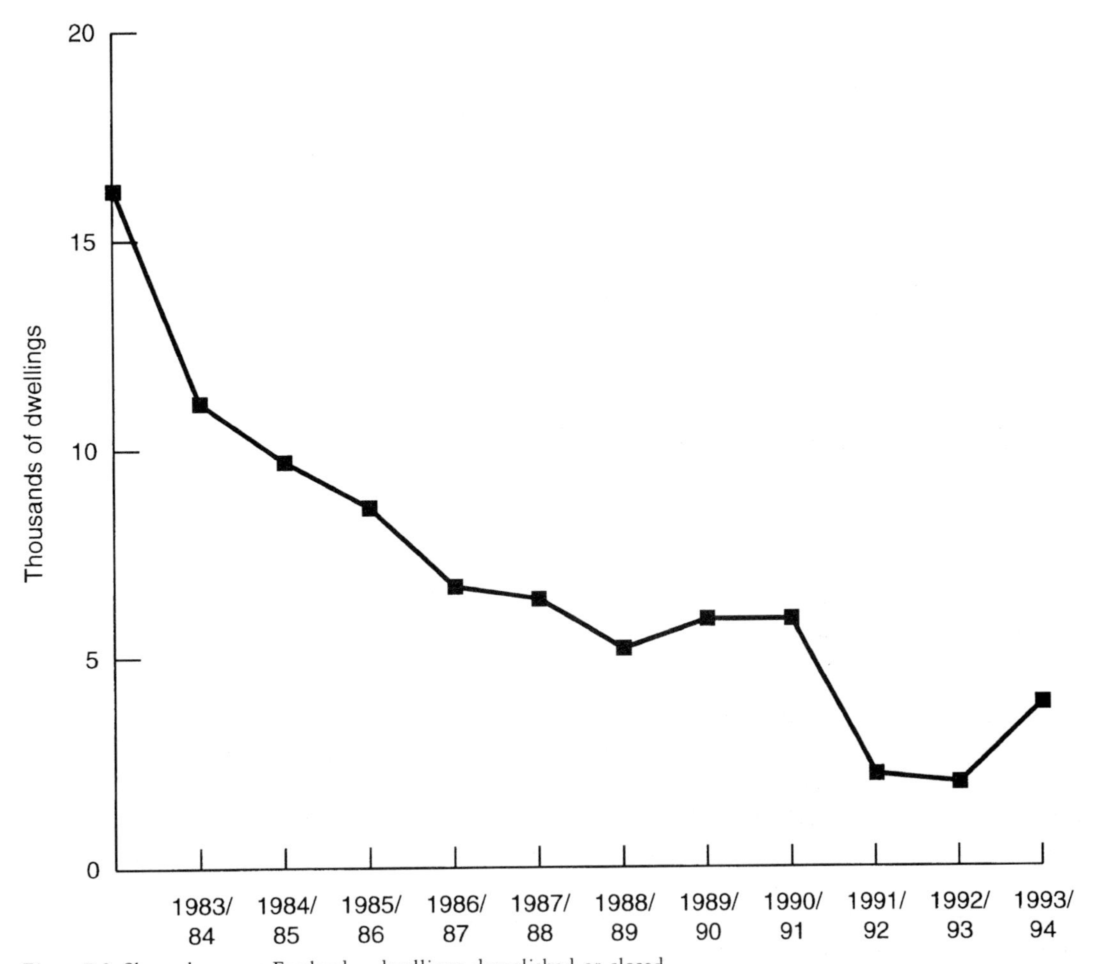

Figure 7.3 Slum clearance: England – dwellings demolished or closed
Source: Department of the Environment, *Housing and Construction Statistics*

recent years there has been increasing emphasis on strategic approaches and the role of area-based renewal in tackling areas of poor housing. These approaches are still developing and can encompass regeneration initiatives linked in with economic development and environmental improvements through to selective clearance and redevelopment of an area.

There are a wide range of factors to consider in declaring a clearance area (Section 289 Housing Act 1985 and DoE *Circular 17/96*). Typically, however, clearance may be chosen as the best course of action to deal with areas where there are concentrations of unfit housing. This action essentially involves demolition of all buildings within the declared area. Detailed processes covering consultation and declaration are laid down in the legislation (Part XI Housing Act 1985 Sections 289–311), and one of the key considerations in this type of approach concerns the housing needs and aspirations of those affected. Residents living within the area are required to vacate their houses and properties can be acquired compulsorily if necessary. Market value compensation is paid to displaced owners in addition to home loss and disturbance allowances which may be awarded, and displaced residents also have a right to rehousing.

The Housing (Grants Reconstruction and Regeneration) Act 1996 introduces new approaches utilising relocation grants to encourage displaced owners to re-invest their equity and compensation into owner-occupied housing. The aim is to help low-income home-owners, affected by clearance, to bridge the gap between compensation payments and the cost of buying another home. By helping such households to remain in the owner-occupied sector, the rehousing burden on local authorities is also effectively reduced. These new forms of grant have been developed from pilot studies in Birmingham, where they proved successful in helping overcome resistance to clearance proposals, as well as preventing the break-up of existing communities (Karn *et al.*, 1995).

In terms of regeneration and renewal initiatives, the declaration of renewal areas has an important role in strategic approaches to private sector housing. The Local Government and Housing Act 1989 enables councils to declare renewal areas and the Housing (Grants Reconstruction and Regeneration) Act has introduced revised provisions covering group repair schemes. Both of these approaches enable authorities to deal with private house renewal and repair on an area basis.

Renewal areas are intended to deal with areas of run-down private housing over a period of up to ten years. Although initial progress was slow, there are currently around ninety renewal areas in England. A key aspect of area-based renewal programmes is the participation of the local community and private sector interests in the area. There is an increasing trend for these approaches to be linked into other funding frameworks such as the single regeneration budget, thus providing additional funding. In order to declare a renewal area the following criteria should be satisfied:

- 75 per cent of the houses in the area must be privately owned.
- 75 per cent of the dwellings must be unfit or in poor repair.
- 30 per cent of the householders must be in receipt of income-related benefit.
- Size of area: around 600–2,000 properties.

Initiatives within a renewal area might include selective clearance, targeting of grant-aided works and environmental improvements, as well as group repair schemes.

Group repair schemes are targeted at the renovation of blocks of housing and utilise a combination of public funding and contributions from owners. The schemes cover external repairs and structural works where dwellings are affected by instability. Typical works include new roofs, boundary walls and rainwater drainage. These schemes are quite different from the other mainstream grants in that the whole process is very much driven by the local authority in consultation with the residents. The authority initiates the scheme, has contractual control of the works, and is responsible for the payment of contractors and collecting any contributions from participants. By tackling groups of properties, these schemes can achieve economies of scale and enable better co-ordination than may be possible with individual renovation grants. An additional benefit is that external improvements can encourage owners to carry out improvements to the interior of their properties and maintain the exterior and so lead to increased confidence in an area.

Until recently, the majority of group repair schemes have occurred in renewal areas. However, the flexibility and application of these schemes have been extended by the 1996 Act and there has been some relaxation of detailed eligibility criteria which had previously provided a barrier in some authorities (Group Repair (Qualifying Buildings) Regulations 1996) (*SI No. 2883*) (DoE, 1996d). These changes, combined with the strategic emphasis of current grant policy, could potentially increase the attractiveness of both renewal areas and group repair as policy options to help arrest decline in deteriorating neighbourhoods.

DISREPAIR AND NUISANCE

In addition to the powers described in the preceding sections, which deal specifically with unfit dwellings, there are wide-ranging provisions to tackle disrepair. The main powers (Section 190 Housing Act 1985) deal with properties in substantial disrepair or where conditions interfere with the personal comfort of the occupying tenant. Unlike the powers to deal with unfit dwellings, these provisions are discretionary and can only be applied to owner-occupied premises in a renewal area. Otherwise, the general administrative processes are similar and are illustrated in Figure 7.2 above.

The Environmental Protection Act 1990 (Part III) provides significant powers enabling local authorities and individuals to take action about matters which are causing a nuisance or which are prejudicial to health. These provisions deal with the effects of defects and cover, among other things, premises where conditions constitute a statutory nuisance. The concept of 'statutory nuisance' (Section 79) is a key aspect of these provisions and is defined as including the following:

a premises in such a state as to be prejudicial to health or a nuisance;
b smoke emitted from premises so as to be prejudicial to health or a nuisance;
c fumes or gases emitted from premises so as to be prejudicial to health or a nuisance;
d dust, steam, smell or other effluvia arising on industrial trade or business premises and being prejudicial to health or a nuisance;
e any accumulation or deposit which is prejudicial to health or a nuisance;
f any animal kept in such place or manner as to be prejudicial to health or a nuisance;
g noise emitted from premises or caused by a vehicle machinery or equipment in the street which is prejudicial to health or a nuisance; and
h any other matter declared by any enactment to be a statutory nuisance.

These powers enable councils to deal with urgent problems with health implications, or which are affecting a third party (they also apply to unfit dwellings). The provisions are also used to deal with noise nuisance where the nuisance is due to inadequate sound insulation (Section 79a) (see *Southwark v. Ince* [1989]), as well as complaints about neighbour nuisance such as the playing of amplified music (Section 79g).

Local authority environmental health departments and public sector housing managers are all experiencing increased levels of complaints relating to noise nuisance. The trends in this area reflect a range of socio-economic factors, including the tensions imposed by differing lifestyles, increased occupation density rates and the legacy of property conversions which fail to meet current standards. Analysis of local authority returns to the Chartered Institute of Environmental Health (CIEH annual reports) indicates that since 1980 domestic noise complaints have increased by as much as 466 per cent! While these figures may overestimate the true incidence of noise nuisance (in that there may be multiple complaints concerning the same source), they do highlight the growing public concern and expectations as to the right to a peaceful environment.

Tenants may take their own action under Section 82 of the Environmental Protection Act. Following a conviction for responsibility for a statutory nuisance the court has power to make a compensation order in favour of the complainant.

Figure 7.4 illustrates the process by which tenants can challenge their landlords. Approaches to these provisions are now a key feature of repairs management strategies in public and private sector housing.

Supplementing the central repairs provisions described above are specific powers under public health legislation to deal with inadequate, defective or blocked drains and private sewers. These are to be found in the Public Health Act 1961, Building Act 1984 and the Local Government (Miscellaneous Provisions) Act 1976. Generally, failure to comply with such notices will result in the local authority carrying out works in default. Other public health powers which are relevant in the context of housing management enable local authorities to undertake cleansing of filthy or verminous premises, articles and people (Public Health Act 1936, Sections 83–86). These provisions may be used to deal with serious cases of hoarding behaviour by occupiers or tenants. Local authorities are also able to secure the restoration of gas, electricity and water services where they have been disconnected or are likely to be disconnected as a result of the landlord's default in making payment (Section 33 Local Government (Miscellaneous) Provisions Act 1976).

While levels of occupation and overcrowding have gradually reduced in more recent times, the health and social problems of overcrowding remain a serious problem for thos households affected. Specific provisions to deal with overcrowding of dwelling houses are contained in Part X of the Housing Act 1985. These provisions are explored in more detail in Chapter 8. The standards for overcrowding are very low and certainly do not meet contemporary expectations; for example, living and dining rooms will be counted in assessing overcrowding.

HOUSES IN MULTIPLE OCCUPATION (HMOs)

Houses in multiple occupation (HMOs) are houses which are occupied by more than one household and can include houses divided into bedsits, shared houses, hostels, as well as self-contained flats in conversions. There is a range of case law which helps determine whether or not dwellings fall into this category and the precedents established under *Simmons v. Pizzey* [1977] and *Barnes v. Sheffield* [1995] are particularly important. It is estimated that there are in England as many as 800,000 HMOs of all types. Of these approximately 350,000 are in the private rented sector, accommodating approximately 2 million people (DoE, 1994). There are a range of provisions covering HMOs, as many of the worst housing problems are concentrated in this sector. With the exception of means of escape in larger HMOs (that is, those of three storeys or more), all powers to deal with this sector are discretionary. As with other enforcement actions, authorities are required to give preliminary notice of their intent to take action, although exceptions are provided where immediate action is considered necessary. The following paragraphs consider some of the main powers enabling authorities to tackle problem HMOs:

Councils are able to require works to make an HMO suitable for a given level of occupation (Section 352 Housing Act 1985) and advice to authorities on typical requirements is provided in government circulars and professional guidance (DoE, 1992 *C12/92* and DoE, 1993c; CIEH 1995b *C12/93*; IEHO, 1994a, 1994b). These powers provide a key underpinning for enforcement roles in this area and cover:

1 Adequate facilities for storage, preparation and cooking of food including cookers, refrigerators and sinks.
2 Adequate personal washing and w.c. facilities; for example, hand basins, baths, showers.
3 Means of escape in case of fire and other fire precautions. These include fire doors, smoke

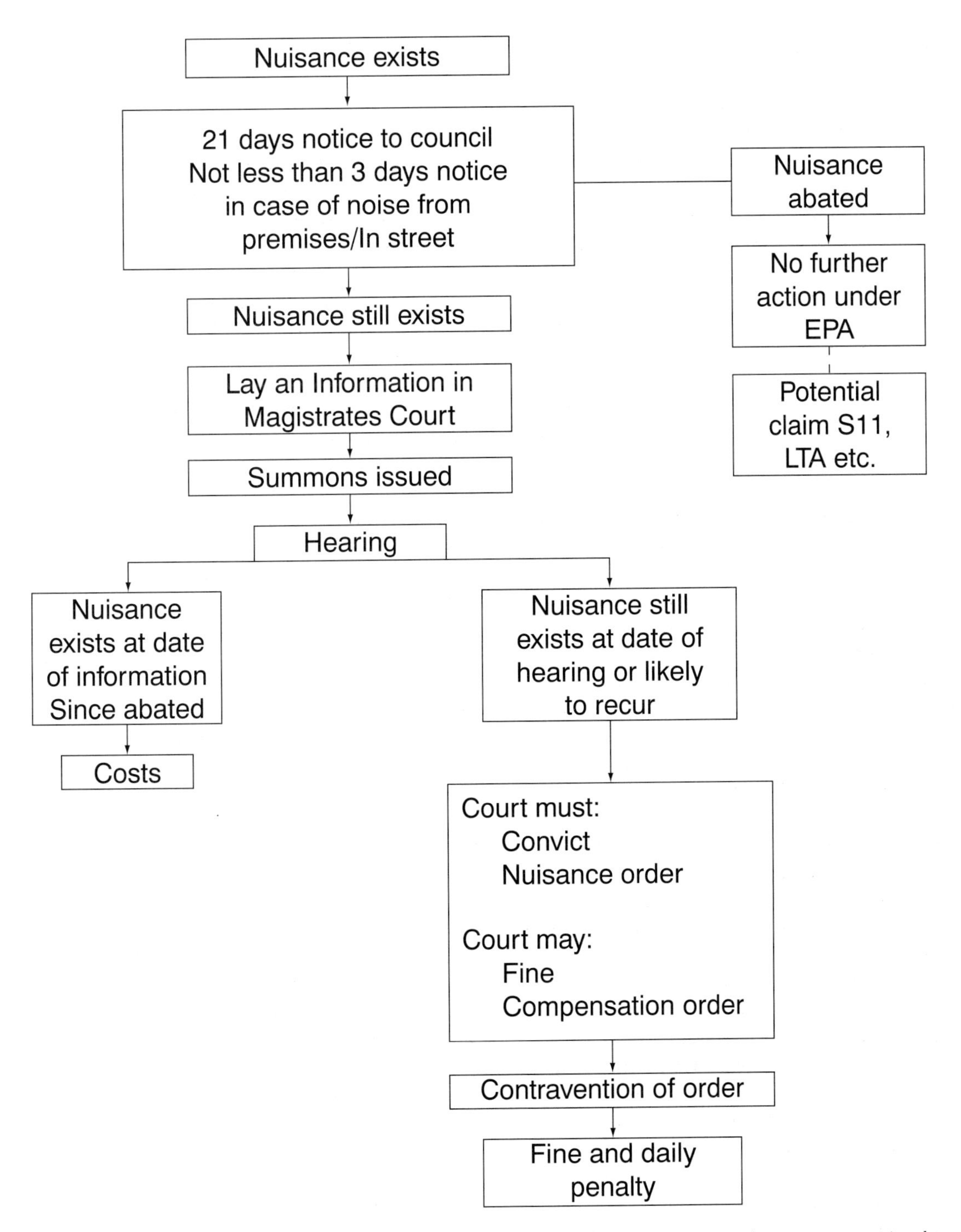

Figure 7.4 Environmental Protection Act 1990: Statutory Nuisance (Section 82) – Action by tenant or aggrieved individual

seals, alarm systems and extinguishers, etc. These fire safety requirements are based on risk assessments (BS 5839 Part VI) and are decided in consultation with the Fire Authority.

Authorities can also restrict the level of occupation in an HMO, having regard to the facilities available to tenants. These controls are enforced by service of a direction (Sections 354–357 Housing Act 1985). Where the HMO is over-occupied and there is no space to put in extra facilities, the direction can require that the level of occupation be reduced so that as tenants leave they are not replaced.

Overcrowding notices (Sections 358–364 Housing Act 1985) allow the authority to control the level of occupation in rooms used for sleeping purposes in HMOs. These provisions are separate from the general overcrowding provisions discussed above and covered in detail in Chapter 8.

Poor standards of management are often found in HMO accommodation. The Housing (Management of Houses in Multiple Occupation) Regulations 1990 apply standards of management to all HMOs. Councils can take action to remedy these deficiencies (Section 372) and failure to comply with the regulations is an offence. These are very useful powers and enable authorities to deal with a wide range of problems arising from the poor maintenance and management of common parts, facilities and installations.

Control order powers are designed to tackle the very worst problems which can occur in HMOs. They can be applied where the conditions are so severe as to pose a threat to the health, safety and welfare of the tenants. These powers enable the council to take over the control of the HMO for a period of up to five years or to acquire the property by way of a compulsory purchase order. Only a handful of control orders are made each year. However, these are key controls to tackle the very worst housing problems which can occur in this sector.

As authorities move towards more strategic and risk-based enforcement approaches in dealing with problematic HMOs, registration schemes can perform a useful role (DoE, 1996c, 1997). Schemes may cover just part or the whole of a local authority's area. The Housing Act 1996 introduced significant changes to registration scheme provisions, introducing three types of registration scheme, of up to five years' duration. The first of these is a notification scheme which simply requires the person responsible for the HMO to apply to the authority for registration. Alternatively, councils may adopt a regulatory scheme which includes control provisions, enabling the revocation of a registration where conditions are not met. Models have been developed for both of these schemes (DoE, 1997). Third, where HMOs are having an adverse impact on the local neighbourhood, special control provisions may be attached which allow an authority to revoke registration or to reduce the level of occupation so that the house is no longer multiply occupied. Clearly, this latter form of scheme is radical in its effect and could potentially lead to tenants being made homeless, and perhaps for this reason schemes using special control provisions are subject to the Secretary of State's approval. Charges may be made for registration of HMOs and there are heavy penalties of up to £5,000 for contraventions of control provisions.

CONCLUSIONS

Since the nineteenth century it has been recognised that poor housing is an important agent in the causes of ill-health. Legislation was thus introduced to alleviate basic problems such as lack of sanitation, contaminated water supplies and gross overcrowding. Currently, the environmental health professional is concerned with the underpinning principles of health, risk assessment and sustainability, together with the alleviation of disrepair, unfitness and an inadequate standard of management in private sector housing.

Clearly, the combination of enforcement approaches and enabling powers discussed in this chapter has a key role to play in strategic approaches in both housing and health policy. Furthermore, as we approach the new millennium the sustainability of our housing stock will be of vital importance in addressing the housing needs of future generations.

Although this chapter explores the above aspects of environmental health control, it is helpful to refer to Chapters 1 and 3, which also consider housing policy and finance in relation to rehabilitation, and to Chapter 6, which examines the causes and rectification of common building defects, and the conversion, adaptation and refurbishment of residential buildings.

QUESTIONS FOR DISCUSSION

1 Briefly discuss a scenario which you may have encountered which could be used to illustrate the following key terms:

(a) premises in such a state as to be prejudicial to health or a nuisance (Section 79 Environmental Protection Act 1990);

(b) premises which are unfit for human habitation (Section 604 Housing Act 1985);

(c) premises where conditions are affecting the personal comfort of the occupier (Section 190 Housing Act 1985);

(d) a house in multiple occupation which is unfit for the number of occupants (Section 352 Housing Act 1985).

2 How effective is the standard of fitness in meeting current housing needs? Consider the factors you feel should be included in the statutory standard of fitness.

3 Given the problems of an ageing housing stock and the increasing number of elderly home-owners, how might services be developed to meet the housing needs of elderly owner-occupiers?

4 You are a housing manager. You receive a notification of intended action under the Environmental Protection Act 1990, concerning a flat in your area. The notice relates to disrepair and severe condensation which are affecting the dwelling. The notification allows twenty-one days in which to remedy the nuisance. How might you respond?

5 Housing and environmental health professionals are engaged in a range of functions dealing with private sector housing. As we move towards the twenty-first century, what changes do you foresee occurring in these roles?

NOTES

1 At the time of writing, a new public health White Paper was expected as a consequence of the change of government in May 1997.

2 Fuel poverty is defined as the inability to afford adequate warmth.

3 Estimates suggest that cold, damp housing costs the NHS £800,000 each year in increased treatment costs (Boardman, 1991).

4 Provision of a healthy housing stock by the year 2000 was a target established under the World Health Organisation's 'Housing For All' strategy (WHO, 1986).

5 Greenhouse gases contribute to global climate change (or global warming); they include carbon dioxide, methane, nitrous oxides and chlorofluorocarbons.

RECOMMENDED READING

Ambrose (1996) *The Real Cost of Poor Homes*, London: Royal Institution of Chartered Surveyors.

Arden, A. and Hunter, C. (1997) *Manual of Housing Law*, 6th edn, London: Sweet & Maxwell.

Balchin, P. (1995) *Housing Policy*, 3rd edn, London: Routledge.

Barlow, J., Chambers, D. and Forrester, P. (1995) *Local House Condition Surveys in England*, A Report to the DoE (unpublished).

Chartered Institute of Environmental Health (CIEH) (1995) *Environmental Health for Sustainable Development*, London: CIEH.

Leather, P., Mackintosh, S. and Rolfe, S. (1994) *Papering Over the Cracks*, London: National Housing Forum.

Luba, J. (1991) *Repairs: Tenants' Rights*, 2nd edn, London: Legal Action Group.

McManus, F. (1994) *Environmental Health Law*, London: Blackstone.

Thornton, R. (1992) *Property Disrepair and Dilapidations: A Guide to the Law*, London: Fourmat.

REFERENCES

Audit Commission (1991) *Healthy Housing: The Role of Environmental Health Services*, London: HMSO.

Bassett, W.H. (1992) *Environmental Health Procedures*, 3rd edn, London: Chapman & Hall.

Boardman, B. (1991) *Fuel Poverty*, London: Bellhaven Press.

Cox, S.J. and O'Sullivan, E.F. (1995) *Building Regulation and Safety*, Garston: Construction Research Communications.

Davidson, M. and O'Dell, A. (1996) *Owner Occupiers in Poor Condition in England.* Paper presented to the European Network for Housing Research Conference, Helsingør, August 1996.

Department of the Environment (1992) *Circular 12/92. Houses in Multiple Occupation. Guidance on Standards of Fitness Under Section 352 of the Housing Act 1985*, London: HMSO.

—— (1993a) *English House Condition Survey 1991*, London: HMSO.

—— (1993b) *The Future of Private Housing Renewal Programmes: A Consultation Document*, London: DoE.

—— (1993c) *Circular 12/93: Houses in Multiple Occupation – Guidance to Local Authorities on Managing the Stock in Their Area*, London: HMSO.

—— (1994) *Houses in Multiple Occupation: Consultation Document on the Case for Licensing*, London: DoE.

—— (1995) *Housing and Construction Statistics 1984–1994*, London: HMSO.

—— (1996a) *The United Kingdom Environmental Health Action Plan*, London: HMSO.

—— (1996b) *Circular 2/96: Home Energy Conservation Act 1995*, London: HMSO.

—— (1996c) *Circular 17/96: Private Sector Renewal – a Strategic Approach*, London: HMSO.

—— (1996d) *SI No. 2883: Group Repair (Qualifying Buildings) Regulations*, London: HMSO.

—— (1996e) *Houses in Multiple Occupation: Establishing Effective Local Authority Strategies*, London: DoE.

—— (1997) *Circular 3/97: Houses in Multiple Occupation: Guidance on Provisions in Part II of the Housing Act 1996*, London: HMSO.

Department of Health (DoH) (1992) *Health of the Nation: A Strategy for Health in England*, London: HMSO.

—— (1996) *The Health of the Nation – Consultative Document: The Environment and Health*, Cm 3323, London: HMSO.

Department of Trade and Industry (1990) *Home and Leisure Accident Research: 12th Annual Report of the Home Accident Surveillance System*, London: HMSO.

Harrison, C., Humfrey, C.D.N. and Shuker, L.K. (1996) *Institute for Environment and Health Assessment on Indoor Air Quality in the Home: Nitrogen Dioxide, Formaldehyde, Volatile Organic Compounds, House Dust Mites and Bacteria*, Assessment A2, Leicester: IEH.

Home Office (1995) *Fire Statistics United Kingdom 1993*, London: HMSO.

Hopton, J. and Hunt, S. (1996) 'The Health Effects of Improvements to Housing', *Housing Studies*, Vol. 11, No 2.

Institute of Environmental Health Officers (IEHO) (1994) *Amenity Standards For Houses in Multiple Occupation*, London: IEHO.

Karn, V., Lucas, J. *et al.* (1995) *Home Owners and Clearance: An Evaluation of Rebuilding Grants*, London: HMSO.

Local Government Library (various) *Encyclopedia of Housing Law and Practice*, London: Sweet & Maxwell.

Local Government Management Board (LGMB) (1995a) *A Framework for Local Sustainability*, Luton: LGMB.

—— (1995b) *Sustainability Indicators Research Project: Consultants' Report of the Pilot Phase*, Luton: LGMB.

Lomas, P.R., Green, B.M.R., Miles, J.C.H. and Kendall, G.M. (1996) *Radon Atlas of Great Britain*, Didcot: NRPB.

Lowry, S. (1991) *Housing and Health*, London: British Medical Journal.

Luba, J. (1991) *Repairs: Tenants' Rights*, 2nd edn, London: Legal Action Group.

Miles, J.C.H., Cliff, K.D., Green, B.M.R. and Dixon, D.W. (1992) 'Preventing Excessive Radon Exposure in UK Housing', in *Conference Proceedings of 29th Hanford Symposium on Health and the Environment – 1990*, Richland, Wash.: Battelle Press.

Office of Population Censuses and Surveys (OPCS) (1989) *Mortality Statistics Accidents and Violence*, London: HMSO.

Ratcliffe, P. (1996) *Race and Housing in Bradford*, Bradford Housing Forum.

Raw, G.J. and Hamilton, R.M. (1995) *Building Regulation and Health*, Garston: Construction Research Communications.

World Commission on Environment and Development (1987) *Our Common Future* (The Brundtland Report), Oxford: Oxford University Press.

World Health Organization (1986) *Healthy Cities: Action Strategies for Health Promotion*, First Project Brochure, Copenhagen: WHO.

8

LEGAL STUDIES, PROPERTY AND HOUSING LAW

Mark Pawlowski

No one who is studying[1] housing law can do so successfully without understanding the legal system within which it operates. In this chapter, we look initially at how English law is made and how it is applied to the settlement of disputes. The former question involves some knowledge of case law, legislation and EU (European Union) law. The latter requires an understanding of the civil court system and the legal personnel of the law (solicitors, barristers and judges). We shall also look in outline at the law of contract and tort (the law of negligence, trespass, nuisance and occupiers' liability) in order to provide a foundation for studies in housing law. Subsequently, the chapter provides the reader with an introductory text on the subject of property and housing law. It is not intended to be exhaustive and the reader is referred to the Recommended Reading at the end of the chapter. Specifically, this chapter examines:

- Legal studies:
  - The development of English law.
  - Sources of law.
  - Civil court structure.
  - Personnel of the law.
  - Legal personality.
  - Law of contract.
  - Law of tort.
- Property and housing law:
  - Freehold and leasehold estates.
  - The essentials of a valid lease.
  - The formalities of a lease.
  - Types of leases and tenancies.
  - Enforceability of covenants in leases.
  - Express covenants.
  - Implied covenants.
  - Assigning and sub-letting.
  - Leasehold dilapidations.
  - Termination of leases.
  - Statutory protection of residential tenants.
  - Leasehold enfranchisement.
  - Homelessness.

Case studies are used to test some of the above aspects of law.

LEGAL STUDIES

The development of English law

English law has developed gradually over a long period and has drawn on many different sources, the most important being common law and equity.

Common law

Before 1066, there was no law common to the whole of England. Justice was administered in local courts and local customary laws were applied which often varied considerably from one part of the country to another. After the Norman Conquest (in 1066), William I began the process which eventually led to a unified system of law. The growth of the common law (as it became known) depended on the basic principle that a legal right only existed if there was a procedure for enforcing it (that is, no remedy, no right). Common law procedure was based on the writ system. All cases had to be commenced by means of a writ (a command from the king to the defendant to submit to the court's jurisdiction, setting out the cause of action). The system was inflexible and caused injustice if a particular claim could not be brought within an existing writ. This (and other) shortcomings led to the development of a body of law known as equity.[2]

Equity

Persons unable to obtain justice in the common law courts would petition the king (the fountain of all justice) who passed on these petitions to the Lord Chancellor. The petitions became so numerous that in time a new court, the Court of Chancery, was created in order to deal with them, and a new system of law known as 'equity' was developed in this court. Equity means fairness, and the Court of Chancery concentrated on looking at the merits of a dispute and doing what was right between the parties, rather than on applying a rigid set of legal procedures. All equitable remedies (for example, the injunction[3] and specific performance) originated from the Court of Chancery and are discretionary in nature. The common law remedy of damages, on the other hand, is awarded as of right.

By virtue of the Judicature Acts 1873–1875, the old common law courts and the Court of Chancery were abolished and a new Supreme Court was established in their place. This consisted of an Appeal Court and a High Court.[4] Any judge within this new structure could administer both common law and equity so that litigants could obtain both kinds of remedies in one court. With the administrative fusion of common law and equity, the rules of procedure were simplified and much improved.

It is important to bear in mind that the Judicature Acts fused only the *administration* of common law and equity. They did not fuse the substantive rules, so that the distinction between these two historic branches of English law is still very relevant to this day. Equity is seen essentially as a body of rules which were created to mitigate the rigours of the common law. A good example is to be found in the law of forfeiture of leases. At common law, a lease may be forfeited by the landlord if the tenant breaks any of the covenants in the lease regardless of the seriousness of the breach. Equity, however, steps in to mitigate the harshness of this common law remedy by allowing the tenant in default to apply to the court for relief from forfeiture.[5]

Sources of law

The expression 'sources of law' can mean at least two different things. First, it can refer to the historical origins from which the law has come (that is, common law and equity) discussed above. Second, it can refer to the body of rules which a judge will draw upon in deciding a case. In this second sense, there are three main sources of English law, namely legislation, case-law (or precedent) and EU law.

Legislation

The Queen in Parliament is the supreme law-making body in the United Kingdom and is recognised as such by the courts. The Queen in Parliament means the Queen, House of Lords and House of Commons. Before a legislative proposal (a bill) can become an Act of Parliament (a statute) it must be approved by both Houses of Parliament and receive the royal assent. In the absence of any special provision, the new Act will operate from the date of assent.

A *public* Act of Parliament covers the whole of England and Wales, and also Scotland and Northern Ireland unless there is a provision to the contrary. Such Acts include government bills (introduced by the government of the day) and private member's bills (introduced by private members of Parliament). About twelve Fridays in each session are reserved for the introduction of private member's bills and members ballot for the privilege of introducing such bills. A *private* Act of Parliament, on the other hand, does not affect the public at large, but generally confers some additional powers on some person or some public body. Most private bills are sponsored by local authorities which, for example, may require additional borrowing powers or additional powers for the compulsory purchase of land.

A legal adviser must know how to interpret the law. In order to do this, he or she needs to be able to read legislation with a basic knowledge of the rules on statutory interpretation. The Interpretation Act 1978 defines words commonly used in statutes. It contains, for example, provisions stating that 'male' includes 'female' and 'plural' includes 'singular', unless a contrary intention is expressed. In a complex Act, there will be an interpretation section defining terms used in the particular Act (see, for example, Section 152(1) of the Rent Act 1977). There are also a number of general rules evolved by the judiciary used to assist in the interpretation of words and phrases in statutes. The 'literal rule', for example, dictates that the task of the court is to give the words construed their plain, ordinary or literal meaning regardless of whether the result is sensible or not. Under the 'golden rule', however, the wording of a statute is construed in such a way as to avoid absurd results. The 'mischief rule' requires the judge to consider what mischief or weakness in the law the statute was intended to correct. The court may also have regard to Hansard, the printed report of debates in Parliament, as an aid to construing ambiguous or obscure legislation.[6]

Some of Parliament's legislative functions are delegated to subordinate bodies that (within a limited field) are allowed to enact rules. For example, local authorities are permitted to enact by-laws but they can only do this because an Act of Parliament has given them power to do so. This is known as delegated legislation.[7]

Case-law (or precedent)

Despite the growth of the importance of legislation as a means of making and changing the law, the bulk of English law is still case-law. This is law which has not been enacted by Parliament, but has been developed through the centuries by judges applying established rules of law to new situations and cases as they arise. Case-law is based on the doctrine of judicial precedent or *stare decisis* (standing by things decided). Once a court has stated the legal position in a given situation, then the same decision will be reached in any future case where the material facts are the same. The modern doctrine of binding precedent did not develop until the mid-nineteenth century because it requires (1) accurate reporting[8] of judicial decisions and (2) a settled hierarchy of courts. In 1875, the Incorporated Council of Law Reporting was established to produce officially authorised reports, prepared by barristers and revised by the judges themselves, so that their accuracy is virtually guaranteed. In addition, several private firms publish similar series of reports (for example, the All England Reports and Housing Law Reports). Reports of cases are also found in *The Times* and *Independent* newspapers, and in legal magazines such as *Estates Gazette*. Whether a court is bound by a previous decision depends on the court which gave the previous decision. Generally speaking, a lower court must follow the decision of a superior court but a superior court is not bound to follow the decision of a lower court.

In any given case, the decision of the court will be made up of the following elements: (1) findings by the court of material facts; (2) statements of the principles of law applicable to the problem disclosed by the material facts of the case; and (3) a judgment by the court based upon (1) and (2) above. The principle of law contained in the decision of the court is known as the *ratio decidendi* (the reasoning vital to the decision). It is this part of the court's judgment which is binding on future courts. Other statements are regarded as *obiter dicta* (things said by the way). A statement may be classified as an *obiter dictum* if it is a statement of law based upon facts which were not found to be material to the actual decision.

It is not always easy to identify the *ratio decidendi* of a given case. The problem is increased when considering the *ratio* of a Court of Appeal decision where, generally, three judgments are given. In the House of Lords, usually five speeches may be given. The *ratio* of a given case is found (if the opinions of the judges are not unanimous) by identifying the majority viewpoint. This may not always be straightforward since, although there may be a majority decision, the reasoning used by each member of the majority may differ. The role of the lawyer when analysing a case is, therefore, to extract the principle(s) of law from the decision of the court.

EU law

Membership of the European Union means the incorporation of Community law into a state's internal legal system.[9] The primary sources of Community law are the three Foundation Treaties of Paris and Rome. To these have been added subsequent treaties, including the Treaty of Merger 1965 and the Brussels Treaty of Accession 1972. These treaties set out in broad terms the objects to be attained and leave many of the detailed means of achieving them to the European Council and the Commission, which have law-making powers. These may take the form of (1) *Regulations*, which are directly applicable in all member states without the need for further legislation; (2) *Directives*, which require a particular member state to make such changes in its own national law as are necessary to bring it into line with Community requirements; (3) *Decisions* of the Council or Commission, which are binding upon those to whom they are addressed. The decision may sometimes be addressed only to a single firm or group of companies. Unlike regulations, they do not have general legislative effect; and (4) *Recommendations* and *Opinions*, which merely express the views of the Council and the Commission on Community policies and are persuasive only and not binding on member states.

There are now two courts, the European Court of Justice and the Court of First Instance, whose function is to ensure observance of the treaties and the various other forms of law made under it. These courts have power over member states and other EU institutions. The United Kingdom signed the Treaty of Accession in Brussels in January 1972. The effect of the European Communities Acts 1972–86 is to require the British courts to give effect to European Community law and to enforce, where necessary, any decisions of the European Court of Justice. The impact of Community law on English law is concerned primarily with economic and financial matters. Commercial and labour law has also been affected.

Civil court structure

Civil cases are heard in either county courts or the High Court of Justice. County courts were created in 1846. They are now governed by the County Courts Act 1984. There are about 400 such courts in England and Wales, presided over by circuit judges. The majority of civil cases are, in fact, disposed of in the county courts by judges sitting alone without a jury. Solicitors and barristers have rights of audience. There are limits as to the court's jurisdiction based to some extent on the amount claimed. The distribution of business between the county courts and the High Court is dependent on a number of criteria:

1. the financial value of the case;
2. the nature of the proceedings;
3. the parties to the proceedings;
4. the degree of complexity; and
5. the importance of any question involved in the case (for example, whether an important point of law is involved).

So far as the financial value of the case is concerned, there is a presumption that claims for less than £25,000 should be brought in the county court. In respect of claims for more than £50,000, there is a presumption that the proceedings should be commenced and tried in the High Court. For claims in the middle range (i.e. between £25,000 and £50,000), there is no presumption as to the court to be used. Cases are allocated to whichever court is appropriate having regard to the criteria listed above (especially the criterion of complexity). The administration of each county court is carried out by the District Judge, who is a solicitor of at least seven years' standing. He (or she) also performs a judicial function, having his own court which deals with relatively straightforward civil cases. He also hears cases under the county court arbitration scheme in respect of small claims limited to under £3,000 (the small claims court).

The High Court forms part of the Supreme Court of Judicature. As mentioned earlier, the Judicature Acts 1873–1875 reorganised the system of civil courts into a Court of Appeal and the High Court of Justice, the latter now having three divisions: Queen's Bench Division, Chancery Division and Family Division (Figure 8.1). Cases are assigned to the division specialising in that particular type of action.[10]

The High Court also exercises certain appellate jurisdiction. For this purpose, two or three judges sit together and constitute a divisional court. Appeals from county courts (except bankruptcy), the High Court and certain tribunals (for example, the Lands Tribunal) are normally heard by three judges of the Civil Division of the Court of Appeal. A further right of appeal exists from the Court of Appeal to the House of Lords (Figure 8.1). The House of Lords, when it sits as a judicial tribunal, differs in constitution from the legislative body which forms the House of Parliament. The judges are known as 'Law Lords' or 'Lords of Appeal in Ordinary'. A minimum of three judges (but in practice five) will sit to hear a case. By the Administration of Justice Act 1969, a 'leap-frog' procedure was introduced whereby the Court of Appeal can be avoided and the appeal can go direct from the trial judge to the House of Lords. The trial judge must grant a certificate that the case is suitable for an appeal direct to the House of Lords on the grounds that it involves a point of law of general public importance which involves statutory interpretation or is a case in which the judge is bound by a previous decision of the Court of Appeal or House of Lords. The parties must consent to the leap-frog. Also, the House of Lords must grant leave for the direct appeal. Finally, reference must be made to the Judicial Committee of the Privy Council, which has the right to hear appeals made to the Crown from certain overseas territories. Decisions of this court have only strong persuasive force in this country. Most Commonwealth states, however, have now removed their rights of appeal to the Privy Council, Singapore being a notable exception.

Personnel of the law

In England and Wales, the legal profession is divided into two branches and the expression 'lawyer' may be used to refer to either a solicitor or a barrister. By analogy with the medical profession, the solicitor may be viewed as a general practitioner, the person to whom the individual first turns for legal advice. The barrister, on the other hand, is the consultant, whose specialist advice is sought on a particular legal problem. Although solicitors have now been given wider rights of audience (in the High Court), it is still customary to brief a barrister to conduct the case in court. After a number of years' practice, a barrister may apply to the Lord Chancellor for recommendation to the Crown to be appointed a Queen's Counsel – he (or she) then 'takes silk' and exchanges a stuff gown for a silk one. Normally, a Queen's Counsel appears in court with a junior barrister and is restricted to more important cases. In due course, a Queen's Counsel may be appointed a circuit or High Court judge. Although much has been written about the possible fusion of the two branches of the legal profession,[11] it is likely that the fundamental distinctions between barrister and solicitor will remain intact, at least for the foreseeable future.

Most judgeships are selected from the ranks of practising barristers. A circuit judge is appointed by the Crown on the advice of the Lord Chancellor from barristers of at least ten years' standing or Recorders who have held office for at least five years. A Recorder is a part-time judicial appointment who may be a practising barrister or solicitor of at least ten years' standing. High Court judges are appointed by the Crown on the advice of the Lord Chancellor from barristers of at least ten years' standing. A High

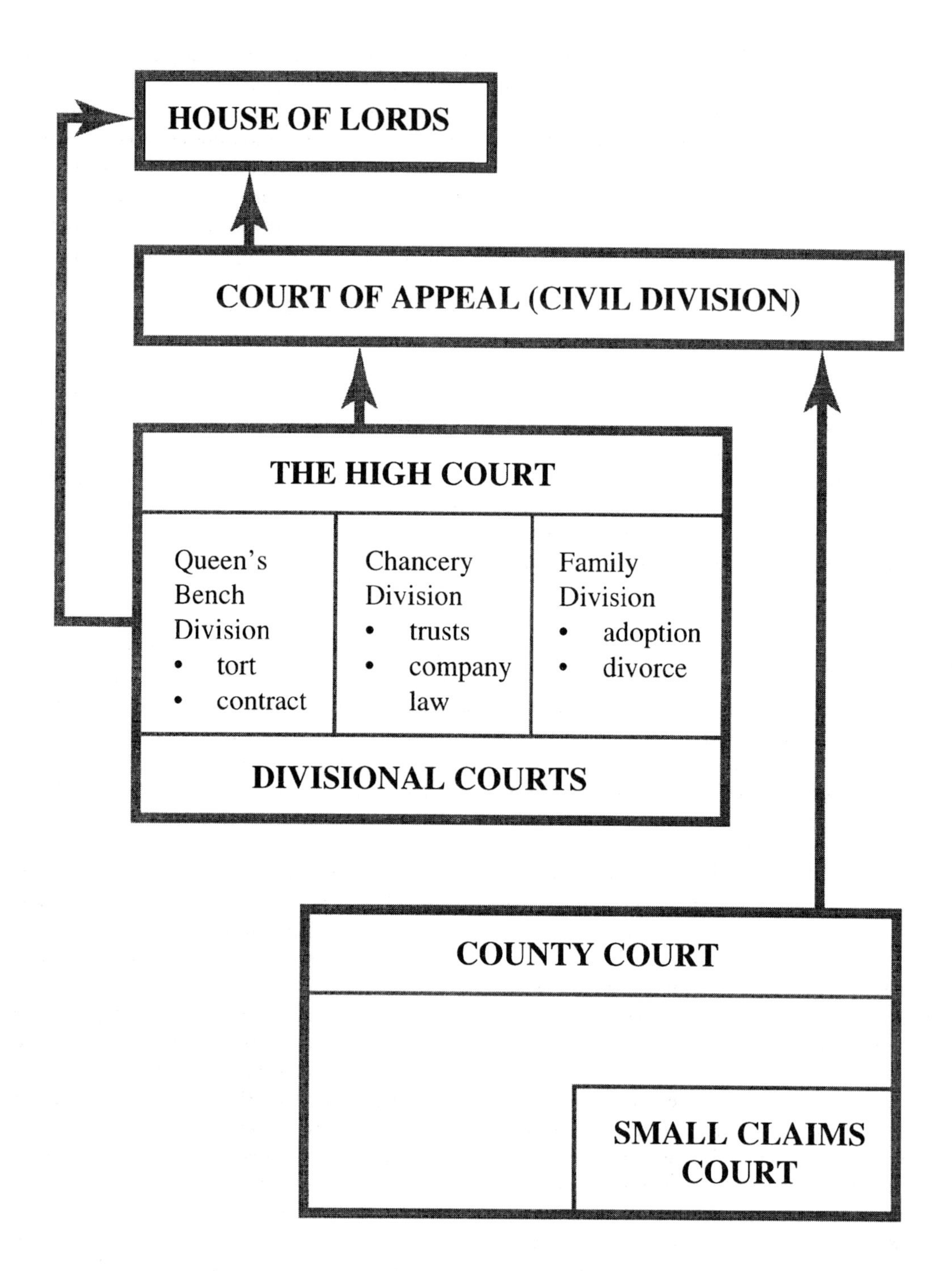

Figure 8.1 The civil courts

Court judge will be assigned to one of the three divisions of the High Court and will receive a knighthood upon appointment. Higher judicial appointments (judges of the Court of Appeal and House of Lords) are made by the Crown on the advice of the prime minister.

In this country, a prospective candidate for judicial office receives little or no formal training in judicial skills prior to taking up appointment on the bench. By contrast, in continental European countries the roles of judge and lawyer are seen to require different skills and training. A lawyer may actually start his legal career trained as a small-time judge and move up the judicial ladder as he or she becomes more proficient. English judges, on the other hand, are chosen from the ranks of the senior bar (and, as we have seen, to a limited extent from solicitors). The system works on the assumption that a good barrister (or solicitor) will also make a good judge.

Legal personality

A legal person is any person or thing recognised by the law as possessing rights and duties. Where legal personality attaches to a human being, it normally begins with birth and ends with death. However, before birth, a foetus's life may be protected by the criminal law, property rights may attach to it and there may be an action for negligence for injury to the pregnant mother affecting the child.[12] The child's civil law rights will be exercised later if the child is actually born alive.

Upon death, legal personality will continue in the sense that outstanding rights and liabilities pass to the personal representatives for the purpose of winding up the deceased's estate.

The only exception to the rule that legal personality attaches to a human being arises where the law grants personality to a group of persons together and creates an artificial person. This process is known as incorporation. The corporation (company or other corporate body) possesses a personality entirely separate from its members with many of the powers of a natural person (it can own property, enter into contracts, and so on, in its own name).[13] Persons may also combine to further a common interest *without* creating an independent legal person. The law regards such an association (for example, a club or society) as a number of individual persons. Any property of the unincorporated association is the joint property of the members and not the property of the association itself. A partnership[14] is a form of unincorporated association which 'subsists between persons carrying on business with a view to profit' (Section 1 of the Partnership Act 1890). Since the partnership is not a legal person, it only exists in the persons of its members. In contrast to shareholders of a company, the liability of the partners is generally unlimited, so that each partner is liable to the full extent of his or her own property for the debts of the partnership.

Law of contract

A contract is a legally binding agreement. The law of contract is concerned primarily with providing a framework within which business can operate. If an agreement is broken, the law intervenes and makes the person who broke the contract pay compensation (damages) to the other party. In order to create an agreement which the law will recognise and enforce, the following conditions must be satisfied:

1 There must be an intention to create legal relations. Unless the court is satisfied that the parties intended the agreement to be legally binding, it will not be enforceable. Many agreements are, clearly, never intended to have legal force. In the case of an agreement of a friendly, social or domestic nature, there is a strong presumption that the parties did not intend to create a legal relationship. On the other hand, there is a strong presumption that commercial agreements are intended to create legal relations, unless there is a clear statement to the contrary.
2 The court must be satisfied that the parties have reached an agreement and are not still in the process of negotiation. An agreement is marked by the unconditional acceptance of an offer. An

offer is a statement of the terms on which the offeror is willing to be bound. If the offer is accepted as it stands, then agreement is reached. It is to be contrasted with a mere invitation to treat, which is simply an indication that a person is willing to enter into negotiations but not that he is yet willing to be bound by the terms mentioned (for example, goods displayed on a supermarket shelf[15]). A mere advertising boast, which no one would take seriously, will not be treated as a firm offer. Much, however, will depend on the circumstances of the case. In *Carlill v. Carbolic Smoke Ball Co*,[16] for example, the defendants advertised that they would pay £100 to anyone who caught influenza after using their smoke balls and that, as evidence of their sincerity, they had deposited £1,000 with a bank. Mrs Carlill followed the instructions and used the smoke balls but still caught influenza. She claimed the £1,000. One of the defences was that the advertisement was not an offer. The Court of Appeal rejected this submission, holding that a reasonable person would take the advertisement seriously and assume that the advertiser intended to be bound on the terms stated.

As a general rule, acceptance of an offer must be communicated to the offeror (that is, there is no contract until the offeror knows that his (or her) offer has been accepted). One exception arises where the offeror dispenses with the requirement of communication and indicates that the offeree should, if he wishes to accept, simply carry out his side of the bargain without informing the offeror (for example, if a customer writes ordering goods and, without further communication, the goods are delivered in accordance with the order). Another exception arises where the so-called 'post-box' rule applies. Under this rule, a letter of acceptance, properly addressed and stamped, is effective from the moment of posting, even if it never arrives![17]

3 A contract is essentially a two-sided bargain (that is, a mutual exchange of promises). If A merely promises to give something to B, the law gives no remedy if A breaks his promise. But if B promises to do something in return, this reciprocal element turns the arrangement into a contract. In legal terminology, A's promise or action is the 'consideration' for B's and vice versa. Consideration, therefore, may be described as something given, promised or done in exchange.

   Although a promise made without consideration cannot be sued upon and will not amount to a contract, it may have limited effect as a defence to an action. If A promises to do something and the promise is intended to be acted upon and is, in fact, acted upon by B to his detriment, then A may be 'estopped' (prevented) from doing something which is inconsistent with that promise. This equitable defence is called promissory estoppel.[18]

4 Certain types of contract are only valid if made in a particular form (in writing). For example, a contract for the sale of land must be made in writing, be signed by the parties and incorporate all the essential terms, in order to be valid.[19]

5 It must be possible for the court to ascertain what the parties have agreed upon. If the terms are so vague as to be meaningless, the law will not recognise the agreement.

6 Certain types of agreement are so plainly contrary to public policy that the law will not recognise them (for example, an agreement to commit a criminal offence).

Vitiating factors

Where a person is driven into a contract by duress or by undue influence, the contract is voidable and may be set aside by the court. Mistake may also vitiate a contract. Thus, a mutual mistake as to the identity of the subject-matter will render the contact void. A mutual mistake will occur where the parties are, unknown to each other, thinking about different things. Neither is right nor wrong – they are simply at cross-purposes and have never really agreed on anything. In this situation, there is no *consensus ad idem* (meeting of the minds). For example, in *Raffles v. Wichelhaus*,[20] a cargo of cotton was described as being on the SS *Peerless* sailing from Bombay. There were, in fact, two ships of this name sailing from Bombay with an interval of three months between them. The seller intended to put the cargo on the second ship and the buyer expected it on the first. The contract was held void for mutual mistake.

A fundamental common mistake about the subject-matter of the contract will also render it void. A common mistake occurs when both parties are under the same misapprehension – both are wrong (for example, where, unknown to both parties, the subject-matter of the contract does not exist).

In exceptional cases, a party to a contract may avoid liability under a contract by pleading that the document was not his deed (*non est factum*[21]). A mistake by one party as to the identity of the other may also sometimes invalidate the contract.[22]

In general, all persons have full legal power to enter into any contract they wish but a few groups of persons do not have this power and are said to be under incapacity. The main groups concerned are minors, persons lacking mental capacity, and drunkards!

Misrepresentation

The conclusion of a contract is often preceded by negotiations, in the course of which one party will make statements of fact intended to induce the other to enter into the contract. If any such statement is false, it is called a 'misrepresentation' (that is, a false statement of fact made by one party to the contract to the other before the contract is made with a view to inducing the other to enter into it). The statement must be a representation of fact (as opposed to law or opinion).

Many misrepresentations also amount to promises which are incorporated into the contract itself. In this case, the party misled will normally sue for breach of contract rather than for misrepresentation because, once a breach of contract is proved, damages will automatically be awarded. Where a mere misrepresentation is proved, the party liable may still have a defence to an action for damages if he can prove that he reasonably believed himself to be telling the truth.[23]

The remedies for misrepresentation are damages and rescission of the contract. Any misrepresentation will give the innocent party a right to rescind the contract (that is, put an end to it if he so wishes). Each party, however, must be restored to his original position (for example, the goods must be returned to the seller and the price to the buyer). The right to rescind is lost if it becomes impossible to return the parties to their original pre-contact position (for example, if the goods have been consumed or destroyed).

Terms of the contract

The terms of the contract may be divided into (1) express terms and (2) implied terms. Express terms are those which are specifically mentioned and agreed by the parties at the time of the contract. Contractual terms, whether oral or written, differ in importance and may be classified into:

1 conditions;
2 warranties; and
3 innominate terms.

A condition is an important term which is vital to the contract so that non-observance of it will affect the main purpose of the agreement. Thus, breach of a condition will give the injured party a right to terminate the contract at common law. Alternatively, he (or she) may (if he so wishes) treat the contract as continuing but recover damages for his loss as a result of the breach of condition.

A warranty, on the other hand, is a less important term, non-observance of which will cause loss but not affect the basic purpose of the contract. Breach of warranty will only give the injured party the right to sue for damages.

Many express terms are difficult to classify in advance and can only be classified by reference to the nature of the breach as and when it occurs. Thus, a minor breach of the term might only be a breach of warranty, whereas a serious breach of the same term might be a breach of condition. This type of term is called an innominate (or indeterminate) term.[24]

Any term may be implied into a contract by the court or by statute. Where the parties have not made express provision on some point, the court will sometimes imply a term to give 'business efficacy' to the contract (give business effect to the parties' intentions). The parties obviously intended such a term (but failed to mention it) and the contract makes no commercial sense without it.[25]

In some types of contract, detailed terms are implied by statute. A good example of this is Section 11 of the Landlord and Tenant Act 1985, discussed in more detail below,[26] which implies obligations to repair on the part of a landlord in respect of residential lettings for under seven years.

Breach of contract

A breach of contract can occur in several different ways. For example, one party may expressly repudiate his liabilities and refuse to perform his side of the bargain. This can happen either at or before the time when performance falls due. If a party renounces his obligations in advance, this is known as 'anticipatory breach' and entitles the innocent party to sue immediately for damages for breach without waiting for the date of performance. A person can also impliedly renounce his obligations by rendering himself incapable of performing them (for example, if he has contracted to sell a specific item to A, he would impliedly renounce by selling to B). Alternatively, a party may simply fail to perform all (or some) of his obligations under the contract.

Every breach of contract will give the injured party the right to recover damages for any loss he has sustained as a result of the breach. If the breach is sufficiently serious (as in a breach of condition), it will also give the injured party a right to terminate the contract at common law (see above).

The basic principle is that damages should put the injured party in the position he would have been in had the contract been performed (loss of bargain damages). The loss can be financial, damage to property, personal injury or even distress (as where a holiday firm defaults on its obligations[27]). The innocent party cannot, however, be compensated for *all* of the consequences which might logically result from the defaulting party's breach, otherwise there would be no end to liability. Some losses will be too remote. In *Hadley v. Baxendale*,[28] it was held that damage or loss treated as resulting from the breach should only include (1) such damage as may 'fairly and reasonably be considered as arising naturally' (according to the usual course of things) from the breach or (2) such other loss as may 'reasonably be supposed to have been in the contemplation of both parties at the time they entered into the contract', so that the defaulting party in effect accepted responsibility for it.

The innocent party is also under a duty to mitigate (or minimise) his damages (that is, take all reasonable steps to reduce them). Thus, an employee who is wrongfully dismissed must attempt to find other employment. A loss arising from a failure to take such reasonable steps is not recoverable.

Apart from damages, the injured party may seek other types of remedy for breach of contract. He may, for example, seek an order for specific performance (if damages are an inadequate remedy) compelling the party in breach to carry out his obligations under the contract. This is an equitable remedy granted at the court's discretion. In some situations, a claim for damages may not be an appropriate financial remedy. Instead, the innocent party may seek a *quantum meruit* for the value of work already done under the contract.[29]

Discharge of contracts

A contract may be discharged (brought to an end) in four main ways: (1) performance; (2) breach; (3) new agreement; and (4) frustration.

The basic rule is that each party must perform completely and precisely what he has contracted to do, although the court will ignore minor deviations under the so-called *de minimis* rule. Thus, for example, in *Shipton, Anderson & Co v. Weil Bros & Co*,[30] the court ignored a deviation of 55 lb in a consignment of 4,950 tons of wheat!

We have already looked briefly at breach of contract (see above) and seen that an acceptance of a repudiatory breach will discharge the contract. There are also a number of ways a contract may be discharged by agreement. First, the parties may have made provision for discharge in the contract itself (for example, the parties may have agreed that the contract should terminate automatically on some determining event or on the expiration of a fixed time). A good example is a lease which will end on the expiry of its term.[31] The contract may also provide that one or both parties may terminate if they so wish (as in a break-clause in a lease). Second, the contract will end if it is replaced by a binding new agreement. Third, one party can simply release the other unilaterally from his obligations under the contract. A contract may also be discharged by frustration. If some extraneous event occurs, for which neither party is responsible and which makes future performance of the contract impossible or impracticable, then the contract will be discharged by frustration. A good illustration is *Taylor v. Caldwell*,[32] where a music hall was hired for a series of concerts but was burnt down before the date for the first performance. This was held to frustrate the contract because there was no longer any hall to hire. Frustration automatically brings the contract to an end and, as a general rule, all sums paid by either party in pursuance of the contract before it was discharged are recoverable and all sums not yet paid cease to be due.[33]

Law of tort

The word 'tort' means a civil wrong, which entitles a person who is injured by its commission to claim damages for any loss sustained. A tort, like a breach of contract, is a civil wrong but there the similarity ends. Whereas contractual duties are imposed by the parties to the contract themselves, the duty to refrain from committing a tort is imposed by the general law independently of the wishes of the parties.

Tort of negligence

A plaintiff must establish three elements to succeed in an action for negligence: (1) that the defendant owed the plaintiff a legal duty of care; (2) that the duty was broken; and (3) that damage was suffered as a result.

There are many situations where a legal duty of care is recognised as existing, and which are constantly being extended and added to by judicial decisions. The 'neighbour' principle enunciated by Lord Atkin in the celebrated case of *Donoghue v. Stevenson*[34] provides the general test for determining whether or not a duty of care exists. In this case, a woman drank some ginger beer which had been bought for her by a friend. The beer was in an opaque bottle and, when the last of it was poured out, it was found to contain what was thought to be the decomposed remains of a snail! The woman suffered nervous shock and became ill. The House of Lords held that a manufacturer owed a duty of care to the consumer of its products when they are marketed in the form in which the consumer will receive them. The snail was contained in an opaque bottle and there was no reasonable possibility of its being discovered between leaving the manufacturer and reaching the consumer.

The neighbour principle requires that reasonable care should always be taken to avoid injury to your 'neighbour' (that is, any person closely affected by your conduct and who you should reasonably foresee might be injured by it). Under this principle, a professional adviser (such as a solicitor, surveyor or accountant) owes a duty of care not only to his own client who employed him but also to other persons who he knew were relying on his skill and advice.[35]

Despite this universal principle, there are many recognised exceptions to it. For example, a barrister owes no duty of care to those for whom he acts as an advocate in court.[36] This exception is based on considerations of public interest: a barrister should be immune from liability when conducting a case in court so that 'he may do his duty fearlessly and independently as he ought and to prevent him being harassed by vexatious actions'.[37] Generally speaking, there is also no duty of care as regards purely financial or economic loss, save in circumstances where the defendant has made a negligent statement[38] (as opposed to committing a negligent act).

In deciding whether or not a duty of care has been broken, the standard against which the defendant's conduct is measured is that of the so-called 'reasonable man' (or woman). In other words, there will be a breach of duty if the defendant did not act in a reasonable manner in the circumstances of the given situation. The plaintiff must also show that he has suffered loss as a result of the defendant's breach of duty of care. Loss can involve personal injury, damage to property, and financial loss (in the case of a negligent statement[39]). The resulting damage may, however, be too remote to be recoverable because it could not reasonably have been foreseen by the defendant at the time when he acted negligently.[40]

The plaintiff may also be held to be contributorily negligent and his award of damages reduced by an appropriate percentage to reflect his own negligent conduct. The Law Reform (Contributory Negligence) Act 1945 provides that the court may reduce the damages by an amount proportionate to the plaintiff's share of responsibility. For example, the courts will reduce damages resulting from a motor accident if the driver has failed to wear a seat-belt.[41]

Tort of trespass

The tort of trespass is committed by direct interference with the person or property of another without lawful justification. It can take several forms, namely, trespass to the person (the tort of assault), trespass to goods,[42] and trespass to land.

In regard to trespass to land, the plaintiff must establish a direct interference with his land. The most common example is where a person enters upon land without the owner's permission. Land, in a legal sense, comprises the surface of the earth together with air space and buildings above and the ground below. A trespass is actionable *per se* (the act alone is sufficient to give rise to a cause of action without the need to prove that damage has resulted from the act).

The remedies for trespass to land include damages for any loss suffered, an injunction to restrain the trespass, and the ancient remedy of distress damage feasant. This latter remedy entitles a landowner to take prompt action to prevent or stop damage to his land (or anything on it) by seizing and impounding any 'trespassing chattel' until its owner claims his chattel and tenders appropriate compensation for the damage done. A good example is where a football is kicked into a neighbour's garden, damaging a greenhouse. Its applicability to the growing practice of wheel-clamping cars parked unlawfully on private land has recently been questioned by the courts.[43]

Tort of nuisance

Nuisance involves the unreasonable interference with a person's use or enjoyment of his land (for example, by noise, smoke or smell), or of some right over it, or in connection with it (for example, blocking a right of way or interfering with a fishing right). It differs from trespass to land in that the injury is indirect and usually nuisance requires some continuous interference. An action is possible only if the interference is unreasonable. In deciding whether or not the interference is substantial enough, the court will consider all the circumstances of the case (the character of the neighbourhood, for example, and the defendant's motives, may be very relevant). Most private nuisances are not actionable *per se*; some loss of amenity or enjoyment of the land is essential. The primary remedy is an injunction to put an end to the unlawful interference, rather than merely compensation.

A *public* nuisance will be actionable if there is an unlawful interference which affects the public generally. A plaintiff may only sue, however, if he can show that he has suffered special damage over and above that suffered by the public at large. For example, an obstruction of a public highway will be a criminal offence and will affect anyone who wishes to use it. If the plaintiff is particularly affected (for example, because the entrance to his shop is blocked), he will be entitled to bring a civil action for public nuisance.

Occupier's liability

The duty owed by an occupier of premises towards people coming on to his premises is now largely governed by statute; in particular, the Occupier's Liability Act 1957,[44] which governs the liability of an occupier to his lawful visitors, and the Occupier's Liability Act 1984,[45] which covers persons other than his lawful visitors (trespassers).

Under the 1957 Act, an implied duty of care is imposed upon those in physical occupation of premises to all visitors who enter with the express or implied permission of the occupier. The duty concerns the state of the premises and things done (or omitted to be done) on them. The duty of care is to take such care as, in all the circumstances of the case, is reasonable to see that the visitor will be reasonably safe in using the premises for the purpose for which he is invited or permitted by the occupier to be there. The occupier may discharge his legal obligation by giving adequate warning to visitors (for example, by displaying an appropriate notice[46]) of specific dangers facing them on the premises. The duty of care under the 1957 Act cannot be excluded by contract in so far as death or personal injury is concerned. The occupier may, however, contract out of liability for other loss or damage (for example, to property), provided he can show that it is reasonable to do so.

In relation to trespassers, the 1984 Act states that a duty of care will arise if the occupier is aware of a 'danger' on his land (or has reasonable grounds to believe that it exists) and he knows (or has reasonable grounds to believe) that another person is in the vicinity of the danger or may come into the vicinity of it. The danger in question must be one that creates a risk against which, in all the circumstances of the case, the occupier 'may reasonably be expected to offer some protection'. The duty is to 'take such care as is reasonable in all the circumstances of the case' to see that the other person does not 'suffer injury on the premises by reason of the danger concerned'. In appropriate cases, the duty may be discharged by giving adequate notice of the danger or by discouraging persons from incurring the risk. The occupier's duty to the trespasser under the 1984 Act is more limited in scope than that owed to a lawful visitor. Apart from the fact that it extends only to personal injury (whereas the duty to lawful visitors includes the protection of their property), the 1984 Act duty arises only where the occupier knows, or has reasonable grounds to believe, both that the danger in question exists and also that the trespasser may come into its vicinity. No duty is owed to a person who willingly accepts the risk in question, and so the general maxim of *volenti non fit injuria* ('to a willing person, no legal injury can be done') applies to both lawful visitors and trespassers.

PROPERTY AND HOUSING LAW

Landlord and tenant law has been described as a 'labyrinth of technicality, complexity and difficult concepts'.[47] It is true that many students who embark on the study of this subject do so with a sense of dread or, at best, unease. The student is faced with the double hurdle of mastering not only the basic common law rules but also highly complex statutory codes which govern most forms of letting. We begin with a short analysis of the nature of freehold and leasehold ownership.

Freehold and leasehold estates

An estate in land may be defined as a certain legal status for a period of time. The doctrine of estates is, therefore, to do with time – for how long land is held. By virtue of Section 1 of the Law of Property Act 1925, only two estates in land are capable of subsisting or of being conveyed or created in law, namely, a 'fee simple absolute in possession' (commonly known as a freehold estate) and a 'term of years absolute' (commonly referred to as a leasehold estate). We are here concerned predominantly with the law governing leaseholds (that is, the relationship of landlord and tenant).

Freehold ownership

The 'fee simple absolute in possession' is the greatest estate known to English land law. The word 'fee' means that it is an estate capable of inheritance and 'simple' signifies that it is inheritable by heirs generally. Thus, a fee simple estate will end only if there cease to be any heirs of X (X being the current owner). If this happens, the land will revert to the Crown as *bona vacantia* since, under the doctrine of tenures, all land in England and Wales ultimately belongs to the Crown (or Duchy of Lancaster or Duke of Cornwall, as appropriate).

The holder of a freehold estate is entitled to unrestricted use and enjoyment of the land, subject to (1) any third party rights created over the land (for example, an easement of a right of way over the land); (2) statutory restrictions and duties (for example, under the Town and Country Planning Act 1990); and (3) the general law of nuisance and trespass (for example, the rule in *Rylands v. Fletcher*[48]). A freeholder also enjoys various 'rights of alienation' in respect of the land: he is entitled to lease, mortgage, gift and sell it, if he so wishes.

Leasehold ownership

A 'term of years absolute' is also a legal estate (since 1925) but it is less enduring than a fee simple because it is for a defined period of time. The 'term' may be for just one year, less than a year, or a number of years. A lease may also be periodic, which may be weekly, monthly, quarterly or yearly.

The leaseholder's rights and obligations are determined by the general law but, more importantly, by the terms of the lease itself. Thus, the tenant will usually covenant to pay the rent, repair the premises, use the property only for the purposes designated by the lease, and not assign, sublet or part with possession of the land without the landlord's prior consent. Essentially, there are two types of leases: (1) fixed term and (2) periodic. A fixed term will come to an end by expiry of time (when the term expires). A periodic tenancy will end upon the expiration of a notice to quit served by either the landlord or the tenant on the other party.

The essentials of a valid lease

There are a number of essential prerequisites to the creation of a valid lease. These may be listed as follows:

1. The landlord and tenant, respectively, must be different persons. In other words, a landlord cannot grant a lease of land to himself.[49] A statutory exception[50] allows A to grant a lease to A and B. Similarly, A and B can grant a lease to A (or B).
2. The subject-matter of the lease must be land.[51] Land has an extended meaning for this purpose and includes not just the ground itself but any buildings on the land, as well as 'incorporeal hereditaments' (rights over land). It is, therefore, possible to have a lease of fishing rights, or mining rights. Livestock and chattels cannot form the subject-matter of a lease; they may be hired out under a 'hiring agreement' governed by the law of contract.
3. The grant must be of a lesser estate. If a tenant divests himself of everything he has got (that is, of the leasehold estate), which he must do if he transfers to his sub-tenant an estate as great as his own, the relationship of landlord and tenant cannot exist between him and the so-called subtenant.[52] In effect, in this situation, the tenant will have *assigned* (parted with) all his interest in the property.
4. The term or duration of the lease must be certain. Thus, a tenancy 'for the duration of the war' has been held invalid as being for an uncertain term.[53] The doctrine of certainty of term applies to both fixed and periodic tenancies. In regard to the former, the maximum duration of the term must be ascertainable at the outset. In relation to a periodic tenancy, it may be said that the ultimate length of this kind of tenancy cannot be determined at the outset – it could go on forever. But, so long as each separate period is definite (a week, month, quarter or year), the test of cer-

tainty of duration is satisfied.[54] Moreover, a term totally precluding the landlord from determining a periodic tenancy is treated as repugnant to the nature of such a tenancy and void.[55]

5 A term which is expressed to commence more than twenty-one years from the date of the lease creating it is void under Section 149(3) of the Law of Property Act 1925. Subject to this qualification, a landlord may wish to grant a 'reversionary lease' whose term is expressed to commence at some date after its execution (not being more than twenty-one years). For example, the premises may be subject to an existing lease which is not due to expire until some specified date in the future.

6 The lease must grant the tenant exclusive possession of the demised premises (the property conveyed by a lease). The test whether an occupancy of residential (or commercial) premises is a tenancy or a licence is whether, on the true construction of the agreement, the occupier has been granted exclusive possession of the premises for a fixed or periodic term at a stated rent.[56] There may, however, be special circumstances which negative the presumption of a tenancy: (1) where the occupier is a 'lodger' in the sense that the landlord provides attendance or services which require the landlord (or his employees) to exercise unrestricted access to and use of the premises;[57] (2) where, from the outset, there is no intention to create legal relations;[58] (3) where possession is granted pursuant to a contract of employment;[59] and (4) where the relationship between the parties is that of vendor and purchaser.[60] The retention of a key by the landlord is not by itself decisive in determining whether an agreement for the occupation of residential premises is a tenancy or a licence.[61] Also, since *Street*,[56] it has been suggested[62] that the reservation of a rent is not strictly necessary for the creation of a tenancy.

An employee of the landlord will be a licensee if he is genuinely required to occupy his employer's premises for the better performance of his duties.[63]

The most important practical effect of categorising the occupier as a licensee (as opposed to a tenant) is that he will have little or no statutory protection (security of tenure) under the Rent Act 1977 or Housing Act 1988 (see below).

The formalities of a lease

Section 52(1) of the Law of Property Act 1925 provides that all conveyances of land or of any interest therein are void for the purpose of conveying or creating a legal estate unless made by deed. A lease is a 'conveyance' for this purpose and, hence, the general rule is that a lease must be made by deed to be legally valid. One important exception to this rule is contained in Section 54(2) of the 1925 Act, which provides that no writing is required for a lease taking effect in possession (that is, the tenant is entitled to immediate occupation) for a term not exceeding three years (which includes all periodic tenancies) at a full economic rent. In other words, such a lease will be a valid legal lease despite the absence of writing.

Where the parties enter into a *contract* for the grant of a lease in the future, such contract must be made in writing in order to comply with Section 2(1) of the Law of Property (Miscellaneous) Provisions Act 1989.

If a tenant wishes to *assign* his tenancy, he must use a deed. The requirement of a deed applies to the assignment of any tenancy, even a weekly tenancy[64] which was created orally pursuant to Section 54(2) (see above).

Where a tenant has a right to occupy premises as a residence in consideration of a rent payable weekly, the landlord is obliged to provide a rent book (or other similar document) for use in respect of the premises.[65] A failure to comply with this requirement does not, however, excuse the tenant from paying the rent.[66]

Types of leases and tenancies

Leases (or tenancies) are categorised as being either specific (fixed term) or periodic. Periodic tenancies have already been mentioned earlier. Fixed-term tenancies may take a number of different forms:

Tenancy determinable with a life

It is possible to create a tenancy which is expressed to determine automatically with the life of the tenant or landlord, or some other person. By virtue of Section 149(6) of the Law of Property Act 1925, such a lease is converted into one for a fixed period of ninety years, if it is granted at a rent.

Concurrent lease (or lease of the reversion)

This arises where the landlord grants a lease to T1 and subsequently grants a lease to T2 for a term to commence before the expiry of the lease in favour of T1. So long as the leases are concurrent, the disposition in favour of T2 operates as a part assignment of the landlord's reversion entitling T2 to the rent reserved in the previous lease and the benefit of the covenants given by T1. If T1's lease is prematurely determined before T2's lease has expired, T2 will be entitled to possession of the premises. However, forfeiture (see later) by the landlord of T2's lease cannot affect T1, for although T1's lease is akin to a sub-tenancy, it is not derived out of T2's lease. It is possible for any number of concurrent leases to exist. Often such leases are created so as to give a lender of money security in relation to the property.

Perpetually renewable lease

If a lease contains an option to renew in favour of the tenant under which the new lease is to include, *inter alia*, the option to renew, a perpetually renewable lease will exist and, in accordance with Schedule 15 to the Law of Property Act 1922, will be converted into a 2,000-year term determinable by the tenant by ten days' notice expiring on any date upon which the original lease would have expired if it had not been renewed.[67] From the landlord's standpoint, the consequences of a perpetually renewable lease are very severe. With perpetually renewable leases, the original rent payable becomes the rent of the 2,000-year term!

Apart from specific and periodic tenancies, there are also a number of other types of tenancy:

Tenancy at will

A tenancy at will exists where land is occupied 'at the will' of the landlord (that is, the occupation of the land is carried on with the landlord's consent). However, because the permitted occupation is for an uncertain period, it is not a 'term of years absolute' and, hence, may be determined at any time by a demand for possession. If possession is demanded by the landlord, a tenant at will must leave immediately. It is interesting to note that Section 5 of the Protection from Eviction Act 1977 (which provides for the service of a notice to quit in respect of residential dwellings) does not apply to tenancies at will.[68] Such tenancies may arise expressly,[69] or by implication of law[70] from the conduct of the parties where, for example, a tenant holds over rent free at the end of the lease with the landlord's permission.

An express tenant at will is protected by the Rent Act legislation.[71]

Tenancy at sufferance

This denotes the relationship of owner and occupier where the tenant holds over on the expiry of his lease and the landlord has neither consented nor expressed objection. The legal effects of this relationship will depend on subsequent events. Thus, if the landlord requires the tenant to quit, the tenant becomes a trespasser. If, on the other hand, the landlord signifies his consent (for example, by a demand for rent), the tenant becomes a tenant at will and, if rent is paid with reference to a particular period, the tenant becomes a periodic tenant. Alternatively, if the landlord simply acquiesces, he will be statute barred from reclaiming possession if he fails to reassert his title within twelve years from the date that the tenant's possession ceased to be lawful.[72]

Tenancy by estoppel

The doctrine of estoppel precludes parties, who have induced others to rely upon their representations, from denying the truth of the facts represented. For the purposes of landlord and tenant law, this means that neither landlord nor tenant can question the validity of the lease granted, once possession has been taken up. Thus, even if the landlord is not the legitimate owner of the estate out of which the tenant is granted his lease, the tenant cannot deny any of his leasehold obligations to the landlord by arguing that the grant was not effectively made. Similarly, the landlord cannot set up his want of title as a ground for repudiating the lease.

Enforceability of covenants in leases

In relation to leases created *prior* to 1 January 1996, an original tenant is, as a matter of contract, legally liable to the landlord on all the covenants of the lease for the duration of the term, even where he subsequently assigns his interest. The original tenant's liability will also continue despite an assignment of the landlord's reversionary interest in the premises. With regard to leases (both residential and commercial) created *after* this date, the position is now governed by the Landlord and Tenant (Covenants) Act 1995. The basic principle under the Act is that, on assignment, a tenant will be automatically released from his liabilities and only the tenant for the time being will be liable to the landlord.

An important provision which applies to all leases, including those entered into before 1 January 1996, requires that if the landlord wishes to sue a former tenant for arrears of rent or service charges, he must give notice of his intention to do so within six months of the liability arising. Failure to do so means that the landlord loses the right to recover from the former tenant to the extent that the sums fell due more than six months prior to any notice served.

It is frequently overlooked that this rule, known as privity of contract, applies to landlords just as it does to tenants. Under the 1995 Act, if the landlord wishes to be released when he assigns the reversion, he must serve notice on the tenant giving details of the proposed assignment and indicating that he wishes to be released. The tenant then has four weeks within which to object to the proposed release. If he does not object within that period, then, on completion of the assignment, the landlord will be released. If there is an objection, the landlord can apply to the county court for a declaration that it is reasonable for him to be released.

Express covenants

Express covenants are obligations which arise as the result of an express agreement between the parties and are invariably set out in the lease or tenancy agreement itself. Examples include the tenant's covenant to pay rent,[73] the landlord's covenant for quiet enjoyment (discussed below), the tenant's covenant against assigning, sub-letting or parting with possession of the demised premises (also discussed below), the tenant's covenant against the making of alterations and improvements to the premises, the tenant's covenant restricting the user of the premises, the covenant to repair (also discussed below) and the covenant to insure.

Implied covenants

In the absence of an express covenant dealing with the point, the law will automatically imply certain covenants into the lease on the part of the landlord and the tenant. The most important implied covenants on the part of the landlord concern quiet enjoyment, non-derogation from grant and liability to repair the demised premises.

Landlord's implied covenants

Quiet enjoyment

As its name suggests, this covenant is designed to ensure that the tenant may peacefully enjoy his land without interruption from the landlord or persons claiming under him (for example, other tenants in the same building). Most commonly, a breach of the covenant arises when the landlord seeks forcibly to evict the tenant from the demised premises. In such circumstances, exemplary damages may be awarded for the unlawful eviction where the facts constitute the tort of trespass[74] or nuisance.[75] Exemplary damages are awarded not to compensate the tenant for any loss he may have suffered as a consequence of the unlawful eviction but simply 'to teach the landlord a lesson'.[76]

A landlord will also be liable under the covenant if he causes physical interference with the tenant's land, irrespective of whether he has actually intruded on to the demised premises.[77] Thus, the landlord's act of cutting off the tenant's gas and electricity supply to the premises from outside will constitute a breach even in the absence of any direct physical interference with the premises.[78]

In addition to claiming exemplary damages, the tenant may seek compensation for any damage to (or loss of) belongings and the cost of alternative accommodation while wrongfully kept out of the premises. Damages may also be awarded to compensate for inconvenience and discomfort but not, it seems, for injured feelings and mental distress.[79] Also, the provisions of Section 27(3) of the Housing Act 1988 impose a statutory liability to pay damages on a landlord who personally, or through his agents, has committed acts which amount to the offences of unlawful eviction or harassment under Section 1 of the Protection from Eviction Act 1977. This liability is incurred by acts occurring after 9 June 1988 and arises irrespective of whether or not there has been a conviction for the offence. By virtue of Section 27(6), the landlord has a complete defence either if the tenant is actually re-instated in the premises before the proceedings are finally disposed of or if a court makes an order re-instating the occupier. Section 28(1) provides that the measure of damages for the purposes of Section 27(3) is the difference between the value of the landlord's interest subject to the residential occupier's right, and its value in the absence of the occupier's right. On this basis, the tenant in *Tagro v. Cafane*[80] was awarded £31,000 damages for the loss of her accommodation.

Non-derogation from grant

A landlord is subject to an implied covenant not to derogate from his grant. Essentially, if a landlord lets land for a particular purpose, he must not do anything which prevents its being used for that purpose. The covenant differs from quiet enjoyment in three important respects. First, non-derogation only operates when the landlord lets part of his land and retains the other part – the covenant limits the use that can be made of the retained part. Second, if the tenant intends to use the premises for some special purpose, this must be made known to the landlord at the commencement of the letting in order to render the latter liable under the covenant.[81] Third, the tenant need only show that the landlord has rendered the premises less fit for the purpose for which they were let.[82]

Liability to repair

In relation to furnished lettings, there is at common law an implied condition that the premises will be fit for human habitation at the commencement of the tenancy. The following have been held to render premises unfit: (1) infestation with bugs; (2) defective drains; (3) condensation dampness; and (4) previous occupant suffering from a contagious disease. If the condition is not fulfilled on the day the tenancy commences, the tenant is entitled to treat the letting as discharged, quit the premises and sue for damages.

Where the essential means of access/common parts to units in a building in multiple occupation are retained in the landlord's control, then, unless the tenancy expressly deals with the matter, the landlord is impliedly obliged to maintain those means of access/common parts to a reasonable standard.[83] An obligation to repair may also be implied on the landlord in order to match a correlative obligation on the part of the tenant.[84]

There are also implied obligations to repair

imposed on the landlord by statute. Section 8 of the Landlord and Tenant Act 1985, which applies to lettings for human habitation at a rent not exceeding £80 per annum in Inner London or £52 per annum elsewhere, implies a condition on the part of the landlord that the premises will be fit for human habitation at the commencement of the letting and also an undertaking to keep them in that condition throughout the term of the tenancy. Fitness for human habitation is determined having regard to the condition of the house in respect of the following matters: (1) repair; (2) stability; (3) freedom from damp; (4) internal arrangement; (5) natural lighting; (6) ventilation; (7) water supply; (8) drainage and sanitary conveniences; and (9) facilities for storage, preparation and cooking of food and for the disposal of waste water. The very low rent limits imposed by Section 8(3) result in most lettings falling outside the section.

Section 11 of the Landlord and Tenant Act 1985, which applies to most residential lettings under seven years, implies a covenant on the part of the landlord to keep in repair the structure and exterior of the dwelling-house (including drains, gutters and external pipes) and to keep in repair and proper working order the installations in the dwelling-house for the supply of water, gas, electricity and for sanitation (including basins, sinks, baths and sanitary conveniences) and installations for space and water heating. Section 11 has been extended by Section 116 of the Housing Act 1988, but only in respect of tenancies entered into after 15 January 1989. Section 116 provides that the landlord will be obliged to keep in repair the structure and exterior of any part of the building in which the landlord has an estate or interest, provided that the disrepair is such as to affect the tenant's enjoyment of the dwelling-house or of any common parts which the tenant is entitled to use. The landlord's statutory obligation in relation to installations has been extended along similar lines in that he is obliged to keep in repair and proper working order an installation which, directly or indirectly, serves the dwelling-house and which either (1) forms part of any part of a building in which the landlord has an estate or interest, or (2) is owned by the landlord or is under his control. Here again, failure to repair or maintain in working order must affect the tenant's enjoyment of the dwelling-house (or any common parts which he is entitled to use).

Tenant's implied covenants

There are a number of covenants implied by law on the part of the tenant. These include the obligation to pay rent, not to commit waste and use the premises in a tenant-like manner. The latter obliges the tenant to keep proper care of the demised premises and 'do the little jobs about the place which a reasonable tenant would do'[85] (change the light bulbs, mend the fuse, unblock the sink, and so on).

Assigning and sub-letting

Whether or not a tenant can assign or sub-let his premises depends essentially on the form of the lease. There are five possible situations: (1) the lease may contain no restriction, in which case the tenant may freely assign or sub-let without obtaining the landlord's prior consent or licence; (2) the lease may contain an absolute covenant against assigning or sub-letting. A simple prohibition unqualified by any words requiring the consent or licence of the landlord entitles the latter to withhold his consent in all circumstances and to impose what conditions he likes. By way of statutory exception, a covenant purporting to restrict prospective assignees or sub-tenants on grounds of colour, race, ethnic or national origins or sex is deemed to be qualified;[86] (3) the lease contains a qualified covenant not to assign or sub-let without obtaining the consent or licence of the landlord. Such a qualified covenant brings into operation Section 19(1) of the Landlord and Tenant Act 1927 which provides that, notwithstanding any provision to the contrary, such covenant shall be deemed to be subject to the proviso that such licence or consent shall not be unreasonably withheld; (4) the lease may contain an express proviso that consent shall not be unreasonably withheld. In such a case, the tenant is in the same position as (3) above; and

(5) the lease contains a covenant by the tenant to offer a free surrender of the premises to the landlord before assigning or sub-letting. Such a covenant is not invalidated by Section 19(1) of the 1927 Act.[87]

The burden of proof lies on the landlord to show that consent has not been unreasonably withheld. Indeed, the Landlord and Tenant Act 1988 imposes a duty on the landlord to consent to a tenant's application to assign or sub-let unless he has good reason for not doing so. It is, in each case, a question of fact, depending upon all the circumstances, whether the landlord's consent is being unreasonably withheld. Thus, for example, it is reasonable for a landlord to refuse his consent to an assignment of the lease where the proposed assignee would, unlike the assignor, become entitled to purchase the freehold under the enfranchisement legislation[88] (see later).

The landlord must show reasonable grounds for any delay in communicating his decision whether or not to grant or refuse consent to a proposed assignment or sub-letting: Section 1(6) of the 1988 Act.[89]

Leasehold dilapidations

The meaning of repair

Most well-drafted leases will contain a covenant on the part of the landlord or tenant to repair the demised premises during the term of the lease. Three criteria have been established over the years for determining whether particular works constitute repair, as opposed to improvement, of the premises: (1) whether the works go to the whole or substantially the whole of the structure, or only to a subsidiary part; (2) whether the effect is to produce a building of a wholly different character from that which had been let; and (3) the cost of the works in relation to the previous value of the building and the effect on the value and life of the building.[90]

It was once thought that 'repair' did not include the remedying of an inherent defect in the design or construction of the premises. This view no longer represents the law and the question, in all cases, is one of fact and degree.[91] The essential issue is whether, having regard to all the circumstances of the case, the proposed remedial works can fairly be regarded as 'repair' in the context of the particular lease.[92]

It is important to bear in mind that disrepair connotes a deterioration from a former *better* condition, so that where there is no evidence before the court demonstrating that the condition of the property has in any way deteriorated since it was first built, no breach of the repairing covenant will be held to have occurred.[93]

Landlord's liability to repair

The landlord's liability under a repairing covenant in respect of the demised premises only arises when he has notice of the defect in question.[94]

If the landlord is obliged to repair the premises, it is common for the lease to require the tenant to pay a service charge, which represents a proportion of the landlord's maintenance and repair costs. Sections 18–30 of the Landlord and Tenant Act 1985 (as amended by the Landlord and Tenant Act 1987 and Housing Act 1996) impose important duties on landlords with respect to service charges.

As we have seen, in the absence of express provision, the landlord may be under an implied obligation arising by virtue of the common law or statute to repair the premises (see 'Landlord's implied covenants', pp. 245–7).

Also, a landlord who designs and/or builds premises owes a common law duty to take reasonable steps to ensure that the premises are reasonably safe and habitable.[95] Reference must also be made to Section 4 of the Defective Premises Act 1972, which imposes on a landlord, who has covenanted to repair, an obligation to the tenant (and third parties) to keep them 'reasonably safe from personal injury or from damage to their property' caused by defects in the state of the premises.

Finally, Sections 79–82 of the Environmental Protection Act 1990 contain provisions to deal with 'statutory nuisances' as defined in Section 79(1) of the 1990 Act. Section 79(1)(a) includes in the definition of a statutory nuisance 'any premises in such a state as to be prejudicial to health or a nuisance'. Section 80 entitles a local authority to serve an abatement notice on the person responsible for the nuisance and Section 82 empowers a magistrates' court, on a complaint made by a person on the ground that he is aggrieved by the existence of a nuisance, to make an order requiring, *inter alia*, the defendant to abate the nuisance within a time specified in the order and to execute any works necessary for that purpose. These provisions have been used successfully by tenants against local authorities in respect of council accommodation suffering from condensation dampness.[96]

Tenant's liability to repair

The tenant may be liable for the condition of the premises under a tenant's repairing covenant in the lease and under the tort of waste, in nuisance or under the Occupier's Liability Act 1957.

Remedies for breach

It will be convenient to consider separately the remedies of the landlord and the tenant.

Landlord's remedies

Apart from claiming damages for breach of covenant, the landlord may seek to forfeit the lease under a proviso for re-entry contained in the lease. In virtually all cases (other than non-payment of rent) a prerequisite for forfeiture is the service by the landlord of a notice under Section 146(1) of the Law of Property Act 1925. He must, in the notice, specify the particular breach complained of and, if the breach is capable of remedy, require the tenant to remedy the same within a reasonable time. In addition, where appropriate, the notice should refer to the landlord's claim for compensation.

Where the lease in question was granted for seven or more years and three years or more remain unexpired at the date of the Section 146 notice, the landlord's remedies of forfeiture and damages are limited by the provisions of the Leasehold Property (Repairs) Act 1938. Where the Act applies, the landlord cannot proceed without first serving a Section 146 notice, which must inform the tenant of his right to serve a counter-notice claiming the benefit of the 1938 Act. If the tenant does serve such a counter-notice within twenty-eight days, no further proceedings by action or otherwise (by initiating proceedings for possession or by physically re-entering upon the demised premises) can be taken by the landlord without leave of the court establishing a case on the balance of probabilities[97] that one or more of the five grounds set out in the Act have been fulfilled (see Section 1(5), grounds (a) to (e) of the 1938 Act). As an alternative to damages and/or forfeiture, the landlord may have the benefit of a clause in the lease enabling him to inspect the state and condition of the demised premises and serve notice on the tenant requiring him to execute necessary repairs. If he fails to do so, the landlord may carry out the work himself and recover the cost from the tenant. In these circumstances, an action by the landlord to recover the cost of repairs is, by reason of the express terms of the covenant, either a claim for a debt or rent due under the lease rather than a claim for damages for breach of covenant within the meaning of the 1938 Act. Accordingly, the landlord does not require leave under the Act to bring the debt or rent action against the tenant.[98]

It appears that a landlord cannot compel the tenant by mandatory injunction or specific performance to perform his repairing obligations.[99]

Tenant's remedies

Where the landlord is in breach of his repairing obligations (express or implied), the tenant has essentially five remedies available to him, namely: (1) damages for breach of covenant;[100] (2) specific performance;[101] (3) a set-off against rent; (4) termination of the contract of letting;[102] and (5) the appointment of a receiver or manager by the court.[103] In addition to the foregoing, the tenant may be able to invoke action by the local authority in extreme cases of disrepair. Where a local authority is satisfied that housing is unfit for human habitation, there are a number of possible courses of action which it can take under the Housing Act 1985, namely, serving a repair notice (under Section 189), making a closing order (under Section 264), making a demolition order (under Section 265), or declaring the area in which the building is situated to be a clearance area (in accordance with Section 289). There are also numerous provisions which confer on local authorities a range of powers which may be exercised in relation to housing which is unsatisfactory, although not unfit for human habitation. For example, in regard to premises which are in a state of disrepair, a local authority may serve a repair notice (under Section 190) or, with respect to houses in multiple occupation, it may require works to be carried out (under Section 352). If the living conditions in an area within its district are unsatisfactory, it may declare that area to be a renewal area (under Section 89 of the Local Government and Housing Act 1989).

Termination of leases

There are a number of different ways in which a lease or tenancy agreement may validly be brought to an end.

A landlord may, for example, be entitled to forfeit the lease under a proviso for re-entry (a forfeiture clause) contained in the lease for breach of covenant on the part of the tenant. In order for a landlord to bring about an effective forfeiture, he must take some positive step to signify to the tenant his intention of treating the lease as at an end as a consequence of the tenant's breach. This may be done either by actually (physically) re-entering on to the demised premises or by suing for possession. If the landlord adopts the former method, he must be careful not to infringe the provisions of Section 6 of the Criminal Law Act 1977, which prohibit the landlord from using or threatening violence for the purposes of securing entry on to the premises. Moreover, actual re-entry is not available to a landlord where the premises are let as a dwelling and while any person is lawfully residing therein.[104] In view of these difficulties, the more usual course in practice is for the landlord to initiate proceedings for possession where the service of the writ (or summons) will operate in law as a notional re-entry.

In cases other than non-payment of rent, the landlord cannot forfeit unless he has first served notice (under Section 146(1) of the Law of Property Act 1925) on the tenant specifying (1) the particular breach complained of; (2) if the breach is capable of remedy, requiring the tenant to remedy the breach; and (3) in any case, requiring the tenant to make compensation for the breach.

Instead of electing to forfeit, the landlord may decide to treat the lease as continuing and waive the forfeiture. The acceptance of rent, for example, with knowledge of the breach, will amount to a waiver.

Even if a ground of forfeiture exists (and has not been waived), it does not necessarily follow that the landlord will be successful in his claim to recover possession of the premises. Both equity and statute law have intervened so as to provide the tenant with the right to apply for relief from forfeiture of his lease.[105]

A tenancy may also be terminated by the acceptance of a repudiatory breach of its terms[106] and by frustration.[107] A fixed-term lease may also entitle the landlord or tenant (or both) to determine the term early (prior to its expiry date) by means of an express power to determine. An option to determine (sometimes referred to as a break-clause) must be exercised strictly within any time limits laid down for its exercise. Many leases entitle the landlord to break the term for redevelopment or reconstruction. In relation to a periodic tenancy, the common law has long accepted that such a tenancy may be determined by a notice to quit of appropriate length served by either landlord or tenant on the other party. The period of notice required (in the absence of statutory or express provision) is that which corresponds in time to the length of the particular periodic tenancy in question (for example, a weekly periodic tenancy requires a notice to quit of one week). However, the termination of a yearly tenancy provides a significant exception to this common law rule in so far as a tenancy from year to year requires six months' notice to quit. These common law rules must also be read in the light of statutory intervention. Thus, under Section 5 of the Protection from Eviction Act 1977, in relation to premises let as a dwelling, a notice to quit is not valid unless it is in writing, given not less than four weeks before the date on which is to take effect, and contains (in relation to a landlord's notice to quit) certain prescribed information regarding the tenant's legal rights.

A lease may also be brought to an end by surrender, which is the process by which a tenant gives up his leasehold estate to his immediate landlord. The lease is essentially destroyed by mutual agreement. A surrender may arise expressly (in which case a deed is required[108]) or by implication (by operation of law). A common example of the latter type of surrender is where the tenant abandons the premises and the landlord accepts his implied offer of a surrender by changing the locks and reletting the premises to a third party.[109] The insolvency of the tenant, whether an individual or a corporate body, provides a further possibility for the termination of the lease by virtue of disclaimer under Sections 172–182 of the Insolvency Act 1986.

Statutory protection of residential tenants

Residential tenants are given a large measure of security of tenure (and rent control) in respect of their premises. The two main statutes are the Rent Act 1977 and Part I of the Housing Act 1988, which apply to private sector tenants.[110] Public sector tenants (and licensees) are afforded protection under Part IV of the Housing Act 1985.

The Rent Act 1977

Where a dwelling-house has been let as a separate dwelling under a tenancy entered into prior to 15 January 1989, that tenancy, if falling within the Rent Act 1977, will be a protected tenancy. A tenancy of a dwelling-house granted on or after this date, however, is subject to Part I of the Housing Act 1988 and will be either an assured or an assured shorthold tenancy (see below).

A protected tenancy is defined as (1) a tenancy (as opposed to a licence) (2) of a dwelling-house[111] (3) which is let as a separate dwelling (or a part of a house being a part so let) (4) unless the tenancy falls within one or more of the exclusions. Excluded tenancies include (1) tenancies of a dwelling-house with a high rateable value or high rent (exceeding £25,000 per year); (2) tenancies at a low rent or no rent (the annual rent for the time being must exceed £1,000 in Greater London or £250 elsewhere); (3) tenancies where the rent includes *any* payment in respect of board[112] or any *substantial* payments in respect of attendance;[113] (4) lettings to students; (5) holiday lettings;[114] (6) tenancies granted by resident landlords;[115] (7) agricultural holdings; and (8) business tenancies.

When a protected tenancy ends, the tenant will automatically become a statutory tenant of the dwelling-house so long as he remains in residence. As a statutory tenant, he will be entitled to the benefit of all the terms and conditions of the original tenancy. Although a statutory tenancy is personal to the tenant (he has no estate or interest in the land but merely a statutory privilege to remain in occupation), the 1977 Act (as amended by the Housing Act 1980) does permit certain rights of succession. Essentially, a succession may be claimed by a surviving spouse or, if there is none, a member of the deceased tenant's family subject to certain residence requirements. In relation to post-1988 deaths, a cohabitee may claim to succeed as a surviving spouse.[116]

A landlord who wishes to claim possession against a protected or statutory tenant must establish (1) the effective termination of the contractual tenancy (for example, by expiry of time, notice to quit, or forfeiture); and (2) that one or more of the statutory grounds for possession apply (see Schedule 15 to the 1977 Act as amended by Section 66 and Schedule 7 of the Housing Act 1980). It follows that a common law forfeiture of a fixed-term tenancy will only destroy the protected (contractual) tenancy and the tenant will remain a statutory tenant of the premises. However, security under the 1977 Act (and, therefore, the need to establish a statutory ground) is removed entirely where the occupier is guilty of the offence of overcrowding under Part X of the Housing Act 1985.[117]

A landlord relying on any of the so-called 'discretionary grounds' for possession (see below) must, in addition to establishing the ground in question, prove that it is reasonable to make the order for possession. In considering reasonableness, the court may consider the widest range of circumstances (an example would be the fact that the landlord has committed perjury).

The discretionary grounds, which are contained in Schedule 15, Parts I, III and IV, are as follows: unpaid rent or obligation of tenancy broken (case 1); nuisance or conviction for illegal/immoral user (case 2); deterioration by waste or neglect (case 3); ill-treatment of furniture (case 4); notice to quit given by tenant (case 5); assignment or sub-letting of entire premises without consent (case 6); new employee in place of old (case 8); premises reasonably required as a residence for the landlord or his family (case 9); and sub-letting at an excessive rent (case 10). The requirement of reasonableness also applies where possession is sought on the ground of the provision of suitable alternative accommodation: Part IV of Schedule 15.

The so-called 'mandatory' grounds for possession, which are not subject to any overriding consideration of reasonableness, are contained in Schedule 15, Part II. A landlord who establishes any of the mandatory grounds can claim possession as of right provided that (1) the dwelling-house was let on a protected tenancy, and (2) not later than 'the relevant date' the landlord gave the tenant written notice that possession might be recovered under the ground in question. For most tenancies, the relevant date is the commencement of the tenancy. The mandatory grounds are: premises required by former owner-occupier, etc. (case 11); premises required for retirement home, etc. (case 12); holiday homes let out of season (case 13); student accommodation let out of term-time (case 14); premises required for minister of religion (case 15); premises required for agricultural employee (cases 16, 17 and 18); termination of protected shorthold tenancy (case 19); and letting by a member of the armed forces (case 20).

In addition to the protected and statutory tenancy, the Rent Act 1977 also provides for restricted contract occupiers.[118] Examples of contracts within this category include furnished contractual licences, tenancies or licences where the rent includes *insubstantial* payments in respect of board,[119] tenancies or licences where the rent includes substantial payments in respect of attendance,[120] furnished tenancies at a low rent, tenancies (regardless of furniture or services) which are not protected tenancies solely because the tenant shares some essential living accommodation with his landlord,[121] and tenancies subject to the resident landlord exclusion.[122] A restricted contract occupier has no security of tenure other than: (1) protection from harassment and unlawful eviction under Section 1 of the Protection from Eviction Act 1977; and (2) the right to apply to a rent tribunal for deferment of a notice to quit for up to six months at a time in respect of a tenancy granted before 28 November 1980 or, alternatively, to apply to the court for deferment of the order for possession for up to three months subject to the imposition of conditions with regard to payment by the tenant of arrears of rent (if any) and rent or mesne profits, in respect of periodic or fixed-term contracts created on or after 28 November 1980.[123]

As well as security of tenure, tenants falling within the 1977 Act have the benefit of rent control. The rent payable for a protected tenancy is regulated by Part VI of the 1977 Act, which provides for the registration of fair rents. An occupier with a restricted contract is entitled to apply to a rent tribunal in order to register a reasonable rent.[124]

Part I of the Housing Act 1988

The assured tenancy scheme (assured and assured shorthold tenancy) replaced that of the Rent Act 1977 in respect of tenancies created on or after 15 January 1989. One of the basic aims of the new scheme was to enable landlords to grant tenancies to residential tenants at a market rent with the ability more easily to regain possession of the premises at the expiry of the lease than is possible in the case of a Rent Act protected or statutory tenant. Section 1 of the 1988 Act defines an assured tenancy as 'a tenancy under which a dwelling-house is let as a separate dwelling'. To this extent, Part I reiterates the terminology of the Rent Act 1977. However, the tenancy will only qualify as an assured tenancy if and so long as the tenant is: (1) an individual and (2) occupies the dwelling-house as his only or principal home. If there is more than one tenant, at least one of them must occupy the premises as his only or principal home. Part I, therefore, requires more than just simple residence. There are a number of exclusions (broadly similar to those under the Rent Act 1977) in which a tenancy will not qualify as an assured tenancy.[125] One significant difference is that a tenancy granted by a resident landlord falls wholly outside Part I.

So far as security of tenure is concerned, the Act treats periodic and fixed-term tenancies in different ways to ensure that both categories of tenant have the right to remain in the premises at the end of the contractual term. In the case of an assured fixed term, Part I provides that the tenancy will not end automatically by expiry of time but instead a new tenancy will come into existence known as a statutory periodic tenancy on substantially the same terms as the original contractual tenancy. In the case of an assured periodic term, the 1988 Act provides that a notice to quit given by the landlord is ineffective with the result that the original tenancy will continue on its terms. A landlord must obtain a court order for possession on the grounds specified in the 1988 Act if he wants to evict the tenant. In the case of a periodic assured tenancy, a landlord may bring proceedings at any time, but in the case of a fixed-term assured tenancy, proceedings may only be brought to evict when the lease has expired (unless there is a power to forfeit the lease earlier). As a preliminary to seeking possession, the landlord must serve a notice (under Section 8 of the 1988 Act) setting out the grounds for possession under which the landlord intends to proceed. The notice must inform the tenant that proceedings will not start before a certain date, which in most cases will be two weeks from the service of the notice. In some cases (where the landlord intends to proceed on grounds 1, 2, 5–7, 9 or 16) the time will be two months from the service of the notice. The notice must also inform the tenant that the proceedings will not begin any later than twelve months from the service of the notice. These notice requirements may be waived by the court if it considers it 'just and equitable' to do so (Section (8)(1)(b)). As under the Rent Act 1977, the grounds for possession are divided between those where the court has a discretion as to whether to make a possession order and those where its grant is mandatory. The grounds are similar to those under the 1977 Act, but there are also some additional grounds which are entirely new (for example, two more grounds concerned with non-payment of rent and a new re-development ground).

In addition to the assured tenancy, Part I of the 1988 Act has also created the assured shorthold tenancy, which enables landlords to regain possession of the premises as of right at the end of a fixed term without the need to establish any grounds. Since an assured shorthold tenancy is a species of assured tenancy, all the qualifying criteria for the creation of an assured tenancy also apply to assured shortholds. In addition (prior to the enactment of the Housing Act 1996), the tenancy had to be for a fixed term with a minimum duration of at least six months and the landlord must have served a notice on the proposed tenant before the tenancy began, informing the tenant that the tenancy was to be a shorthold. The court had no power to waive this requirement. Thus, if the requisite notice was not served, the tenancy would be an assured tenancy enjoying the full benefit of security under Part I. The 1996 Act has now changed the law by providing (in Section 96) that an assured tenancy post-dating the commencement date of the section (28 February 1997) is to be automatically treated as an assured shorthold, unless it falls within any of the stated exceptions. The exceptions are narrowly defined to maximise the scope of shortholds and contain obvious exclusions such as, for example, where it is explicitly stated (either before or during the assured tenancy) that it is not to be an assured shorthold. The upshot of these amendments is that a landlord is no longer required to serve any notice on the tenant, nor is there any requirement that the tenancy is of a fixed term. However, the landlord is not entitled to an order for possession until at least six months has elapsed from the grant of the tenancy (Section 99). Under Section 97 of the 1996 Act, there is an obligation on the landlord to provide, if requested by the tenant, a statement of the terms of any post-Housing Act 1996 tenancy.

During the term of an assured shorthold, the landlord may be entitled to evict using the grounds for possession available against any fixed-term assured tenancy. Alternatively, he must give the assured shorthold tenant at least two months' notice (under Section 21) of the fact that he requires possession of the premises. The notice must specify a date of termination which cannot be earlier than the end of the fixed term. If the tenant holds over after a fixed term has expired, he will do so as a statutory periodic tenant whose tenancy may also be brought to an end by two months' notice at any time (under Section 21), but the date for possession must correspond to the date on which notice to quit could have been stated to expire at common law.

The assured shorthold tenancy has become a popular mechanism for avoiding residential security by landlords.

Part IV of the Housing Act 1985

By Section 79(1) of the Housing Act 1985, a tenancy under which a dwelling-house is let as a separate dwelling is a secure tenancy at any time when both the landlord and tenant conditions are satisfied. The landlord condition is that the interest of the landlord belongs to one of the authorities or bodies listed in Section 80 of the 1985 Act (a local authority, a new town corporation, an urban development corporation, the Development Board for Rural Wales, the Housing Corporation, a housing trust which is a charity, and certain housing associations and housing co-operatives). The tenant condition is that the tenant is an individual and occupies the dwelling-house as his only or principal home or, where the tenancy is a joint tenancy, that each of the joint tenants is an individual and at least one of them occupies the dwelling-house as his only or principal home (Section 81[126]). The provisions of Part IV apply also in relation to a licence to occupy a dwelling-house (whether or not granted for consideration) as they apply in relation to a tenancy (Section 79(3)). However, for the licence to be secure, there must be exclusive possession of the premises.[127]

A landlord can only regain possession of a secure tenancy by a court order on one of a number of statutory grounds listed in Schedule 2 to the 1985 Act.[128] Interestingly, the Housing Act 1996 has brought in a new ground (ground 2A) for possession based upon domestic violence (see Section 145 of the 1996 Act).

If the secure tenancy is a fixed-term tenancy, a periodic tenancy will arise automatically when it ends, unless a further secure tenancy is granted. The procedure for terminating a secure tenancy commences with a termination notice (in the prescribed statutory form) served by the landlord specifying the ground of possession relied on.

Where the tenant dies, the tenant's spouse or another member of the tenant's family may be entitled to succeed to the tenancy (Section 87).[129] Only one right of succession is permitted. Under Part V of the 1985 Act,[130] a secure tenant has the right to buy his dwelling-house from the landlord on favourable terms coupled with a right to a mortgage (Section 118).

Part V of the Housing Act 1996 has introduced a new type of tenancy called the 'introductory tenancy' for local authorities (and housing action trusts) that decide to put their new tenants 'on probation' for twelve months. Where a local housing authority (or housing action trust) elects to operate an introductory tenancy regime, no periodic tenancy (or licence) of premises entered into by the authority (or trust) will be secure. Instead, such a tenancy will be an introductory tenancy, provided the tenancy would otherwise be a secure tenancy. The need for the tenancy to otherwise be a secure tenancy means that the landlord and tenant conditions (see earlier) must be satisfied and the tenancy must not come within any of the exceptions to secure tenancy status. The tenancy remains an introductory tenancy until the end of the one-year 'trial period', when it will become either secure or an ordinary contractual tenancy. If the landlord wishes to bring proceedings for possession against an introductory tenant, it must serve a notice on the tenant. The tenant then has fourteen days to request the landlord to review the decision to seek possession. After the review has been completed, the landlord is obliged to notify the tenant of its decision, and if the decision to evict stands, then the landlord must give reasons for the decision.

The Housing Act 1996 has also given local authorities new powers to obtain injunctions against members of the public who use or threaten violence against people living in, or visiting, their housing stock (Part V, Chapter III). The court may attach a power of arrest to an injunction which it intends to grant (Section 152(6)).

Leasehold enfranchisement

Section 1 of the Leasehold Reform Act 1967 confers on a tenant of a leasehold *house*, occupying the house as his residence, a right to acquire on fair terms the freehold or an extended lease (that is, for fifty years) of the house. In order to qualify for enfranchisement, the tenancy must be a long tenancy (exceeding twenty-one years) at a low rent (as defined in Section 3(1)). It should be noted that Chapter III of Part I of the Leasehold Reform, Housing and Urban Development Act 1993 makes significant changes to the 1967 Act in relation to the qualifying conditions for enfranchisement and the basis of valuation for price determination. (See also, Part III, Chapter III of the Housing Act 1996.)

Part I of the 1993 Act also gives most owners of long leases of *flats* a right either to collective enfranchisement (a collective right to buy the freehold of a block of flats) or an individual right to acquire a new ninety-year lease.[131]

Moreover, Part I of the Landlord and Tenant Act 1987 confers on qualifying tenants of residential flats a right of first refusal where the landlord disposes of his interest in the premises. Under Part III of the 1987 Act, tenants of residential flats held on long leases (for a term exceeding twenty-one years) also have the right to acquire compulsorily the landlord's interest. This procedure, however, only applies where the landlord is in unremediable breach of his covenant to repair.

Homelessness

Homelessness has been the subject of increasing public concern, and the statistics available indicate that the number of people who are homeless in the United Kingdom has risen considerably in recent years.

The Housing (Homeless Persons) Act 1977 introduced a new statutory scheme which imposed on local housing authorities an obligation to assist and secure accommodation for the homeless. The relevant statutory provisions are now contained in Part III of the Housing Act 1985, as amended by Part VII of the Housing Act 1996.

Essentially, a person is 'homeless' if he has no accommodation in England, Wales, Scotland or elsewhere (that is, outside the United Kingdom). A person is treated as having no accommodation if there is no accommodation which he is entitled to occupy by virtue of an interest in it or an order of a court, or has a licence to occupy, or occupies as a residence by virtue of any enactment or rule of law giving him the right to remain in occupation (or restricting the right of another person to recover possession) (Section 58 of the 1985 Act).

A number of persons are designated as having a 'priority need' for accommodation: (1) pregnant women; (2) a person with dependent children; (3) a person who is vulnerable as a result of old age, mental illness or handicap or physical disability; and (4) a person who is homeless or threatened with homelessness as a result of an emergency (such as a flood, fire or other disaster) (Section 59 of the 1985 Act).

A person becomes intentionally homeless if he

deliberately does or fails to do anything in consequence of which he ceases to occupy accommodation which is available for his occupation and which it would have been reasonable for him to continue to occupy (Section 60(1) of the 1985 Act). The overwhelming majority of homelessness cases have involved the issue of intentionality. For a person to be intentionally homeless, the homelessness must result from a deliberate (voluntary) act. Thus, to fail deliberately to pay the rent with the result that the landlord takes possession of the premises is intentional homelessness. By contrast, where a family loses accommodation on account of the woman becoming pregnant, this does not amount to intentional homelessness.[132] Equally, a woman who leaves her accommodation because of domestic violence is, generally speaking, not to be regarded as intentionally homeless.[133] Section 177 of the Housing Act 1996 now expressly provides that it is not reasonable for a person to continue to occupy accommodation if it is probable that this will lead to domestic violence against him (or her), or against (1) a person who normally resides with him as a member of his family, or (2) any other person who might reasonably be expected to reside with him. In this context, the expression 'domestic violence' means violence from a person with whom he is associated, or *threats* of violence from such a person which are likely to be carried out (Section 177(1)).

The local authority must also consider whether it would have been *reasonable* for the applicant to continue to occupy the accommodation which the applicant has given up. It has been held that a local authority should not confine itself to consideration of housing circumstances, but should have regard to all the reasons for leaving the accommodation which the applicant has ceased to occupy.[134]

The local authorities' duties in relation to homeless persons are set out in Sections 62–69 of the 1985 Act. The authority does not have a duty to provide accommodation for everyone who is homeless. However, the provisions do require local authorities to make enquiries in any case where they have reason to believe that the applicant is homeless (or threatened with homelessness) (Section 62). Then, depending on the circumstances, it may have recourse to a number of different courses of action. These may range from the provision of advice and assistance to the securing of permanent accommodation for the applicant (especially if he or she has a priority need).

NOTES

1 For a good introduction to legal studies, see, Mainly for Students, [1984] 272 E.G. 894. See also, 'Law Exam Techniques', Mainly for Students, [1993] E.G. 9313, 103.

2 See further, 'Common Law and Equity', (Mainly for Students), [1989] E.G. 8907, 87.

3 See further, 'Injunctions', (Mainly for Students), [1990] E.G. 9017, 87.

4 See further, pp.232–3, below.

5 See further, p.250, below.

6 *Pepper (Inspector of Taxes) v. Hart* [1993] 1 All E.R. 42, (H.L.) and *Three Rivers District Council v. Governor and Company of the Bank of England (No. 2)*, *The Times*, January 8, 1996.

7 See further, 'Secondary Legislation: Definitions', (Mainly for Students), [1996] E.G. 9603, 118.

8 See, 'Law Reports and Law Reporting', M. Pawlowski, (1986) *The Valuer*, 132.

9 See further, 'European Courts and English Law', (Mainly for Students), [1990] E.G. 9033, 55 and 'EC Law and the English Legal System', (Mainly for Students), [1990] E.G. 9045, 47.

10 See further, 'Civil Proceedings – Terminology', (Mainly for Students), [1996] E.G. 9607, 135 and 'Civil Action – Pleadings', (Mainly for Students), [1990] E.G. 9009, 83.

11 See e.g., *The Bar on Trial*, ed. R. Hazell (1978), Quartet Books, pp.169–178 and 'Solicitors, Barristers, or Just Plain Lawyers', G. Bindman, *New Law Journal*, December 7, 1990, 1712.

12 See, the Congenital Disabilities (Civil Liability) Act 1976, which makes provision for civil liability in the case of a child being born disabled in consequence of some person's fault.

13 See further, 'Private Companies', (Mainly for Students), [1991] E.G. 9149, 86.

14 See further, 'Partnerships', (Mainly for Students), [1992] E.G. 9211, 111.

15 See, *Pharmaceutical Society of GB v. Boots Cash Chemists (Southern) Ltd* [1953] 1 Q.B. 401, (C.A.).

16 [1893] 1 Q.B. 256, (C.A.).

17 *Adams v. Lindsell* (1818) 1 B. & Ald. 681; *Household Fire Insurance Co Ltd v. Grant* (1879) 4 Ex. D. 216;

Entores Ltd v. Miles Far East Corporation [1955] 2 Q.B. 327, (C.A.) and *Byrne v. Van Tienhoven* (1880) 5 C.P.D. 344.

18 See, *Central London Property Trust Co Ltd v. High Trees House* [1947] 1 K.B. 130.

19 Section 2 of the Law of Property (Miscellaneous Provisions) Act 1989.

20 (1864) 2 H. & C. 906.

21 *Saunders v. Anglia Building Society* [1971] A.C. 1005, (H.L.); *Avon Finance Co Ltd v. Bridger* [1985] 2 All E.R. 281, (C.A.) and *Norwich and Peterborough Building Society v. Steed (No. 2)* [1993] 1 All E.R. 330, (C.A.).

22 *Phillips v. Brooks* [1919] 2 K.B. 243; *Lewis v. Averay* [1972] 1 Q.B. 198, (C.A.); *Ingram v. Little* [1961] 1 Q.B. 31, (C.A.); *Cundy v. Lindsay* (1878) 3 App. Cas. 459, (H.L.) and *Citibank NA v. Brown Shipley & Co Ltd* [1991] 2 All E.R. 690.

23 See, Section 2(1) of the Misrepresentation Act 1967.

24 See e.g., *Hong Kong Fir Shipping Co Ltd v. Kawasaki Kisen Kaisha Ltd* [1962] 2 Q.B. 26, (C.A.).

25 See e.g., *Liverpool City Council v. Irwin* [1977] A.C. 239, (H.L.), referred to at p.246, below.

26 See, p.247, below.

27 See e.g., *Jarvis v. Swan Tours Ltd* [1973] 2 Q.B. 233, (C.A.) and *Jackson v. Horizon Holidays Ltd* [1975] 1 W.L.R. 1468, (C.A.).

28 (1854) 9 Ex. 341. See also, *Victoria Laundry (Windsor) Ltd v. Newman Industries Ltd* [1949] 2 K.B. 528, (C.A.).

29 See e.g., *Planche v. Colburn* (1831) 5 C. & P. 58, where the plaintiff was commissioned by a publisher to write a book for £100. After he had done the necessary research and written part of the book, the publisher repudiated the contract. It was held that the plaintiff could recover £50 on a *quantum meruit* (i.e., for the value of the work already done on the book).

30 [1912] 1 K.B. 574.

31 See further, p.250, below.

32 (1863) 3 B. & S. 826.

33 See, the Law Reform (Frustrated Contracts) Act 1943.

34 [1932] A.C. 562, (H.L.).

35 See e.g., *Hedley Byrne and Co Ltd v. Heller and Partners Ltd* [1964] A.C. 463, (H.L.), (firm of advertising agents) and *Smith v. Eric S. Bush* [1990] 1 A.C. 831, (H.L.) (a surveyor employed by a building society).

36 *Rondel v. Worsley* [1969] 1 A.C. 191, (H.L.).

37 *Ibid.*, [1967] 1 Q.B. 443, 501–504, (C.A.), *per* Lord Denning M.R.

38 *Caparo Industries plc v. Dickman* [1990] 2 A.C. 605, (H.L.) and *Murphy v. Brentwood District Council* [1991] 1 A.C. 398, (H.L.).

39 See e.g., *Hedley Byrne and Co Ltd v. Heller and Partners Ltd* [1964] A.C. 465, (H.L.).

40 *Overseas Tankship (UK) Ltd v. Morts Dock and Engineering Co Ltd, The Wagon Mound* [1961] A.C. 388, (H.L.).

41 See, *Froom v. Butcher* [1976] 1 Q.B. 286, (C.A.).

42 See, Torts (Interference with Goods) Act 1977.

43 See, *Arthur v. Anker* (1995) 146 N.L.J. 86.

44 See further, 'Occupier's Liability', (Mainly for Students), [1986] 280 E.G. 903 and 1155; 'Occupier's Liability', (Mainly for Students), [1985] 273 E.G. 87 and 'All Fall Down – Occupier's Liability Round-up', (Legal Notes), [1990] E.G. 9021, 171.

45 See further, 'Occupiers' Liability Act 1984', C.M. Brand and D.W. Williams, (1984) 270 E.G. 394.

46 See further, 'Occupiers' Liability and Notices', J. Murdoch, [1984] 271 E.G. 170.

47 B. Mitchell, *Landlord and Tenant Law*, (1987), BSP Professional Books, Preface.

48 (1868) L.R. 3 H.L. 330, (H.L.), which states that, if a person for his own purposes brings on to his land and keeps there anything likely to do mischief if it escapes, he must keep it in at his peril, and if it escapes he is *prima facie* liable for all damage which is the natural consequence of its escape. Things 'escaping' include gas, explosions, chemicals, electricity, etc.

49 *Rye v. Rye* [1962] A.C. 496, (H.L.).

50 Section 72(2), (3) and (4) of the Law of Property Act 1925.

51 See, Section 205(1)(ix) of the Law of Property Act 1925, for the definition of land.

52 See e.g., *Milmo v. Carreras* [1946] K.B. 306, (C.A.).

53 *Lace v. Chantler* [1944] K.B. 368, (C.A.). The Validation of War Time Leases Act 1944 was passed to reverse the actual decision in this case.

54 *Prudential Assurance Co Ltd v. London Residuary Body* [1992] 3 W.L.R. 279, (H.L.).

55 *Centaploy Ltd v. Matlodge Ltd* [1974] Ch. 1.

56 *Street v. Mountford* [1985] A.C. 809, (H.L.). See also, *A.G. Securities v. Vaughan/Antoniades v. Villiers* [1990] 1 A.C. 417, (H.L.); *Aslan v. Murphy (No. 1)/Aslan v. Murphy (No. 2)/Duke v. Wynne* [1990] 1 W.L.R. 766, dealing with the problem of multiple flat-sharing arrangements.

57 See e.g., *Marchant v. Charters* [1977] 1 W.L.R. 1181, (C.A.).

58 *Booker v. Palmer* [1942] 2 All E.R. 674, (C.A.).

59 *Norris v. Checksfield* [1991] 1 W.L.R. 1241, (C.A.).

60 *Sharp v. McArthur and Sharp* [1987] 19 H.L.R. 364, (C.A.). Contrast, *Bretherton v. Paton* [1986] 1 E.G.L.R. 172, (C.A.).

61 *Family Housing Association v. Jones* [1990] 1 W.L.R. 779, (C.A.).

62 *Ashburn Anstalt v. Arnold* [1989] Ch. 1, (C.A.), *per* Fox L.J.

63 *Norris v. Checksfield* [1991] 1 W.L.R. 1241, (C.A.). See also, *Torbett v. Faulkner* [1952] 2 T.L.R. 659, (C.A.).

64 See, *Crago v. Julian* [1992] 1 W.L.R. 372, (C.A.).

65 Sections 4 and 5 of the Landlord and Tenant Act 1985.
66 *Shaw v. Groom* [1970] 2 Q.B. 504, (C.A.).
67 See e.g., *Caerphilly Concrete Products Ltd v. Owen* [1972] 1 W.L.R. 372, (C.A.), *per* Lord Russell.
68 *Crane v. Morris* [1965] 3 All E.R. 77, (C.A.).
69 *Manfield & Sons Ltd v. Botchin* [1970] 2 Q.B. 612; *Hagee (London) Ltd v. A.B. Erikson and Larson* [1976] Q.B. 209.
70 *Wheeler v. Mercer* [1957] A.C. 416, (H.L.).
71 *Chamberlain v. Farr* [1942] 2 All E.R. 567, (C.A.).
72 See, Section 15 of the Limitation Act 1980.
73 The rent payable by a tenant to his landlord, under an express covenant to pay rent, must be calculable with certainty at such time as when payment becomes due: *Greater London Council v. Connolly* [1970] 2 Q.B. 100, (C.A.).
74 *Drane v. Evangelou* [1978] 1 W.L.R. 455, (C.A.).
75 *Guppy's (Bridport) Ltd v. Brookling* [1984] 269 E.G. 846, 942, (C.A.), where two tenants were forced to leave their premises due to the extent of disruption caused by the landlords' building works.
76 *Ramdath v. Oswald Daley* [1993] 20 E.G. 123, (C.A.).
77 See e.g., *Owen v. Gadd* [1956] 2 Q.B. 99, (C.A.) and *Kenny v. Preen* [1963] 1 Q.B. 499, (C.A.).
78 See, *Perera v. Vandiyar* [1953] 1 W.L.R. 672.
79 See, *Branchett v. Beaney, Branchett v. Swale Borough Council* [1992] 3 All E.R. 910.
80 [1991] 1 W.L.R. 378, (C.A.).
81 *Robinson v. Kilvert* (1889) 41 Ch. 88, (C.A.), where the tenant used the premises for the storage of paper which was peculiarly susceptible to heat. This was not made known to the landlord at the time of the letting, who commenced a manufacture in the cellar below involving a heating apparatus which damaged the paper.
82 *Harmer v. Jumbil (Nigeria) Tin Areas Ltd* [1921] 1 Ch. 200, (C.A.) and *Aldin v. Latimer Clark, Muirhead & Co* [1894] 2 Ch. 437.
83 *Liverpool City Council v. Irwin* [1977] A.C. 239, (H.L.).
84 *Barrett v. Lounova* (1982) Ltd [1990] 1 Q.B. 348, (C.A.).
85 *Warren v. Keen* [1954] 1 Q.B. 15, (C.A.), *per* Denning L.J.
86 Section 24 of the Race Relations Act 1976 and Section 31 of the Sex Discrimination Act 1975.
87 *Bocardo SA v. S. & M. Hotels Ltd* [1980] 1 W.L.R. 17, (C.A.).
88 *Norfolk Capital Group Ltd v. Kitway Ltd* [1977] Q.B. 506, (C.A.) and *Bickel v. Duke of Westminster* [1977] Q.B. 517, (C.A.).
89 See also, *Midland Bank plc v. Chart Enterprises Inc* [1990] 44 E.G. 68.
90 *McDougall v. Easington District Council* [1989] 25 E.G. 104.
91 *Brew Bros Ltd v. Snax (Ross) Ltd* [1970] 1 Q.B. 612 and *Ravenseft Properties Ltd v. Davstone (Holdings) Ltd* [1980] Q.B. 12.
92 *Holding & Management Ltd v. Property Holding & Investment Trust plc* [1990] 05 E.G. 75, (C.A.).
93 *Quick v. Taff-Ely Borough Council* [1986] Q.B. 809; *Post Office v. Aquarius Properties Ltd* [1987] 1 All E.R. 1055 and *Stent v. Monmouth District Council* [1987] 282 E.G. 705.
94 *O'Brien v. Robinson* [1973] A.C. 912, (H.L.); *Sheldon v. West Bromwich Corporation* (1973) 25 P. & C.R. 360, (C.A.); *Dinefwr Borough Council v. Jones* [1987] 284 E.G. 58, (C.A.) and *British Telecommunications plc v. Sun Life Assurance Society plc* [1995] 4 All E.R. 44, (C.A.).
95 *Rimmer v. Liverpool City Council* [1985] Q.B. 1 and *McNerny v. Lambeth London Borough Council* [1989] 21 H.L.R. 188, (C.A.).
96 See e.g., *Greater London Council v. London Borough of Tower Hamlets* [1983] 15 H.L.R. 54 and *Dover District Council v. Farrar* [1980] 2 H.L.R. 35. The court has jurisdiction to make a compensation order under Section 35 of the Powers of Criminal Courts Act 1973 on the making of a nuisance order under Section 82 of the Environmental Protection Act 1990: *Herbert v. Lambeth London Borough Council* [1993] 90 L.G.R. 310.
97 *Associated British Ports v. C.H. Bailey plc* [1990] 2 A.C. 703, (H.L.)
98 *Colchester Estates (Cardiff) v. Carlton Industries plc* [1986] Ch. 80 and *Jervis v. Harris* [1996] 1 All E.R. 303, (C.A.).
99 *Hill v. Barclay* (1810) 16 Ves. Jr. 402 and *Regional Properties v. City of London Real Property Co* [1980] 257 E.G. 64, 66, *per* Oliver J.
100 See, *Calabar Properties Ltd v. Stitcher* [1984] 1 W.L.R. 287, (C.A.).
101 See, Section 17(1) of the Landlord and Tenant Act 1985 and *Jeune v. Queens Cross Properties Ltd* [1974] Ch. 97.
102 *Hussein v. Mehlman* [1992] 32 E.G. 59.
103 See. Section 36 of the Supreme Court Act 1981 and Part II of the Landlord and Tenant Act 1987, as amended by the Housing Act 1996.
104 Section 2 of the Protection from Eviction Act 1977.
105 See generally, M. Pawlowski, *The Forfeiture of Leases* (Sweet & Maxwell, 1993), ch. 1.
106 See, *Hussein v. Mehlman* [1992] 32 E.G. 59. See generally, M. Pawlowski, 'Acceptance of Repudiatory Breach in Leases', [1995] Conv. 379.
107 *National Carriers Ltd v. Panalpina (Northern) Ltd* [1981] A.C. 675, (H.L.).
108 In accordance with Section 52(1) of the Law of Property Act 1925.
109 See e.g., *R. v. London Borough of Croydon, ex p. Toth*

[1986] 18 H.L.R. 493. The abandonment must, however, be of a permanent and not temporary nature.

110 See also, Part I of the Landlord and Tenant Act 1954 and Section 186 and Schedule 10 of the Local Government and Housing Act 1989, dealing with long residential tenancies at a low rent.

111 The premises will comprise a 'dwelling-house' if they are let for the purpose of the tenant living in them as his home: *Metropolitan Properties Co (FGC) Ltd v. Barder* [1968] 1 W.L.R. 286, (C.A.). The purpose of the letting must be full residential use as one dwelling, which includes all the major activities of life, particularly sleeping, cooking and washing: *Curl v. Angelo* [1948] 2 All E.R. 189, (C.A.).

112 *Otter v. Norman* [1989] A.C. 129, (H.L.), where a continental breakfast was held to constitute board for the purpose of the exclusion.

113 *Nelson Developments Ltd v. Taboada* [1992] 34 E.G. 72, (C.A.), involving *inter alia* daily room cleaning, full weekly laundry and refuse removal.

114 *Buchmann v. May* [1978] 2 All E.R. 993, (C.A.) and *R. v. Rent Officer for London Borough of Camden, ex p. Plant* [1980] 257 E.G. 713.

115 Section 12 of the Rent Act 1977. See also, *Palmer v. McNamara* [1990] 23 H.L.R. 168, (C.A.) and *Jackson v. Pekic* [1989] 47 E.G. 141, (C.A.).

116 Schedule 4, paragraph 2 of the Housing Act 1988.

117 See, Section 101 of the Rent Act 1977 and *Zbytniewski v. Broughton* [1956] 2 Q.B. 673, (C.A.).

118 See, Section 19 of the Rent Act 1977.

119 Such tenancies or licences, although excluded from protection under Section 7 of the 1977 Act, may still fall within the definition of a restricted contract. If *substantial* board is provided by the landlord, the tenancy or licence will fall outside the Rent Act 1977 altogether.

120 Such tenancies or licences, although excluded from protection under Section 7 of the 1977 Act, may still fall within the definition of a restricted contract.

121 See, Section 21 of the 1977 Act.

122 Section 12 of the 1977 Act.

123 Section 69 of the Housing Act 1980, inserting a new Section 106A into the Rent Act 1977.

124 Section 77 of the 1977 Act.

125 See, Schedule 1 to the 1988 Act.

126 See also, *Crawley Borough Council v. Sawyer* [1987] 20 H.L.R. 98, (C.A.).

127 See e.g., *Westminster City Council v. Clarke* [1992] 1 All E.R. 695, (H.L.).

128 See, Section 84 of the 1985 Act.

129 See, *Peabody Donation Fund Governors v. Grant* [1982] 264 E.G. 925, (C.A.).

130 As amended by the Housing and Planning Act 1986.

131 For a good summary of the provisions, see, D.J.W. Greenwish, 'Leasehold Reform, Housing and Urban Development Act 1993', [1994] RRLR Vol. 41/2, 7. See also, P. Matthews and D. Millichap, *Guide to the Leasehold Reform, Housing and Urban Development Act 1993* (Butterworths, 1993).

132 *R. v. Eastleigh Borough Council, ex p. Beattie (No. 1)* [1983] 10 H.L.R. 134.

133 But see, *R. v. Wandsworth London Borough Council, ex p. Nimako-Boateng* [1984] 11 H.L.R. 95; *R. v. Eastleigh Borough Council, ex p. Evans* [1984] 17 H.L.R. 515; *R. v. Purbeck District Council, ex p. Cadney* [1985] 17 H.L.R. 534.

134 *R. v. Hammersmith & Fulham London Borough Council, ex p. Duro-Rama* [1983] 81 L.G.R. 702.

CASE STUDY 8.1

Read the decision in *Guppy's (Bridport) Ltd v. Brookling/Guppys (Bridport) Ltd v. James* (1984) 14 H.L.R. 1; (1984) 269 E.G. 846, 942 and then answer the questions set out below:

1 What were the essential facts of the case?
2 What court had the case before it?
3 Who were the judges?
4 Where had the case come from? What happened in the court below?
5 Who wanted what from the court?
6 Who appeared on behalf of the plaintiffs and defendants?
7 What precedents were relied upon by the court?
8 What was the *ratio decidendi* of the case?
9 What reasons did the court give for reaching its decision?
10 What remedies (if any) were awarded?
11 What legislation was referred to?
12 How could the plaintiffs have avoided the defendants' counterclaim?

Having done this exercise, you may wish to visit a county court and then compile a report (1) indicating the nature of the cases you heard and their outcome; (2) describing the function and jurisdiction of the court; (3) explaining the legal personnel involved in the legal process; and (4) outlining the system of appeal from the court to the higher courts.

CASE STUDY 8.2

Read the Court of Appeal decision in *McNerny v. Lambeth London Borough Council* (1989) 21 H.L.R. 188 and then answer the questions set out below:

1 What was the nature of Mrs McNerny's tenancy?
2 What Act afforded her security of tenure?
3 Why was the landlord held not liable in negligence for the condensation dampness in her flat?
4 Why did not Section 4 of the Defective Premises Act 1972 apply?
5 What reasons were given for Sections 8 and 11 of the Landlord and Tenant Act 1985 not applying to the facts?
6 Was liability imposed on the landlord under the implied common law condition relating to fitness for human habitation?
7 What redress (if any) might Mrs McNerny have had under Sections 79–82 of the Environmental Protection Act 1990?
8 What changes have been proposed by the Law Commission regarding the legal responsibility for the state and condition of leasehold property? (See, M. Pawlowski, 'Tenant's Remedies for Condensation Dampness', [1993] E.G. 9335, 108).

QUESTIONS FOR DISCUSSION

1 What should be the primary aim of a civil justice system? What are the respective roles of the barrister and solicitor within the English legal system? What do you consider would be the main advantages of a fused legal profession in this country?

2 What are the fundamental elements of a legal contract? What remedies are available for breach of a contract?

3 Do you consider that English law draws a fair balance between the interests of landlord and tenant? Give examples to illustrate your answer.

4 Compare and contrast the remedies available to:

(a) a landlord; and

(b) a tenant

for breach of a covenant to repair. To what extent are the landlord's remedies limited by statute and the tenant's remedies extended by statute?

5 You have been asked by a client to draft a short lease for use in respect of residential premises.

(a) Outline the form and contents of your draft;

(b) explain what obligations would be imposed by law (including statute) if your client chose to proceed in the absence of such a lease.

RECOMMENDED READING

Arden, A. and Partington, H. (1994) *Housing Law*, 2nd edn, Sweet & Maxwell.

Arden, A. and Hunter, C. (1994) *Manual of Housing Law*, 5th edn, Sweet & Maxwell.

Barnet, D. (1996) *Introduction to Housing Law*, Cavendish Publish-

ing. Clements, L.M. and Fairest, P.B. (1996) *Housing Law: Test Cases and Materials*, Cavendish Publishing.
Harpwood, V. (1993) *Law of Tort*, Cavendish Publishing.
James, P. (1989) *Introduction to English Law*, 12th edn, Butterworths.
Partington, M. and Hill, J. (1991) *Housing Law: Cases, Materials and Commentary*, Sweet & Maxwell.
Pawlowski, M. and Brown, J. (1995) *Casebook on Landlord and Tenant Law*, Sweet & Maxwell.
—— (1995) *Law Q & A: Landlord and Tenant*, Blackstone Press.
Stuart, A. (1996) *Rethinking Housing Law*, Sweet and Maxwell.
Stone, R. (1994) *Contract Law*, Cavendish Publishing.

9

MANAGEMENT AND ORGANISATION

R. Shean McConnell

INTRODUCTION

This chapter is concerned with the application of the generic principles of management to housing organisations. It is not about the details of everyday 'housing management'. While many of the general principles of management apply across the varied fields of housing there are, of course, differences in application, perception and suitability. It is the case that the management objectives within housing departments in the social rented and private sectors respectively became closer from the 1980s as a consequence of political pressure and economic necessity. Nevertheless the goals, or the long-term aims, mission and purposes, of housing management remain essentially different within the social rented and private sector housing organisations. Moreover there are also differences between housing associations and local authority housing departments. While most private landlords, with their own or their shareholders' capital invested in housing, must have optimisation of profits from satisfying demand as their major goal, the managers in the social rented sectors will be concerned with meeting housing need and providing quality landlord service within the constraints of financial cost-effectiveness. Managers in the public and social rented sectors, like those working for private landlords, are also concerned with maximising the returns from rented property, so that accommodation can be provided, repaired and improved. The idea that housing is a social welfare service has become more fragile in the real world of cost-effectiveness. The truth, however unpleasant, is that many of the decision-making criteria of the market place are now having to be applied to social as well as to private sector housing management. There are no longer enough central government subsidies to support either the tenants or their housing managers in the manner of earlier decades. Tenants have to pay more and managers have to work more cost-effectively. Nevertheless, satisfying the housing needs of the most disadvantaged groups in society must remain as the mission and a goal of social housing organisations, and it is this that will continue to differentiate decision-making in the private and the social sectors. Policies and the criteria for the evaluation of alternative strategies will inevitably be different between the sectors. Specifically the chapter examines:

- The economic and political contexts to housing management.
- Accountability.
- Tenant management of housing.
- What is management?
- Organisations and their structure.
- Governance, corporate planning and management.
- The management of change.
- From resistance to planned change.
- Planning and decision-making.
- Quality leadership.
- Delegation and control.
- Empowerment.
- Management by Objectives (MbO).

- Human resource planning and management.
- Motivation and morale at work.
- Conflict and its resolution.
- Communication.
- Some questions about communication.
- Problems in communication.

THE ECONOMIC AND POLITICAL CONTEXTS TO HOUSING MANAGEMENT

Housing issues are especially subject to the dictation of government and, in turn, to the economic and financial policies of the political party, or parties, that hold the reins of power in government at central and local levels. The amount of public money that is allocated to the Housing Corporation and to local government influences not only the extent of new housing provision, but also the objectives of the housing managers. Housing policies, if they are to be realistic, have to be made in the light of the current economic and financial situation and in relation to the local and, especially, national political preferences. It is of course as probable as otherwise that there may be different views between those in control at national and local levels. But there can always be hopes of change. In the United Kingdom there are also the growing influences of the European Union (EU), and the possibilities of EU funding, or restrictions on policy-making, emanating from Brussels.

As explained in Chapter 10, the contexts to housing management include the organisational hierarchies of local and central government. The way that decisions are made within those hierarchies makes it more or less difficult for housing managers to meet their objectives. Political and economic decisions by councillors, and by those acting on behalf of central government, override housing policy in the social sectors, and influence such policies in the private sector. The history of private and local authority housing has been dominated by Acts of Parliament for more than a century (Balchin, 1995). The relative dominance of private renting, council renting and housing association renting has altered very significantly over the last century (see Chapters 1 and 3). Council housing has now acquired a residual role providing, increasingly, accommodation for the most disadvantaged sectors of society: over time the characteristics of households in the council sector have 'changed from the affluent, employed working class family to a low income, benefit-dependent group including disproportionate numbers of elderly persons and lone parent families' (Malpass and Murie, 1994: 147). Ironically these groups, the elderly and the single-parent families, tend to be uncomfortable or unsuitable neighbours, resulting in housing management crises as some alienated and noise-loving youngsters, often with no responsible father figure, annoy and sometimes terrorise their elderly neighbours. Drug abuse is a related problem, with the elderly tenants suffering as a result of the traumatic experiences of those who are abusing themselves and others. Changes in the labour market have of course contributed to increased social inequality because ill-educated tenants are not able to compete for limited skilled work, and unskilled work is declining. Moreover, such people tend to have children who are as ill-equipped as themselves to compete in a market economy, and who therefore may feel that they have no stake in organised capitalist society as they perceive it. Such social problems provide constant challenges to those with management responsibilities and to those with elected responsibilities. The question of accountability can be a complex one.

ACCOUNTABILITY

Whenever something goes wrong in management the blame will be attributed to someone; some person is usually held to be accountable. The responsibility for an error or for negligence may have to be shared by the person who made the mistake and by his or her superior whose control was inadequate. In private sector housing the issue of accountability will be as in any business, and the consequences may lead to dismissal of the person held to be responsible if the error leads to serious consequences. This is also likely to be the case in a housing association. However, the ethics of professionalism may be construed to make housing staff accountable not only to the management committees of the housing association but also to the tenants who may be termed the consumers of the housing services (Malpass and Murie, 1994: 20). By contrast, the members of housing association management boards and committees are accountable, albeit informally, only to those who appointed them. In local authority housing it is different. The housing officers are accountable to senior officers and to housing committees on which elected councillors make the decisions. These councillors are accountable to their own political parties and to those who elected them. For elected members there may be conflict between accountability to the voters in their own ward, who may have a special housing problem requiring a solution that does not conform to the housing policies of the council or of any political group. Such conflicts in accountability can be sharpened in a negative way when a tenants' association does not want, for example, larger flats to be created for homeless families despite this being newly approved council policy. In such a case the members of the tenants' association are accountable to those residents who elected them and the councillors are accountable to those in the ward who elected them. Such councillors may have to vote against their own party members and in defiance of council policy. Or, even more problematical, they may have to vote against their own conscience in the interest of political expediency. But, hopefully, in very serious issues of moral conflict they will choose the moral rather than the political option (McConnell, 1981: 140–44). For housing officers there may also be a conflict between their own professional integrity and personal morality and the demands of superior officers or of councillors on housing committees. Unions and professional associations may be able to offer advice on such complex situations.

The issues of accountability and responsibility may become more complicated once tenants in the social or private sectors are given delegation powers to manage their own estates.

TENANT MANAGEMENT OF HOUSING

Although this chapter is mostly focused on the larger organisation, there is a movement away from institutional and 'top down' management structures to populist 'bottom up' models in housing management. Co-operative housing is a feature in some countries. It was allowed for in the 1961 and 1964 Housing Acts. Later, the 1986 Housing and Planning Act and the 1988 Housing Act provided some encouragement to co-operative and tenant-led housing management initiatives: 'between 1985 and 1994 a considerable number of initiatives were taken by local authorities to help their tenants set up tenant management organisations' (Price Waterhouse DoE, 1995: 4). Tenant management organisations (TMOs) were supported by a range of Department of the Environment publications in 1994 (*ibid.*: 5). Estate management boards (EMBs) were another innovation.

In 1995 the Price Waterhouse research report *Tenants in Control: An Evaluation of Tenant-led Housing Management Organisations* (HMSO) was published for the Department of the Environment. It dealt with TMOs, local authority tenant management and 'par value' co-operatives and estate management boards. A key conclusion was that 'even the less successful forms of TMO included in this study, the EMBs, are regarded by their tenants as delivering a quicker and higher quality repairs service and as being more effective or fairer in dealing with disputes than the local authority' (*ibid.*: 121). However, because the efforts of co-op members are voluntary, exact management comparisons with local authorities and other organisations are difficult. Nevertheless, the tenant management challenges, albeit of a different or smaller scale than in many more conventional housing organisations, may be solved by consideration of some of the ideas in this chapter. There will also be some application to the skills required by landlords in managing flats, but as Malpass and Murie observed (1994: 270), there is a continuum 'from the worst types of exploitive and private landlordism through participative management approaches to forms of management co-operative or community leasehold. The ends of this continuum have little in common.' However, there are complaints by tenants in the private sector about delays in maintenance work and in communication (*ibid.*: 273). Such complaints are echoed by many tenants in the social sector. Management problems may be more common across the sectors than might be expected. What therefore, it can be asked, is management?

WHAT IS MANAGEMENT?

Management is ultimately the taking of responsibility for the work of other people. But there will be elements of planning, delegation, co-ordination and control, the provision of leadership and motivation and, above all, adequate two-way communication. Managers are accountable for anything that goes wrong. They must be able to take decisions, having obtained adequate information and evaluation of alternatives. They need basic IT skills, but also human relations expertise. As they are promoted, managers will undertake less technical and professional work, and become more and more involved with creating new policies and strategies (Oldcorn, 1989; Stoner and Freeman, 1989). Managers from time to time need to submit themselves to additional training as they age, or they will become obsolescent or obsolete, and early retirement may come more quickly than the pension plan has provided for.

Housing managers are part of a hierarchy. They are human elements in an organisation structure. In a small housing association or company everyone will know everyone else. In a large local authority or housing association, the 'top' people may be known only from their photographs. In a local authority, there will be other directorates, some of whose staff will have a working relationship with some of the housing officers. Housing benefit staff, social service officers, environmental health and other staff such as architects and surveyors may have an important work relationship with housing officers. The link may be individual tenants with problems or individual buildings with problems. A particular challenge in an era of reduced budgets may be the need for a relationship between housing management, community care and competitive tendering (Clapham and Franklin, 1994; Lund and Foord, 1996). Large organisations can be very complex.

There are many approaches to setting out principles of management and organisations. One can start with the individual manager, and how he or she should or could manage, or one can start with the organisation and its context, and then proceed inwards with an analysis of the human components and their interaction. It is this latter approach that seems most appropriate to housing: to study the whole, and then to investigate the elements of the systems and structures to reveal the people whose methods and style of communication may or may not make the organisations function well. The risk in this approach is that the study of organisations may seem to be more abstract, and less interesting, than the investigation of individuals and their management styles and interaction. But readers can always start where they wish, provided that they remember that the structure of the organisation is likely to affect those who work within it; but that the people can change the structure, provided that those with the power agree.

ORGANISATIONS AND THEIR STRUCTURE

The structure of any organisation is a representation of the balance of power within it. It also represents levels of accountability, as already outlined, authority, status, responsibility, decentralisation and specialisation. Individuals at the higher levels tend to have the most power; but this power, and the duties that go with it, are delegated downwards to

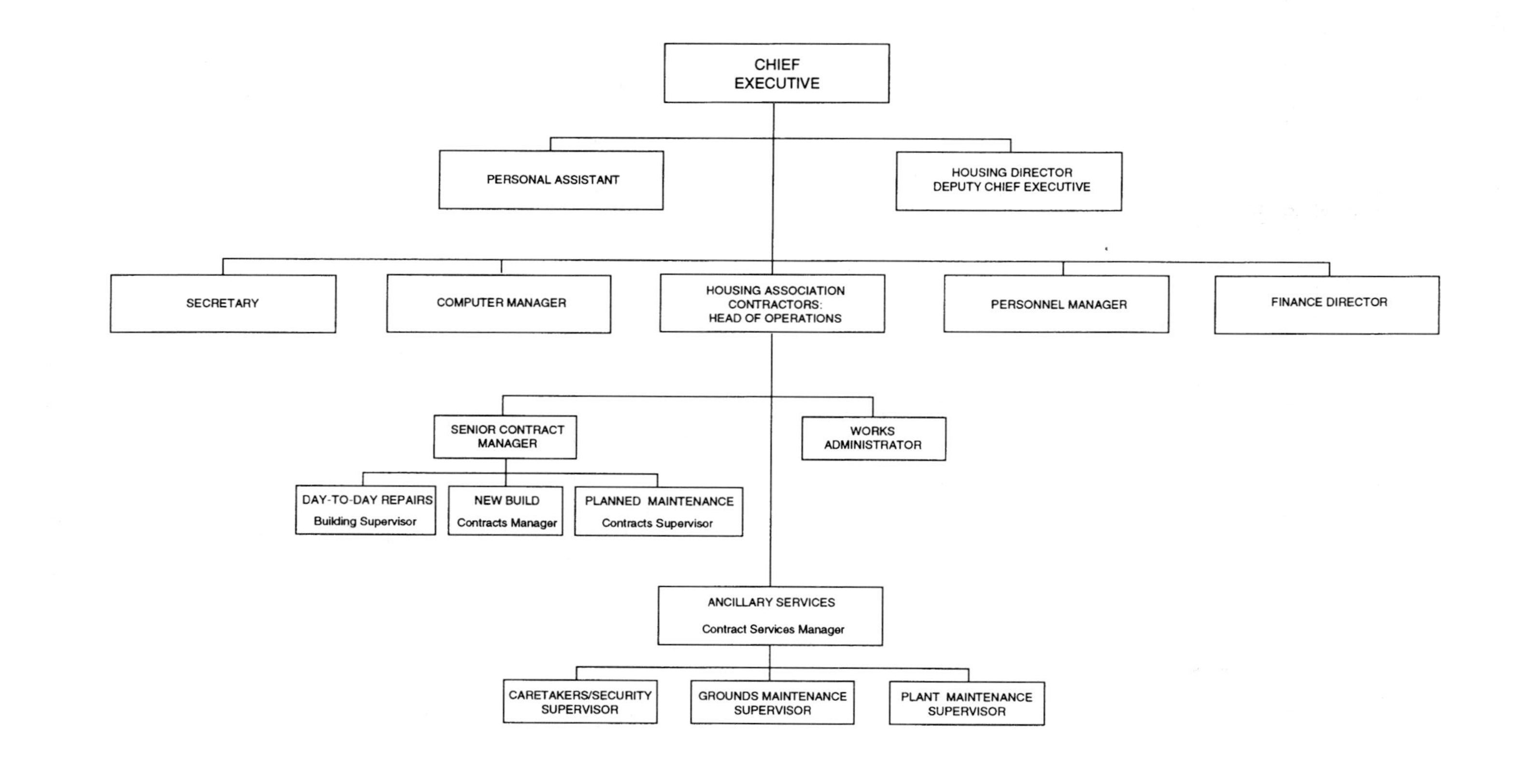

Figure 9.1 A typical housing association organisation structure showing principal personnel and services

subordinates, who in turn may delegate responsibility and accountability to those who work for and with them. Nevertheless, when problems arise, when work is not undertaken adequately and objectives are not met, the responsibility for any failure rises, like an air-filled indicator, upwards to the superior members of the organisation. Thus accountability and responsibility can never be completely delegated away. The 'buck' tends to pass outwards, but eventually upwards, when things go wrong.

The housing organisation structures indicated in Figures 9.1, 9.2 and 9.3 show how many levels of authority there typically may be. Figure 9.1 shows five levels of authority; but that may extend to six levels in the case of ground maintenance staff. Like 9.2, it indicates a line structure, while 9.3 shows a matrix structure.

A tall structure is one in which there are many levels of authority, with relatively narrow spans of control. By contrast a flat organisation has relatively few levels and a wider span of control for each manager. The larger the organisation and the greater the number of staff, the taller the organisational structure is likely to be. It can be theorised that small organisations, with a few hundred employees or less, should have a relatively flatter structure to be cost-effective at the upper levels of salary. The span of control of any one manager is indicated by the number of posts for which he or she is the responsible person. The ideal span of control varies from three to twelve, depending on the level of complexity of the work (Oldcorn, 1989); but there are no absolute rules. A related factor is the number of levels of authority in the organisation. It is currently fashionable to eliminate some of the upper 'Deputy' or 'Assistant' levels in organisations (to save money on top salaries). But a consequence is likely to be an increased level of stress as the top manager's span of control is extended to include more individuals and, perhaps, more specialists.

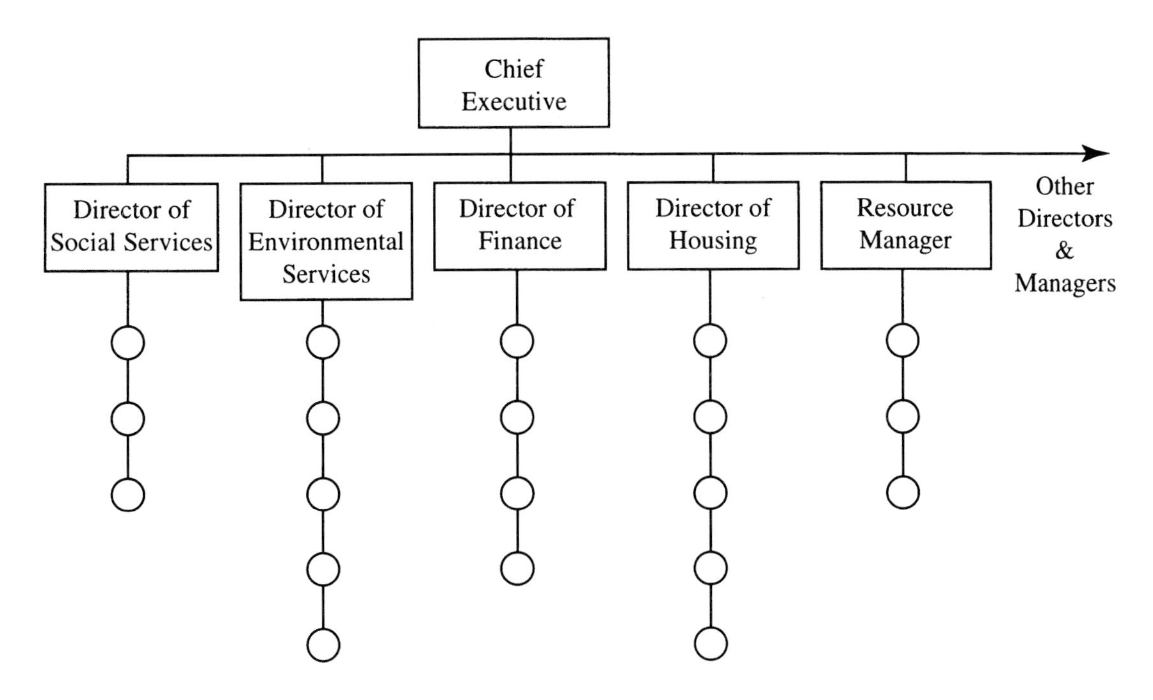

Figure 9.2 Line structure (local authority) showing the vertical but not any horizontal lines of authority

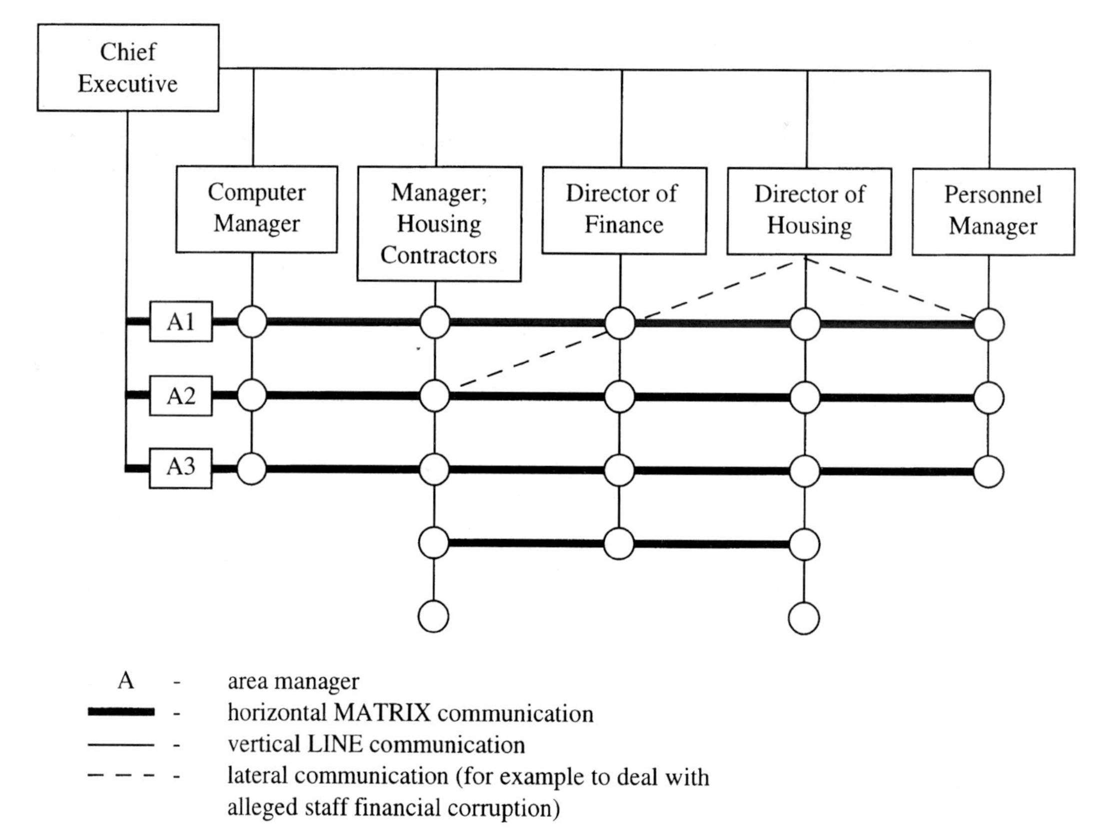

Figure 9.3 Matrix structure (local authority)

Communication within organisational structures can be vertical, horizontal or lateral (Figure 9.3). Organisation structures comprise systems of dividing the activities for which the organisation is responsible into jobs for individuals. But the structure should be based on the work required and not on individuals in the organisation at any one time. Each level of the structure represents levels of delegation of responsibility, accountability and authority. The organisation chart indicates how work is divided by specialism or by expertise. In Figure 9.1, the Chief Executive has seven people reporting to him or her, but one of those is his or her Personal Assistant, who will be responsible for much of the routine day-to-day communication. The Head of Operations (Housing Society Contractors) is responsible for three managers, while the Computer Manager and the Personnel Manager are each directly responsible for only one person next in line of authority. Figure 9.2 is a representation of some of the senior officers of a local authority. Both these figures show the line authority from top to bottom; the chain of command, control and responsibility. There are different chains or lines for each specialist function, for example computing or personnel. The advantages of line structures are, first, that they are easy to understand, the pattern of authority is comprehensible; second, there is a clear responsibility for decision-making; third, co-ordination is straightforward with, fourth, a lucid system of delegation and control. Fifth, individuals' status and roles are unambiguous; and, sixth, discipline is easy to maintain.

There are, however, disadvantages in line structures: individuals' work experience may be very restricted, which may make promotion less easy and make changes in the organisation more unpleasant and job threatening. Second, the influence of each linear boss is strong and this may make organisational change less easy to implement, with a stress on the 'status quo'. Third, and following on from the above, there may be rigidity and lack of innovation and flexibility in responding to changing circumstances. Fourth, line structures, based on staff functional patterns of work and control, may result in poor co-operation across an organisation. This is especially a risk in the formal or classical line structures of local government.

The matrix structure is an alternative to a formal line structure for any organisation that is dependent on good client or customer relations. Admittedly, there is a question as to whether a tenant is as important to a housing manager as a client is to an estate agent or customers are to those selling housing components. Nevertheless, if effective service to tenants is taken as one criterion of housing management, then the matrix organisation is applicable, as shown in Figure 9.3. This shows how the structure is adapted for a territorial re-organisation into three areas, each with a senior area manager, each of whom will need to communicate across the organisation. In practice, it will not be as neat as shown in the figure since such a horizontal framework of communication assumes that the lower-level line staff will be able to carry similar levels of responsibility. The lateral network will be used for any unusual circumstances. Thus matrix structures have vertical authority lines combined with horizontal and lateral lines of day-to-day communication which focus on the requirements of area management. Such a matrix structure can also be focused on corporate client or customer interests, or used more pragmatically for project management. The effect on staff, however, is that many people are subject to two sources of authority: to a specialist or functional boss as well as to the operational area manager (or project manager). This is a dual authority or multiple command system. Each boss may set objectives, and these may be in conflict, not least in terms of urgency.

The advantages of such matrix structures are, first, that there can be effective, rapid team co-ordination and, second, that there is the stimulus of inter-disciplinary co-operation which can motivate staff. Third, staff and other resources can be allocated in accordance with levels of demand, need and urgency which, overall, leads to the effective use of scarce resources and cost rating. Fourth, there should be cross-fertilisation of ideas and initiatives, which also should contribute to job satisfaction. There should, fifth, be consistency of application by all the members of the area or project team. Such an organisational structure should be effective in terms of service delivery as well as rewarding to those working within it. However, there will be problems.

A major disadvantage of a matrix structure may be divided loyalties: staff may not know whether to give priority to the requests of their line boss or to those of their operational matrix manager. Much may depend on the contrasting leadership qualities of the two bosses. A consequence may be that the authority of department heads will be undermined and the management of part of the organisation disrupted. A third and related problem may be that power struggles can ensue between the matrix and line managers. And, fourth, staff morale can be damaged by such problems, especially if the interpersonal skills of the managers are less than satisfactory. A fifth problem may be serious conflict in the bids for allocation of resources, with the problems having to be resolved by the most senior levels of management. In private practice there may also be the temptation for successful area or project managers to set up in practice on their own, in competition with their erstwhile company.

There is no simple conclusion to the arguments outlined above. Generally, a matrix style of organisational management will be more adaptive and flexible in periods of change and will, in the best situations, result in more responsive and better services, staff motivation and prospects of promotion because of growth in personal experience. There is, however, the risk of chaos and staff confusion and alienation. Much depends on the quality of the management, the economic or political environment and the rapidity of change that characterises the organisation. In practice there is no one 'right' way to organise any or every group of housing staff. Because of this it is important to know that the contingency approach in the textbooks emphasises the need for flexibility. Mullins (1993: 337) has explained that 'the "fit" between structure, systems of management, and the behaviour of people will depend upon (be contingent upon) situational variables for each particular organisation'. There will be different organisational influences including the size of the group, the effects of technology in communication and whether or not there are close relationships with other groups. The political and economic 'environments' will change, and with them key influences on the systems of organisation. The contingency approach is derived from systems analysis of organisations. Related influences will be the corporate goals and objectives of the overriding organisation; and how these are interpreted at 'lower' levels. But, also, in the language of urban political analysis, any one organisation is only part of a system of formal and informal bodies and interest groups which are wholly or partly focused on any one activity or set of activities such as, say, housing. There are many interacting systems and sub-systems of 'governance' in which people make decisions that have an impact wider than the original issue or focal point. Moreover, in a large organisation like a local authority or the largest housing associations, estate agencies or property companies, there may be an attempt to plan and to manage all the activities corporatively, with housing as one activity among many.

GOVERNANCE, CORPORATE PLANNING AND MANAGEMENT

'The concept of urban governance includes informal structures and the increasing involvement of private sector interests' (Newman and Thornley, 1996). In housing issues there are the concerns of institutions of central and local government, housing associations and private ownership of housing, not least of flats or houses in estates mostly managed by local authority officers, but with the complication of leasehold interests and active participation of leaseholders in tenant associations.

One way to explain governance is to liken it to '"steermanship" or "decision and control" . . . thus urban government is seen as a part of a more inclusive process, parts which have more informal characteristics, and which we call urban governance' (Diamond and McLoughlin, 1973). The term inter-corporate planning has been used to describe the managing processes whereby the different decision-making networks are linked and co-ordinated; but that is not likely to be a concern of housing managers. By contrast, if they work for a large organisation, they may be involved in corporate decision-making.

Corporate planning has been described as 'systematic and comprehensive long range planning taking account of the resources and capability of the organisation and the environment within which it has to operate, and viewing the organisation as a total, corporate unit' (Eyre, 1993: 55). A five-year view is taken with regular reviews. Another definition is: 'Corporate planning is a continuing process by which the long-term objectives of an organisation may be formulated and subsequently attained by means of long-term strategic actions designed to make their impact on the organisation as a whole. Corporate planning also involves deciding the policy, or code of conduct, of the organisation in pursuit of its objectives' (Cole, 1993: 115). Corporate management, by contrast, is the structuring and control of organisational relationships to ensure that corporate planning takes place and is implemented. It is only possible to manage corporatively when there is a corporate plan.

Corporate planning processes lead to corporate management within the agreed policies and aiming for the agreed objectives. Corporate management also is concerned with the initial structuring of work relationships to ensure that planning can easily take place, and that the plans can be reviewed and implemented. Figure 9.4 shows how the processes of corporate planning commence with goals which have to be agreed by those with the 'power' whether through election, by prescription of government or as a consequence of decisions taken by board members or the equivalent. Within local government, corporate planning is very much wider than that of the housing department, directorate or division. It is a council-wide process aimed at integrating all the services and resource needs of the authority. It is deliberately broader than departmental, and was recommended for British local government by the Bains Committee, 1972, as part of the re-organisation of local government. One major aim was to introduce private sector-type commercial processes of decision-making to local government. The consequence tended to be that a relatively few large directorates replaced many smaller departments, under the overall management of a Chief Executive. At elected member level, a Policy and Resources Committee became the senior committee of the council (Figures 9.4 and 9.5), and the Chief Executive became the senior and highly paid officer. Figure 9.5 shows how the key Policy and Resources committee was inserted above the other committees, between them and the council. Figure 9.4 shows the procedures, and who makes the decisions, or undertakes the work, at each stage of the planning process. This is an applied version of the processes of planning and decision-making shown in Figure 9.7 (p. 277).

SWOT analysis is appropriate to processes of corporate planning and decision-making, and human resource management. The '*s*trengths' and '*w*eaknesses' of internal human resources will be analysed, and then the '*o*pportunities' and '*t*hreats' that the staff and the organisation will face from economic and political changes external to the organisation (Eyre, 1993). An example of SWOT analysis is provided by Catterick (1995).

In many British local authorities such corporate systems were introduced in the late 1970s, but not many survived in the form in which they were set up. The senior officers resented the power of the Chief Executive. Legal and Financial Chief Officers often had been promoted to become the Chief Executive. But in other councils, the Chief Town Planning Officer was given the role. Other Chief Officers were also appointed. Moreover, at member level, the erstwhile autonomy of committees like Housing, Education, Social Services, Technical Services and others were challenged by the then new-found power of the Policy and Resources Committee. Another factor in the practice of corporate planning and management is the relations between officers and members, which vary between local authorities and in any one authority over time, and especially as different parties gain political strength. Skitt wrote perceptively in the early 1970s that in some authorities:

> There is a frank exchange of views, a respect for the others' roles, an acceptance of a joint responsibility to the community and a desire and willingness to work together to fulfil that responsibility. Corporate planning can in those authorities provide the planning and organisational machinery to support that relationship . . .
>
> Elsewhere there is a history of mistrust, of a rigid relationship where officers and members act out long held postures and of contact only in formal situations. Officers openly boast that they can twist members round their little finger, and members attack officers . . . The terms 'corporate planning' and 'corporate management' only have meaning where they embrace a corporateness of purpose, attitude, activity and interaction by members and officers.
>
> (Skitt, 1975: 251)

However, especially in larger local authorities, corporate planning and management structures have survived. But the rate of change that has affected housing staff in recent years has meant that the management of change is perhaps more important for a housing manager to understand than the theories of corporate management.

THE MANAGEMENT OF CHANGE

In any organisation there will be pressures for change. The pressures may be external or internal and may be for economic or political reasons. In local government, change is usually a consequence

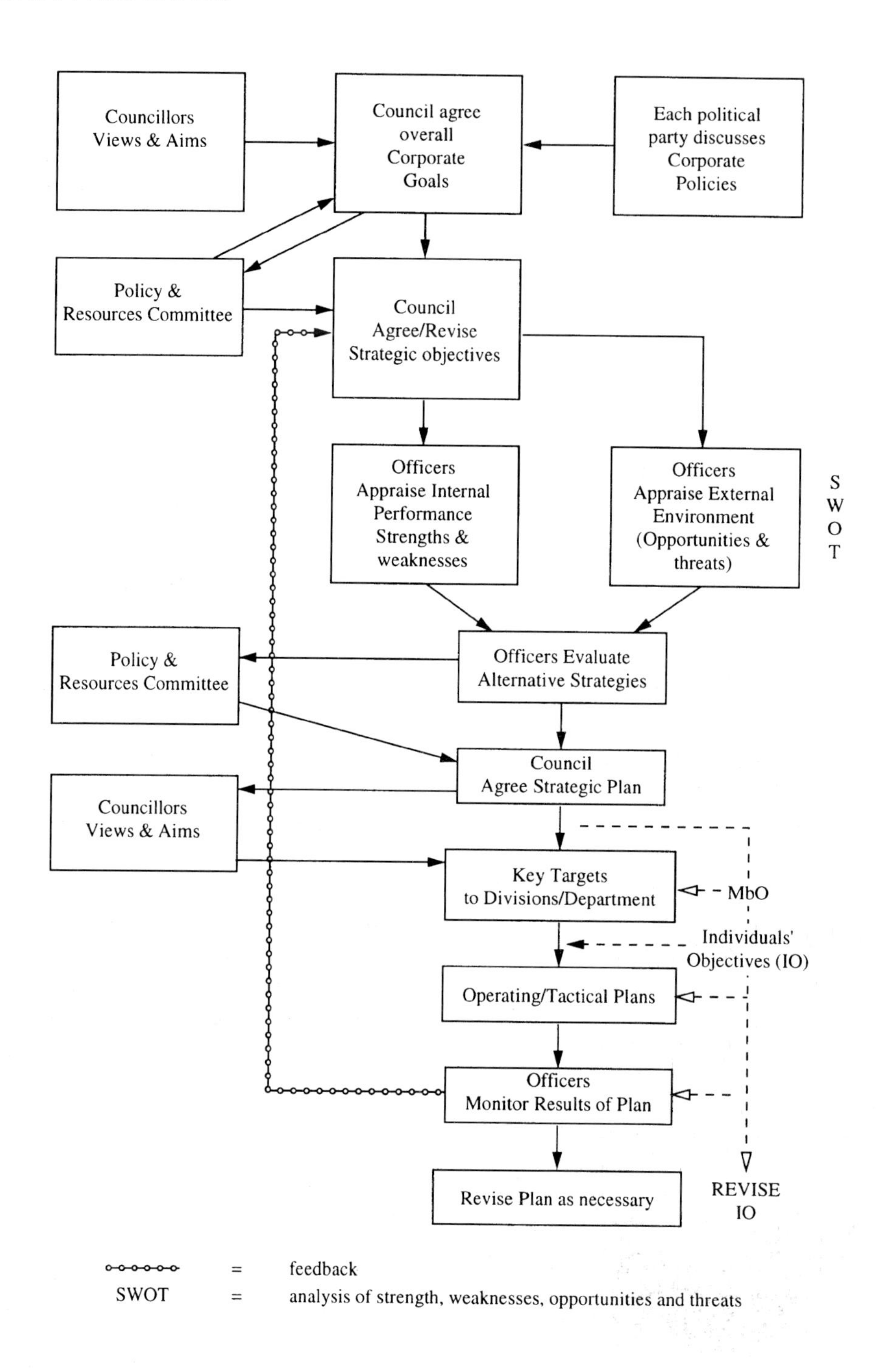

Figure 9.4 Procedures of corporate planning (with management by objectives, MbO) in local government with the agents of decision-making and action

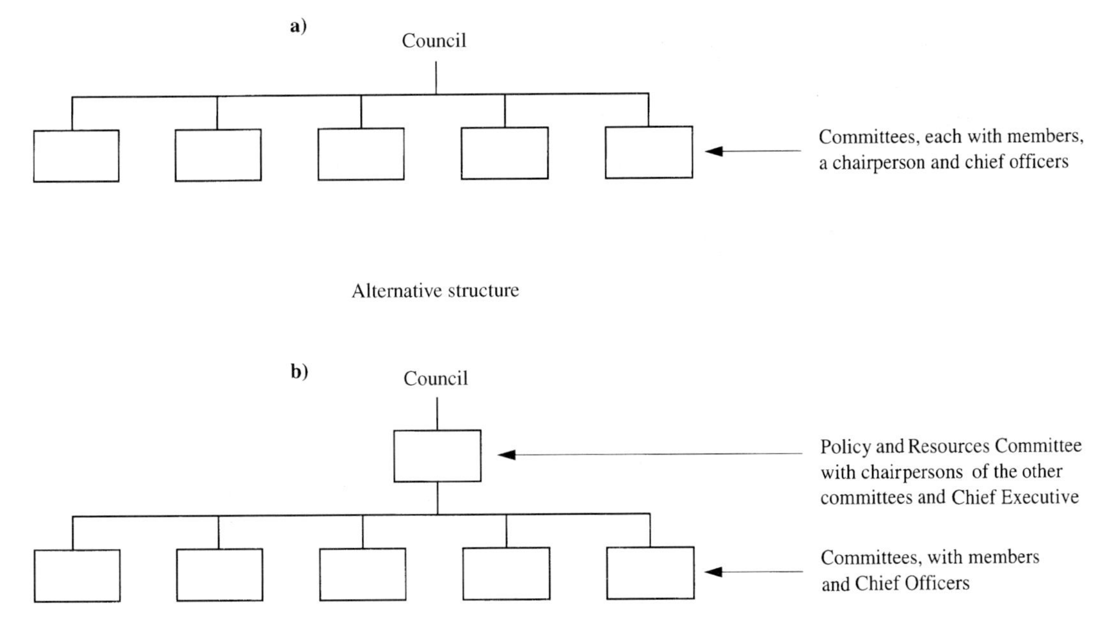

Figure 9.5 Council and committee structures

of the election of people with different political views to those of their predecessors. The main lessons to be learned from Figure 9.6 are that there are positive and negative factors or forces that may cause or prevent great change in an organisation (Oldcorn, 1989). Then, once change is started, there are two fundamental questions: what will be the effect on staff or human resource management? and what will be the effect on the organisation and its structure? These issues are interrelated.

In housing the rate of change in the old public sector has been very great. The main reason has been the government's introduction of competitive tendering for local authority services. This has caused very great changes, and is covered below. In all public organisations the threat and actuality of privatisation has been the cause of many organisational and staff changes. In the private sector, increasing competitiveness and the introduction of systems of information technology (IT), computerisation and more recently globalisation have greatly changed office life for most employees. The public sector has also been subject to these kinds of changes. More part-time rather than full-time employment and short-term contracts, the influence of the European Union directives, and national political uncertainty were other factors creating a climate of uncertainty. There are also more human reasons such as new senior management, with new ideas, or take-overs of one organisation by another. A consequence will be changes that will affect the staff and changes that will affect the organisational structures. These are summarised in Figure 9.6, which shows the split, but also the connections, between human management and 'scientific' management. The terms in the figure are explained in this chapter. There is always a temptation for newly appointed managers to concentrate on the activities shown on right-hand side of the figure rather than those on the left side. If that happens, the change-over from one kind of organisation to another is likely to be less easy, with the possibility of litigation or industrial action and, generally, a low level of morale. Before we examine some of the challenges in managing change, it is appropriate to explain why change has recently been and still is so prevalent in local authority housing organisations.

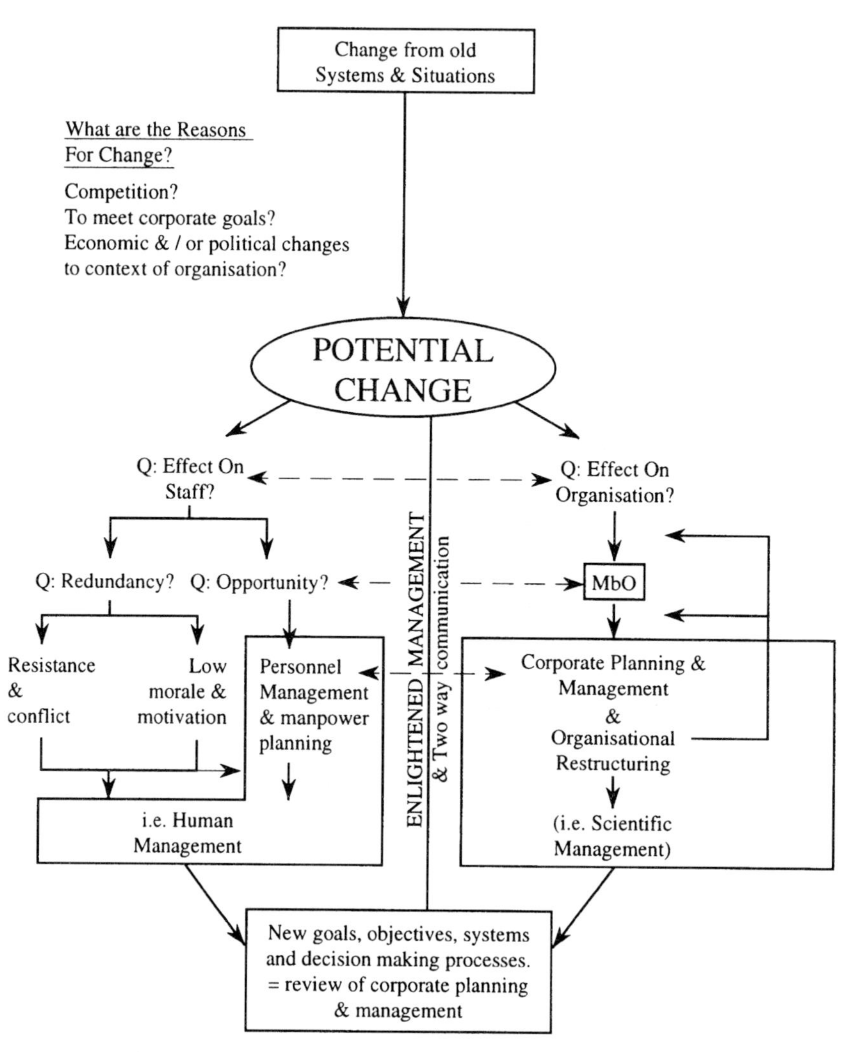

Figure 9.6 The management of change

Changes to allow for competition in housing services

The potential relationships between local authority and the private sector changed in the 1980s, and central government legislation has forced local authorities to adopt a more competitive approach to management (Fenwick, 1995).

In June 1994, the government, under the powers provided in the Local Government, Planning and Land Act 1980, the Local Government Act 1988 and the Local Government Act 1992, approved the regulations to set up compulsory competitive tendering (CCT) for housing management. This is a process by which local authorities have to create a competitive system relating to the management of a range of services. In effect, the client branch of the housing directorate prepares a contract and then invites tenders from the council's own housing management services as part of direct service organisation (DSO) teams, as well as from outside agencies, like private companies and housing associations, which can also submit tenders or bids to manage contract units of the council's housing stock. Fenwick (1995) has observed that the client-side manager pursues objectives related to the clients' (the tenants') interests, while the in-house contractor has objectives concerned with value for money from the contractual perspectives. There may thus be divergences in objectives. The monitoring of contracts is essential (King and Newbury, 1996).

In a large council like the London Borough of Lambeth there are about twenty neighbourhood-based contract areas, with tenders accepted for the management of the northern, central and southern areas for three-year periods, but different starting dates. The following activities must be carried out by each contractor: tenancy and leasehold agreements; management of vacant properties; allocation lettings; collection of rents and service charges (including current arrears, former tenants' arrears and leaseholders' arrears); caretaking and management of repairs and maintenance work (for example, ordering repairs) and managing the performance of repairers and other contractors (for example, grounds maintenance and caretaking services). The contractors must achieve the key targets which are part of the contract, the housing management specification. A local housing office is provided. It is the job of the client section of the housing directorate to monitor the performance of the contractors.

The tenants and leaseholders can attend meetings of a neighbourhood forum, and express their views on the performance of the contractors. Tenants also have the opportunity to apply to form tenant management organisations (TMOs), maybe a tenant management co-op, or estate management boards (EMBs), provided that they meet certain standards. One of the major problems will be the issue of how representative a tenants' association may be of the tenants of different ages and ethnic backgrounds. Another fundamental problem may be apathy of tenants.

From resistance to planned change

Change in an organisation will engender resistance from all those who are not going to gain from it (Oldcorn, 1989). It is therefore important to anticipate the range of possible reactions: from acceptance through indifference to passive or even active resistance (Hunt, 1981). Resistance is concerned with preserving the status quo. It is normal. As Hunt has explained, 'the most frequent cause of resistance is the way the proposed change is introduced' (1981: 277). He recommended face-to-face confrontation with trouble-makers – the games players who are found in every organisation, the rumour mongers – but as he put it, 'in radical change some resistance is inevitable and healthy – it reminds change managers that they have to sell the idea to the people' (ibid.: 278). In changing a housing organisation it will be wise to start with an analysis of its culture (Catterick, 1995).

Resistance will be reduced if the top management is in agreement with the change (not always the case by any means) or if people's income and/or status will be increased as a consequence of the changes (again, by no means always the case). But prior consultation, in the Japanese management style, will pay dividends (Baba, 1990). There must be full communication well in advance of change. Ideally, alternative models and outcomes of change should be debated, but in practice radical change is usually imposed, as with CCT, by governments or it results from irresistible market pressures or changes in ownership in the private sector. In such – the normal – circumstances, the best that can be achieved is to have an implementation plan, with the objectives, performance criteria and dates and processes of change set out, and a communication plan to inform all those who will be affected by the changes how, when and where information will be provided, and by whom. As in all management, the qualities of leadership will be reflected in the levels of staff motivation throughout the processes of change. It may be wise to appoint a transition team (Oldcorn, 1989) to oversee the processes of planned change.

Quality assurance (BS 5750) is a major cause of change in organisations in the 1990s. The British Standards Institution provided rules for 'Quality Assurance' (BS 5750). An organisation can get registered if it proves that it does certain things and is assessed by an independent organisation to check if it meets the demands of quality assurance. It has to prove that it follows its own procedures; keeps procedures up to date; trains its staff; checks what they do; learns from mistakes; keeps important information safe; and listens to its customers. The development of performance measurement in local government is a growing and important part of managing services. It has counterparts in any large housing organisation (King and Newbury, 1996). There are related processes: a performance review which has been defined as 'any systemic attempt to specify what the organisation is trying to achieve' (Fenwick, 1995: 106) and which will be related to devising the goals and objectives of corporate management; performance measurement, which will be the monitoring of the achievement, or otherwise, of the objectives; and performance management, which 'describes the overall process of collecting performance indicators and measures, designing and implementing appropriate management review, and evaluating and acting upon the results of these processes' (Fenwick, 1995: 107). This top-down approach may be opposed to the more person-orientated approach to be described below under the sub-headings of 'Empowerment' and 'Management by Objectives', and may lead to difficulties with union representatives and some councillors in local government. There may indeed be contradictions between corporate management approaches to the management of change and more human management approaches. Resistance will be less if there is full and early communication of intentions and proposed performance indicators, and a well-thought-out planning and decision-making process, with time allowed for adequate consultation.

PLANNING AND DECISION-MAKING

Planning is fundamental to the management of change in any organisation. There are three levels or hierarchies of planning: strategic planning, whereby the mission, goals and overall objectives are determined; tactical or operational planning, whereby the steps needed to attain the agreed objectives across the organisation are agreed; and, finally, project planning, whereby the work activities of the organisation are planned on a task or project basis. There are some key words in planning; and a problem is that they are not used consistently across the management literature or in practice. The following definitions will be used in this chapter:

A *mission* is the purpose or purposes of the organisation; what it is trying to achieve and for whom (Catterick, 1995).

A *goal* is a long-term, ultimate 'end' of a planning process. Unlike its use in football, it is a utopian, ideal state based on values, and therefore may never be attained: for example, in politics and the public sector, 'adequate housing for all'.

An *objective* is a precise, shorter-term, attainable target usually expressed in terms of quantities and time, for example:

1 250 new two-bedroom houses by the year . . .
2 99 per cent of rents collected by . . .

3 all repairs completed within one month of survey.

Setting objectives is a fundamental part of management (Catterick, 1995).

An *aim* is used widely to mean goal and/or objective, but in less quantifiable terms.

A *policy* is a statement of intent which will guide decision-making and use of resources.

The word objective (a target) is one commonly used in housing management, and is fundamental to monitoring, or control, after tenders have been accepted (King and Newbury, 1996). One of the secrets in successful management is to ensure that all published objectives can readily be attained. In an ideal process of planning and decision-making, the scheme shown in Figure 9.7 provides a useful model. In housing, such a model will have its own application. For example, the goals will relate to such matters as tenant satisfaction and to value for money from the viewpoint of the community; the objectives will provide targets for realisation of those goals; the information will be the data required to translate the objectives into action, such matters as the age, household structure and expected income levels of the groups to be housed, and the details of the housing stock. There will also have to be information about resources of all kinds, but especially available finance. The alternative strategies are an

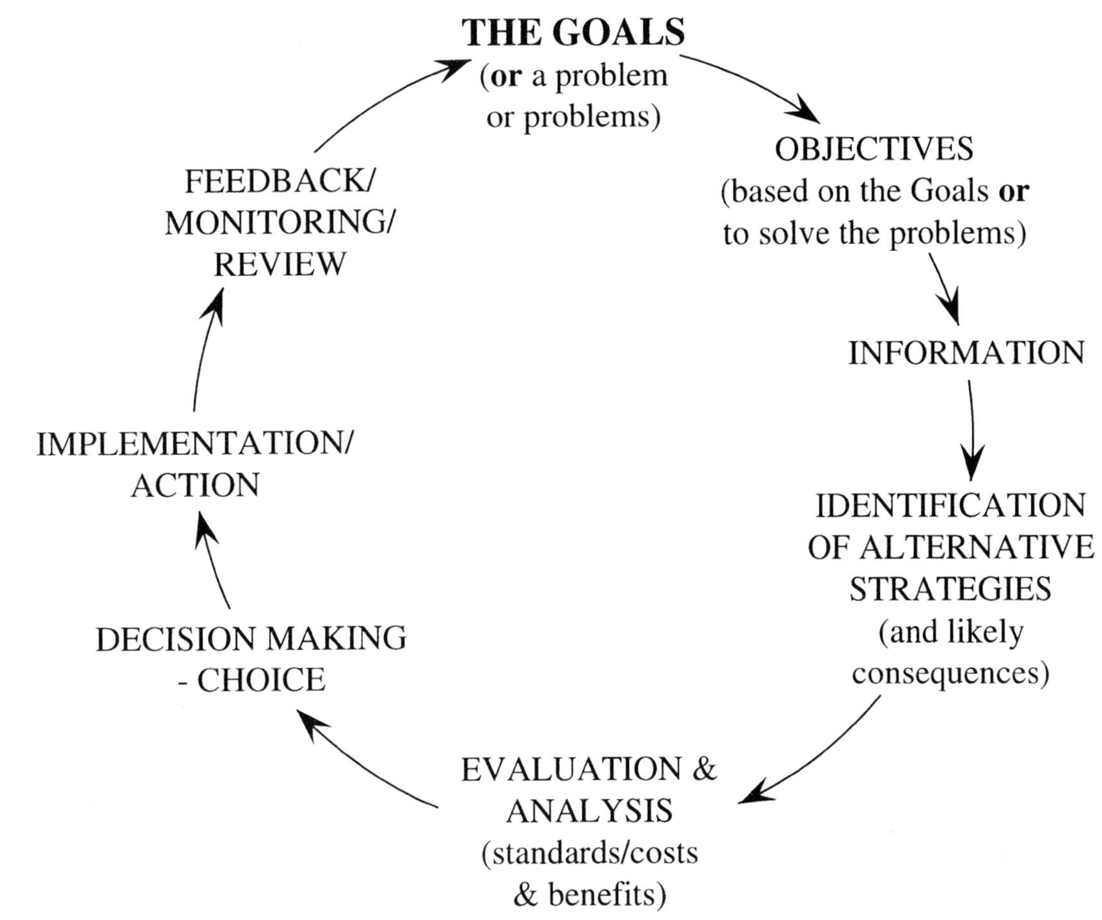

Figure 9.7 Processes of planning and decision-making

important part of any rational decision-making process. They will include different organisational structures to achieve the objectives, for example different numbers of local offices, or different ways to provide central specialist teams to look after, say, the interests of older or disabled people. The possible benefits, consequences and the costs of each alternative will be evaluated in social and financial terms: this is the stage of evaluation and analysis. The decisions will be made by those with the appointed or elected power to decide. The implementation of the chosen strategy will be the responsibility of the officers. Over time there will be feedback and a monitoring review of what has happened.

Alternatively (also shown in Figure 9.7), the decision-making process can originate with a problem, or problems, that need resolution. A problem tends to be something that does not conform with the goals or the objectives, like bad estate cleaning. The problem-solving objective can be stated as washing staircases weekly. Information will be needed about the contract, the staffing and the supervision. The alternatives can be to ensure the contractual terms are met, or to have new supervisors, or to terminate the contract and start the process of getting new cleaners. These alternatives will be analysed from the viewpoints of legal viability, cost and delay; and a decision will be made and implemented. The service will then be monitored regularly.

The evaluation of the alternatives is one of the most technical parts of such a planning process. In broad terms these should be compared under criteria such as these:

1 likelihood of meeting the objectives as specified;
2 financial costs in meeting the objectives
   (a) capital; and
   (b) maintenance terms (revenue);
3 social and environmental costs and benefits to the 'wider' community expressed on a qualitative scale; for example, gain or loss of open space;
4 possibility of harmful or beneficial consequences – expressed as probabilities; or risks, in particular the impact of traffic;
5 Staff costs or buildings required to house extra staff.

Not all of these criteria will apply to every situation, and there are many more possible ways to assess the alternatives. Once the implementation stage is reached, management support services will be useful (Eyre, 1993: 287–93; Cole, 1993: 221). Moreover, from an early stage in the planning process, the use of as much information technology as the organisation can afford, and the staff can learn to operate, will improve the quality of the information, and the decision-making. The implementation procedures will be improved if they are managed with an analytical control system. One method is to use a *critical path analysis* (or *network analysis*) comprising the following stages:

either:

1 deciding what work has to be completed to achieve the objectives;
2 a sequence of activities and a timetable, with dates; and
2 dividing work according to responsibility for doing it with names;

or:

1 allocation of tasks, with targets, to individuals or teams; and
2 regular control and monitoring of the work.

It is a fact that elaborate planning and decision-making processes will not be part of the everyday week of a typical housing organisation. There will often be a senior officer with such strategic-level responsibilities, working with senior management and also with the corporate planning and management experts of the local authority or the company or housing association. However, the related control mechanisms, to be discussed later, will operate throughout the management system. There is also a necessary component to ensure that these processes and mechanisms function well. It is good leadership.

QUALITY LEADERSHIP

Good management is a reflection of the quality of its leaders. High or low levels of staff motivation and morale, especially during and after processes of organisational change, can be attributed to the style of overall leadership. It is, however, true that different approaches to leadership are more, or less, appropriate to different times in the history and in the evolution of an organisation. A new enterprise at its beginning needs the unquestioning energy, commitment and egocentric qualities of an autocratic or strong leader. By contrast, the successful management of change requires a negotiator in charge: someone who can keep in balance the opposing political, economic, institutional, professional and personal pressures.

In housing, with its emphasis often on decentralisation in organisation, some of the definitions of leadership can be seen to be especially applicable. Cole describes leadership as 'a dynamic process in a group whereby one individual influences the others to contribute voluntarily to the achievement of group tasks in a given situation' (1993: 46). Such a level of group commitment can be judged by signs such as whether staff work longer hours, with shorter breaks, than are customary, for the duration of a particular priority situation. Quality leadership is judged by the group behaviour of subordinates. This is an issue considered later in this chapter when staff motivation and morale are discussed.

One type of leader, the one needed to create a new organisation, can be termed an autocratic leader, who tells his or her subordinates what to do. This type of leader can be contrasted with the free-rein or *laissez-faire* leader, who can be found in a long-established organisation: one who is open to endless representations from different groups, the unions or the politicians, the financial experts, the media and more. A characteristic of this second kind of leader is that decision-making is endlessly delayed, unlike with the case of an autocrat, who can often be criticised for making too speedy decisions based on a blend of expediency and ideological arrogance. A third style of leader is the democratic, consultative or participative leader. This is the kind of person who practises two-way communication, listens as much as she or he talks; and is probably the best one for most housing management situations, especially when the hardest decisions have to be made by the elected councillors or members of boards of management. A key contribution to the relationship between leadership and motivation has been made by researchers (Cole, 1993; Eyre, 1993). Moreover, Charsley (1985), writing at a time when British local government was beginning to change, but when many staff were apathetic and uncooperative, advocated a more inspirational kind of leadership (rather than a purely mechanical organisation approach). He stated that leaders should:

- take responsibility rather than duck it;
- promote confidence;
- project a cheerful, enthusiastic and optimistic image;
- avoid fussing, worrying and constant interference;
- 'walk the job' – and be seen;
- set the right example in their personal life off the job;
- think positively;
- accept and capitalise on change;
- show they care for those under them;
- have the normal courage to take the right – sometimes unpopular – decision;
- communicate continuously to all relevant points;
- listen;
- enthuse people;
- take hard decisions, remove passengers and dead wood;
- pick winners;
- avoid needless confrontations.

However, a practical way to judge the effectiveness of the leadership of a directorate or department is to consider how well the delegation and control systems operate.

Delegation and control

> 'Delegation is the act of assigning formal authority and responsibility for completion of specific activities to a subordinate.'
>
> (Stoner and Freeman, 1992: 353)

It can be argued that a good manager delegates so well that he or she has no tasks left, and is free to think about strategic and policy issues. But 'those who delegate still remain accountable' (Rees, 1996: 135). The problems in delegation include the questions as to whether the person to whom the manager delegates (the delegate) is sufficiently able to undertake the tasks that have been delegated, and whether the manager has set in place adequate control procedures. There has to be a proper feedback system. Cole has itemised the range of delegation options:

1. The subordinate is free to act.
2. Action may be taken, but a report is required.
3. The subordinate decides the best action, but before acting consults with manager (the manager can veto).
4. The subordinate examines the problem/issue, outlines alternative solutions and makes recommendations.
5. The subordinate examines the problem/issue, collects all the facts and presents them to manager for his decision.

(Cole, 1993: 183)

The 'correct' option will vary depending on the context of the situation. However, while (1) makes control impossible, the fifth option may be an under-use of the subordinate's experience and intelligence. It is essential for there to be a control mechanism that is explicit, with targets or objectives set out clearly. Verbal directions are not usually adequate. Figure 9.8 shows a methodological approach to the processes of control. A problem in control is that the manager has to protect himself or herself, and the employers, from incompetence or, at worst, fraud (not unknown in housing organisations) – since the manager remains accountable. But at the same time the subordinate must be encouraged to take responsibility, to 'grow' on the job, becoming more confident and ready for more authority. Delegation not only frees the manager

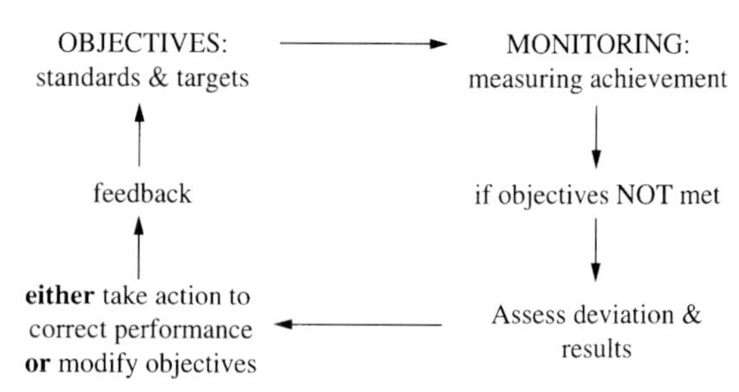

Figure 9.8 Control: processes and methods

for strategic thinking but it improves the morale, motivation and job satisfaction of junior staff if they are allowed to take decisions. Another issue, already discussed, is the span of control. No manager can oversee more than a certain number of people, depending on the complexity of the delegated tasks and the potential consequences of any errors. To conclude, the relationship of delegator and delegatee must be a two-way one, with clearly defined procedures of communication. Compulsory competitive tendering (CCT) in housing 'has the objective of delegation of responsibility for delivering the services for which the client remains accountable' (King and Newbury, 1996). The client should neither interfere in operational matters nor abdicate responsibility (*ibid.*). It may not always be easy.

Empowerment

The term empowerment became more widely used in the mid-1990s. The essence of the idea is that managers should devolve decision-making to those who are most fully involved in any one area of work to empower them to make the decisions, to give them the power to act; and that the consequence will be higher levels of motivation and morale because of more job satisfaction, with high degrees of responsibility. It is an enabling process (Furze and Gale, 1996). The responsibility for control moves from the manager to the team. It is a form of delegation that allows flexibility and innovation in how the housing services are provided. In theory the staff will become more involved and more committed because they have more control over their own work. It is a feature of management associated with flatter organisation structures and decentralisation. However, there are risks in a housing organisation in which staff have power speedily to transfer some tenants and not others (despite the criteria), and to enable improvements to some homes. In the past in some local authorities a few employees may have received financial benefits from some tenants for services rendered. There has to be some control. This dilemma may make true empowerment more difficult for housing managers than for managers in many industrial or commercial organisations. Yet, unless the senior housing staff really believe in empowerment, it will not become effective. Responsibility and control in housing organisations may be a more delicate problem of balance than the textbooks allow for. Nevertheless empowerment is one way to improve staff motivation and morale.

One way to ensure that new practices are working may be through management by objectives. Another is through regular staff appraisal, as outlined later in the chapter.

MANAGEMENT BY OBJECTIVES (MbO)

Management by objectives (MbO) has been defined as 'a principle of management aimed at harmonising individual manager roles with those of the organisation' (Cole, 1993: 118, citing Druckner). (MBO is a different management term meaning Management Buy-Out.) MbO, like corporate planning, to which it is related, is a technique which originated in American industry in the 1950s and was adopted for use by government and other agencies. It requires an organisation to commit itself to corporate aims and a strategic plan (Glendinning and Bullock, 1973). Then the theory is that staff will manage themselves and that this should ensure better motivation. It is a prescription for more effectiveness.

The way in which MbO is introduced depends on the organisation's existing commitment to corporate management; but assuming that the corporate objectives are already established, the idea is that all officers identify their own key tasks and standards of performance in relation to each of their tasks. These will then be discussed with superiors and colleagues within the group to compare perception of roles, duties and relationships in the organisation. The tasks and standards of performance will be agreed for a period of about six months, after which they will be compared and reviewed. As a parallel exercise, the objectives of groups, divisions, sections and departments should also be identified and standards of measurement agreed. In planning there will be decisions to be made about who is responsible for contacts with local groups, other departments and organisations and with the members.

A part of the theory of management by objectives is that the officers will identify their own key tasks and standards, and then, by comparison with colleagues, eliminate any gaps or duplication between job specifications. The idea is self-management, once the job specification is agreed by senior staff – whose own work specification would also be made available for discussion. Lupton (1971: 107) noted that the advantages claimed for MbO were:

1 That as a result of being involved in setting his/her own targets, and in defining his/her own job, the manager will be more highly motivated to improve performance.
2 That because targets are jointly set the superior gains a greater understanding of his subordinate's problems and is therefore better able to assess what changes are needed if these problems are to be minimised.
3 The data from target-setting and appraisal interviews are invaluable for identifying needs for training, education and personal and professional development generally.
4 The data for judging whether an individual is promotable or not are fuller and more systematically comparable under MbO and the criteria for doing so are the same for everyone. Justice is therefore done and seen to be done.

MbO brings together the apparently opposing requirements for discipline on the one hand (which emphasises constraints on the individual) and democracy, participation and consultation on the other (which emphasises opportunities for the individual): 'Because of this, management's authority at all levels is rendered legitimate, and so are the processes of advancement to positions of greater authority' (Lupton, 1971: 108).

MbO seems in theory to be an excellent management technique in which, because an individual's targets are jointly set with the manager, there should be motivation, commitment and job satisfaction within cohesive and effective teamwork. But in practice there are problems, in particular when there are financial pressures on the organisation: tasks may be difficult to define, and, if too rigidly defined, the team may be unable to respond flexibly to new demands and work needs. Another difficulty is that people's identification of their own roles may stress the most responsible, challenging and rewarding aspects, and omit the boring tasks. In the end some of the most mundane tasks may be accepted by no one and may have to be imposed by senior management. Junior officers may find that they are still expected to undertake menial tasks. Thus MbO does not always help in achieving job satisfaction. Another problem is that officers can find that they are judged for promotion on the achievement of their objectives; but this may work against them, if there have been problems. It can be seen that there are relationships between MbO and other management activities, not least corporate management. There can be a positive improvement in the morale within an organisation; but it is the possible link with human resources management, through the medium of staff appraisal, that can be controversial.

HUMAN RESOURCE PLANNING AND MANAGEMENT

The successful management of change is closely related to effective human resource planning and management of the personnel affected by the changes. The older term, manpower planning, as used from the 1960s, was related to mathematical analysis of the predicted staff needs of an organisation based on extrapolation of existing trends and planned developments. But it was unable to anticipate unexpected changes in the external environment, economic and political; and it has generally been replaced by such techniques as human asset accounting and human resource audits (Goss, 1994). These are part of human resource planning.

In the 1980s, the development of human resource management recognised the key importance of relating the psychology of human relations with strategic management theory and the emerging ideas of flexibility and quality management (Goss, 1994). The issues of flexibility included the growing use of part-time or contracted teams of employees, managed by a few full-time employees; and also the idea of multi-skilled staff with specialist as well as wider managerial skills. However, it needs to be recognised that there is no accepted definition of human resource management (Bratton in Bratton and Gold, 1994). But clearly it relates to the overall corporate planning processes and focuses on the personnel element in those processes and in the management of agreed policies or plans. Much of what is traditionally termed personnel management is the face-to-face outcome of the impact of proposals on individuals in the organisation. There may be tension between line and personnel managers in deciding as to the consequences of policy decisions on individual staff members.

Once the human resource planning processes are complete, the personnel (or line) managers have, in crude terms, to get rid of some staff, to persuade other staff to agree to be redeployed and probably retrained, and to recruit new staff. Existing and new staff have to be assessed or appraised, and the organisation's range of jobs evaluated. All the time there will be concerns centred on issues of health, safety, welfare, equal opportunities, grievances and discipline. Industrial relations and negotiation with trade unions may be other important management tasks. Hopefully, there will also be training and development opportunities to be managed, with promotion and career planning, and possibly pay and productivity concerns and the task of implementing policies of redundancy. Communication and counselling will be a nearly daily part of a personnel manager's job. Employees become ill and there is worry about pensions. Employees die and their dependants require help. All these tasks are specialised, some based on evolving statute and case-law. The Institute of Personnel Management seeks to ensure that the standards of personnel managers are maintained. The manager in housing should be aware of the complexity of such personal issues and tasks, and may be able to rely on the personnel experts in the organisation to advise on, or to deal with, any such matters (Cole, 1993; Eyre, 1993; Goss, 1994; Rees, 1996). It is, however, a fact that some local authorities do not now have staff whose job-title is personnel manager, and performance review and staff appraisal is the responsibility of line managers (Catterick, 1995).

MOTIVATION AND MORALE AT WORK

Human relations management theories are usually rooted in psychology or in sociology. They are the antitheses of rational, scientific management principles in which the individual is an element in the systems of the organisation: one who will work harder if 'carrot and stick' techniques are used to give him or her more income for more work or, alternatively, threats of downgrading or unemployment. By contrast, the human relations model emphasises the need that human beings have at work to be seen as belonging to a group, as well as having individual aspirations and abilities. They need to feel secure, and to be esteemed for what they can contribute. In this approach, an employee is viewed as a social person who is responsive to the social forces of the group with whom he or she is working. This is one of the strengths of the unions. They can provide a stronger sense of belonging than do the employers. This can especially be the case in a very large, depersonalised organisation like a major company or a large local authority. The 'good' manager, while recognising the subordinates' social needs and loyalties, will try to provide job satisfaction through two-way communication: listening, encouraging, and thereby providing self-esteem. The 'carrot' of more pay will always be attractive, but may not always outweigh the anxiety that comes with more responsibility and separation from 'the group'. The reality is that every human being is different, and has to be managed differently. This is not good news for a person who believes that information technology can replace personal relations in the workplace.

Japanese management principles are of interest. Baba (1990) has theorised that there are two basic Japanese concepts of employment: employment of the individual as a human being and long-term employment with a company. The role of the unconscious mind in management is recognised. Japanese companies have stressed the life-long prospects of employment, and the social 'togetherness' of the working life, with loyalty to colleagues and to the company, and mutual respect between managers and subordinates. Decision-making has been shared at all levels of the hierarchy, with full consultation and participation. But with growing unemployment in Japan, these principles may be changing. Moreover, the hitherto male-dominated Japanese work culture, with long working hours, does not easily fit the newer world-wide concepts of the role of women in society. Nevertheless, recent Japanese management styles have belonged to the consultative–participative decision-making models in which the human relations factor is emphasised, and that has been a strength of the Japanese approach.

Later in this chapter is a section on communication, which is perceived to be based on either technological or human intercourse. Modes of communication link individuals with groups. A group can be described as 'a collection of individuals contributing to some common aim under the direction of a leader and who share a sense of common identify . . . in the work situation, most tasks are in fact undertaken by groups and teams, rather than by individuals' (Cole, 1993: 53). This may be a less easy mode of working for some 'loners' than for persons of more gregarious instincts. Conflicts inevitably arise. There may be conflicts between the self-esteem of individuals and the demands and needs of the group in deciding how to achieve their targets. 'The level of motivation in the group will be a decisive factor in effectiveness' (*ibid.*: 57). But what happens when there is disagreement?

CONFLICT AND ITS RESOLUTION

The modern view of conflict within and between groups is that it is normal, and can lead to innovation and better management outcomes. 'The task of management is to manage the level of conflict and its resolution for optimal organisational performance' (Stoner and Freeman, 1989: 392). The social scientist may perceive conflict as part of natural political behaviour in which one group seeks more power by forming alliances to achieve changes (Furze and Gale, 1996). Staff meetings can be the forum for conflict. The dynamics of group interaction in a period of conflict tend to increase feelings of uncertainty and a level of anxiety that may block easy solutions (*ibid.*).

'Conflict that has its roots in personal differences is especially difficult to resolve, partly because it is closely tied to values, attitudes and perceptions' (*ibid.*: 318). Discussion may help. However, it has to be accepted that some conflict is due to conflict within the mind of an individual. He or she may be unable adequately to undertake the work that is demanded. This deficiency may not be recognised or admitted, but may result in aggressive behaviour, leading to conflict between individuals, especially between subordinates and managers, or between individuals and groups. Another normal kind of conflict is between groups. There may be a problem of sharing resources, not least space and funds. There may be fundamental differences in goals, values or ideologies. These may have a political dimension. There may be the need for groups to agree priorities in work schedules imposed by higher levels of management: 'work inter-dependence exists when two or more sub-units depend on each other to complete their respective tasks' (Stoner and Freeman, 1989: 394), and that can easily result in disagreements.

There are ways to reduce the growth of conflict: staff responsibilities should be clearly defined; all meetings should be followed by follow-up notes confirming what has been agreed and who is responsible for what, by when; staff must be consulted before major decisions are irrevocably made; specialist staff should be held accountable for failure, with praise for success. There should be open discussion and consultation, with people encouraged to be explicit about their views. It may be necessary to bring in outside arbitrators. Above all, there must be two-way communication.

COMMUNICATION

Communication is the key to all effective management in housing, or in any work or personal situation. It is an activity that permeates all successful management. Communication today is dominated by the rapidly advancing techniques of information technology. But the nature of communication is so basic that it is common to all mammals, and probably in some form exists between all living creatures. It is an activity that bonds, enlightens, nurtures and protects. The word communicate is derived from the Latin word *communicare* (to share): communication is the process of achieving understanding between people (Padget, 1983).

But the receiver must be tuned in to the sender. There must be commonly understood communication media: a language or technological method that all the parties understand. One thing that technologists may overlook is the fact that the human senses used in communication are more than speaking and listening. There is a psychological dimension to communication, a body language, a meeting or non-meeting of eyes, a recognition of the way that the hands and the whole body change as negative or positive messages are given and received (Pease, 1981). Such signals may indeed contradict the words that are uttered and heard: the words may be untrue and consciously or unconsciously intended to deceive. As technological media begin to allow visual as well as oral communication, so the interpretation of body language can be world-wide, as on television. And the erstwhile divisions of communication modes into verbal or non-verbal will come closer in usage.

Some questions about communication

In practice, the objectives of communication should always be determined before the process commences. Some questions should be asked (Padget, 1983: 53–4):

1 Why is communication necessary? Is the purpose to obtain or to give information; to give instructions, or what?
2 What is the exact content? Is it fully and presently available?
3 With whom should the communication be? At what level in the hierarchy should the contact be made; the top person or someone lower in the management structure?
4 What mode of communication is apt? Usually eye-to-eye contact is best, with a follow-up in readable words that are permanent (for legal reasons, not least). But for instant globe-wide communication, the modes increase continually. However, in housing management the desk-top terminal will usually provide the mode, with the telephone and fax. Typed memoranda are still a common mode, not least between officers and members.
5 When should the communication take place? Instantly, or at some strategic time in the future?
6 Where should the communication take place? For communication to be meaningful, the location of the person who is to act on the message is important. The best place is usually in an office or at a meeting. But there may be subtle reasons why a social context is better.
7 How will the communication be followed up? The more impermanent the mode, the more important it is for messages of long-term importance to be in a form that is retrievable (not least as evidence).

In any large organisation, there will be formal, vertical line communications between managers and subordinates, and between directors and managers. These will often be of a functional nature. But there will also be informal communication between people working in the same organisation when they talk at work or socially. There are also horizontal or lateral lines of communication during which information, sometimes gossip, is relayed. Such informal information becomes part of a 'grapevine'. In a matrix structure, the horizontal levels of communication will also be more than functional. There will also be communication, functional and social, between people working in different organisations. Sometimes the impact of informal circuits of communication is greater than that of formal lines of communication. One reason is that in formal communication, there may be 'barriers' between people at work, between the communicators, or other problems.

Problems in communication

There are a range of difficulties that may prevent full communication in the workplace, between people working for the same housing organisation, or between them and those whose housing needs they are trying to satisfy, or between them and the elected members or board members.

1 Lack of forethought is a common problem. The objectives of the communication, as outlined above, may not have been adequately determined. The person receiving the communication may be critical and even hostile. It may be the wrong mode of communication to the wrong person at the wrong time, with an incoherent message.
2 Unclarified assumptions may have been made about the other person. A housing officer may telephone a person to discuss a transfer, only to find that the person who answers the telephone

is a newcomer who has replaced the previous partner of the tenant.

3 Semantic distortion is a use of words that implies something other than the truth to the person who is listening. It may be that the words and seeming promises are a way of keeping the other person happy for a while until the promised action does not materialise. An example is the statement 'The rent will be low' (lower than what?).
4 The body language may contradict the actual words used. The words 'I'll do what I can', said with arms crossed on the chest, and smiling at the next person in the queue, mean that nothing is likely to be done in the near future.
5 Emotional responses, in particular anger, may prevent an effective exchange of views. The converse, when people clearly like each other, and are beginning to use the body language that signifies attraction, may disguise the fact that the housing manager is not in a position to provide what she or he appears to be promising.
6 Hierarchical or social status difference may mean that people are so uneasy with each other, or one with the other, that they cannot express their true views or are unable to listen with belief to the other. People may be scared of, or dislike, anyone with authority or in uniform. Even the wearing of a tie by a man may seem to be a sign of assumed social superiority and intolerance, making timid people more timid. One party or the other may attribute views to the other that are unfounded.
7 Educational or cultural differences involve some of the problems of (3) and (6) above in communication. The 'language' used by one person may not be recognisable to the other. Some of the problems (8) and (9) below may also be a consequence of educational or cultural dissimilarities or conflicts.
8 Ideological differences include the different objectives of people of different political parties or professions. One member may be obsessed by the need for the success of privatisation, while another may be interested only in housing the homeless, irrespective of any social or financial consequences. An architect tends to be design-orientated compared with a quantity surveyor's concern with cost. A housing officer may want a large house subdivided into several small units of accommodation, while a town planning officer may advise against this on the grounds of lack of car parking space on the site and in the road. Ideological barriers can be substantial.
9 Technological differences. These, although related to (7) above, tend to be reflected in the fact that older and less educated people are computer illiterate, and cannot understand mathematical representations of information, nor do they always find it easy to learn how new media operate. The differences will be great between older tenants and younger officers.
10 Geographical separation is a common problem in communication, despite the prevalence of desktop terminals, e-mail, the internet and other modes of communication. There is no perfect human interaction substitute for face-to-face communication. Even offices on different floors, or in different but adjoining buildings, create a peer group reaction to people not part of the group, with the range of behavioural problems in communication that can cause. However, because of IT, geographical separation is not the excuse that it once was for problems in communication.
11 Time separation is a problem across the globe, with time differences. However, another time-related problem is the difficult relationship of historical evidence, sources and decisions on the solution of current problems.
12 Not listening is the final universal problem in effective communication and decision-making. Unless communication is two-way, there is no guarantee that the 'correct', long-term decisions will be made. More seriously, the implementation of such decisions may have to surmount more barriers if those who are responsible for the implementation have not shared in the initial decisions that related to solving the problems.

Every ambitious housing manager has to have a range of personal skills in communication. They need to be able to write grammatically, to be numerate, to understand and be able to use the available range of information technology; to be able to draft reports; to have the personal skills required to communicate with people of different backgrounds from their own; they must learn how to organise and to chair meetings, and to work in groups, as part of teams; and to negotiate with people from other departments and organisations. The effective control of meetings is an essential part of any manager's job. Meetings must have a clear purpose and be well prepared for (Bergin, 1976; Oldcorn, 1989; Rees, 1996; Furze and Gale, 1996).

Really successful housing managers must be superstars, or at least know how to pretend that they are!

CONCLUSIONS

Since the organisation of housing management differs considerably between local authorities, housing associations and private companies, the generalisations that apply to the larger organisations in the social and private sectors are those that are applicable to most larger organisations. Taking the issues in the reverse order in which they have been covered in this chapter, adequate communication in every sense of that word between those who work within an organisation, with other managers and with the tenants (the clients or consumers) is a fundamental component of good human management. As explained, communication problems are likely to be numerous. Much conflict will be avoided if there is adequate communication. Motivation and staff morale will be less likely to be a problem if the levels of communication between all staff at all levels of the organisation are two-way, frequent and comprehensive.

A specialist aspect of management that may have no place within a particular organisation is management by objectives, but with good leadership it can be a way to improve staff motivation and to facilitate empowerment, which is a key part of delegation and control. All management is dependent on there being professionally developed objectives which are the pivotal points of decision-making within the organisation, and fundamental to its control and change. The successful management of change is intimately related to human resource planning and management, which should be conducted as component parts of one operation.

The governance of housing may be part of the wider corporate planning organisation. The latter may be needed because of organisational structures that are not effective and may lead to improved structures as part of managed change overall. Matrix-type organisations tend to be more responsive to the needs and the demands of consumers, the tenants, than do exclusively linear structures. Yet accountability is essential. This chapter has attempted to outline many generic principles of management, some of which will be applicable in all reasonably large housing organisations.

Although this chapter concentrates on an examination of management related to housing, the matters considered are very relevant to the execution of housing policy and finance analysed in Chapter 3, and are particularly important in relation to policy-making and strategy discussed in Chapter 10.

QUESTIONS FOR DISCUSSION

1 Contrast the advantages and disadvantages in management of line structures with matrix structures in organisations.

2 Why are objectives so important in planning, decision-making and control? Discuss, with examples.

3 Explain how different situations may require different styles of leadership.

4 Contrast the ideas of empowerment, delegation and control, and assess any problems that may arise.

5 Which, to you, are the five most important barriers to communication in a housing officer's working life? Explain.

RECOMMENDED READING

Cole, G.A. (1993) *Management Theory and Practice*, DP Publications, London.

Eyre, E.C. (1993) *Mastering Basic Management*, Macmillan (Master Series), Basingstoke.

Fenwick, J. (1995) *Managing Local Government*, Chapman & Hall, London.

Mullins, L.J. (1995) *Management and Organisation Behaviour*, Pitman, London.

Rees, W.D. (1996) *The Skills of Management*, 4th edn, International Thomson Business Press, London.

REFERENCES

Allsop, M. (1979) *Management in the Professions*, Business Books, London.

Baba, K. (1990) *Management by the Unconscious – Reason for Strength of Japanese Management*, Chuoh Keizaisha, Tokyo, quoted by Ng Hui Leng (1996) 'Japanese Management', BSc (Quantity Surveying) dissertation, University of Greenwich.

Balchin, P. (1995) *Housing Policy: An Introduction*, Routledge, London.

Bains, K. (1972) *The New Local Authorities: Management and Structure*, HMSO, London.

Bergin, F.J. (1976) *Practical Communication*, Pitman, London.

Bratton, J. and Gold, J. (1994) *Human Resource Management Theory and Practice*, Macmillan, London.

Catterick, P. (1995) *Business Planning for Housing*, Chartered Institute of Housing, Coventry.

Charsley, W. (1985) 'Morale and Motivation', *Local Government News*, May, pp. 19, 21.

Clapham, D. and Franklin, B. (1994) *Housing Management, Community Care and Competitive Tendering*, Chartered Institute of Housing, Coventry.

Cole, G.A. (1993) *Management Theory and Practice*, DP Publications, London.

Diamond, D. and McLouglin, I.B. (eds) (1973) *Education for Planning: The Development of Knowledge and Capability for Urban Governance*, Pergamon Press, Oxford, vol. 1, part 1 *Progress in Planning*.

Eyre, E.C. (1993) *Mastering Basic Management*, Macmillan (Master Series), Basingstoke.

Fenwick, J. (1995) *Managing Local Government*, Chapman & Hall, London.

Furze, D. and Gale, C. (1996) *Interpreting Management: Exploring Change and Complexity*, International Thompson Business Press, London.

Glendinning, J.W. and Bullock, R.E.H. (1973) *Management by Objectives in Local Government*, Charles Knight, London.

Goss, D. (1994) *Principles of Human Resource Management*, Routledge, London.

Hunt, J. (1981) *Managing People at Work*, Pan Books, London.

King, J. and Newbury, A. (1996) *Managing Housing Contracts: A Good Practice Guide*, Chartered Institute of Housing, Coventry.

Koontz, M. and O'Donnell, C. (1988) *Essentials of Management*, McGraw-Hill, New York.

Lund, B. and Foord, M. (1996) *Towards Integrated Living? Housing Strategies and Community Care*, Policy Press with the Rowntree Foundation, Bristol.

Lupton, T. (1971) *Management and the Social Sciences*, 2nd edn, Penguin Books, Harmondsworth.

McConnell, S. (1981) *Theories for Planning*, Heinemann, London.

Malpass, P. and Murie, A. (1994) *Housing Policy and Practice*, Macmillan, London.

Massie, J.L. (1979) *Essentials of Management*, Prentice-Hall, Englewood Cliffs.

Mullins, L.J. (1993) *Management and Organisation Behaviour*, Pitman, London.

Newman, P. and Thornley, A. (1996) *Urban Planning in Europe*, Routledge, London.

Oldcorn, R. (1989) *Management*, Macmillan (Professional Masters), Basingstoke.

Padget, P. (1983) *Communication and Reports*, Caswell, London.

Pease, A. (1981) *Body Language*, Sheldon Press, London.

Price Waterhouse for the Department of the Environment (1995) *Tenants in Control: An Evaluation of Tenant-led Housing Management Organisations*, Housing Research Report, HMSO, London.

Rees, W.D. (1996) *The Skills of Management*, 4th edn, International Thomson Business Press, London.

Skitt, J. (ed.) (1975) *Practical Corporate Planning in Local Government*, Leonard Hill (International Text Book Co.), Leighton Buzzard.

Stoner, J.A.F. and Freeman, R.E. (1989 and 1992) *Management* (two edns), Prentice-Hall International, Englewood Cliffs.

Torrington, D. and Hall, L. (1991) *Personnel Management: A New Approach*, Prentice-Hall, Hemel Hempstead.

10

POLICY-MAKING AND POLITICS

Maureen Rhoden

Politics is the way society or a nation manages conflicts and disagreements. It is difficult to exclude value judgements from the study of politics. It is important, however, to discover and describe what actually happens. The purpose of this chapter is to review the executive, legislative and judicial functions of central government, the local policy-making process, theories of local government politics and the policy-making process in housing associations. The chapter will begin with an examination of central government and the functions which it performs. It will then investigate the role of local government in practice and in theory and will end with a study of housing associations' structure and purpose. The chapter focuses specifically on:

- The executive, legislative and judicial functions of central government:
  - the executive, civil service and legislature;
  - private members' bills;
  - scrutiny of government by committees, and questions in the House;
  - interest and pressure groups;
  - Scotland and Wales;
  - central and local government relations;
  - sub-central structures;
  - inter-governmental relations;
  - central and local government relations during the 1980s.
- Local policy-making processes:
  - Elected members;
  - Local interest groups;
  - Individuals within the policy-making process.
- Theories on local government politics:
  - localist theory;
  - public choice theory;
  - dual state thesis;
  - local state theory.
- Housing associations:
  - structure;
  - board/committee members.
- Conclusions.

EXECUTIVE, LEGISLATIVE AND JUDICIAL FUNCTIONS OF CENTRAL GOVERNMENT

The executive

The Prime Minister

The function of the Prime Minister is to lead the majority party – a function requiring the loyalty of his or her party and the command of the House of Commons. However, the loyalty of the party is not absolute and revolts do occur from time to time, such as the continuing debate within the Conservative Party regarding whether or not Britain should remain a member of the European Union, and if so, what limits should be set in order to remain a member.

The Prime Minster has the power to appoint and dismiss and decides who will be a member of the

Cabinet (see below). In deciding on appointments to the Cabinet the Prime Minister has to ensure that senior colleagues and important factions within the party are not offended.

The Prime Minister's power of appointment stretches beyond the Cabinet and also includes positions such as hereditary and life peers, heads of Royal Commissions and Permanent Secretaries within the civil service. The Prime Minister is expected to 'encourage, even inspire; he must co-ordinate, preserve balance, keep the convoy moving steadily' (Madgwick, 1984: 58–9).

The Prime Minister chairs the Cabinet and regulates the agenda and its debates. He or she controls the Cabinet committees and how the information is disseminated.

The Prime Minister also has a wide range of public duties at home such as formal speeches, receptions and broadcasts which can enhance the high public image which the position has. When abroad the Prime Minister is expected to represent the nation.

The Prime Minister has an important role to play in the House of Commons, being expected to be the main spokesperson and to lead the party on the floor of the House. He or she must also participate in major debates, make statements and answer questions once a week in Prime Minister's question time. The performance of the Prime Minister in Prime Minister's question time is crucial in terms of the public image which is conveyed in terms the skills which he or she has of debate and repartee. The Prime Minister also has a duty to maintain contact with the governing party's backbenchers (Members of Parliament who are not Ministers) through formal or informal meetings/sessions.

The Cabinet

The Cabinet consists of approximately twenty persons who include Members of Parliament and members of the House of Lords holding offices in the government mostly bearing the title of Secretary of State. The size of the Cabinet reflects the need to ensure that the interests of sections within the government are adequately represented. Other members of the Cabinet may not have the title Secretary of State but have an important role in terms of providing advice or co-ordination at a high level within both Houses or may be special appointments such as Lord Privy Seal and Chancellor of the Duchy of Lancaster.

The function of the Cabinet includes taking decisions on matters of general policy, especially government expenditure; taking decisions on sensitive matters which often have a high media profile, such as the debate on whether to ban the sale of handguns in the wake of the tragic events in Dunblane during 1996; acting as an arbitrator between departments or ministers who are unable to agree on an important issue; and overseeing and co-ordinating the operation and administration of government.

The Cabinet works through a system of sub-committees with the assistance of a secretariat and so the work of the Cabinet is principally conducted outside of the Cabinet. This system allows the Cabinet to meet once or twice a week and to be in a position to take important decisions.

Cabinet committees usually consist of two or three Cabinet senior ministers, other relevant ministers concerned with the subject under discussion and possibly high-level civil servants. The committee is usually formed to discuss issues such as future legislation and to look at details which the Cabinet does not have time to investigate. It is chaired by the Prime Minister or another senior minister.

The Cabinet committee can include any non-Cabinet ministers and civil servants to debate the issue under debate. This system helps to prepare junior ministers for more Cabinet responsibility and helps to improve communication and co-ordination. Any decisions reached by the sub-committee are passed to the Cabinet for approval and the Cabinet has the right of rejection; however, in practice this is unlikely as the committee would have attempted to clarify any issues which would cause the Cabinet to disagree with their conclusions. The position of the Prime Minister remains central within the system as he or she often chairs or liaises with the chairs of the sub-committees to keep informed of the direction which a committee takes on an issue.

The Cabinet has the duty to co-ordinate government policies and so ensure that no policy is inadvertently compromised by another, for example by ensuring that any proposals for expenditure fit the government's overall budget. Madgwick (1984) stated that co-ordination takes place through the following means:

1 the Prime Minister;
2 non-departmental ministers;
3 the Treasury;
4 the Secretaries of State in main departments such as the Ministry of Defence and the Department of the Environment;
5 the committee system.

Ministers and their departments

Members of Parliament are usually expected to prove themselves in Parliament before gaining office and then to prove themselves in office before gaining appointment to the Cabinet. This provides the Member of Parliament with the opportunity to be inducted by means of a system of training and selection. On average most Members of Parliament will have spent fourteen years in Parliament before reaching the Cabinet. Few ministers spend more than two years within an office and they are moved on frequently.

There are, however, situations where a few exceptional people will pass through the system very quickly or where a political party has been in opposition for a long period of time and may be forced to appoint relatively inexperienced ministers to the Cabinet. An example of such a situation occurred with the Labour Party when it won the general election on 1 May 1997. The party had been in opposition for the previous seventeen years and so many of its potential ministers were inevitably inexperienced.

A minister must try to maintain good and productive working relationship with the Prime Minister, the Cabinet, ministerial colleagues and the House of Commons and must ensure that the Prime Minister is kept informed of any major developments within the minister's department and of any potential problems which may arise.

The minister is responsible for managing an important organisation while pursuing duties in or towards the Cabinet and Parliament. The minister is ultimately responsible for the work of his or her department, although ministers cannot know everything that is happening within the department and are unlikely to remain within the department longer than two years on average.

Departments of State

There are approximately twenty major departments which are managed by Cabinet ministers. Some departments rise and fall over time, such as the Department of the Environment, which was created in 1970 and consisted of the previous Ministries of Housing and Local Government, Transport and Public Buildings and Works. Thus, the aim is often to unify departments and create one major department or to seek to highlight the significance and autonomy of an area of work, as with the formation in 1974 of the Department of Prices and Consumer Protection.

Departments vary in size, with approximately 600 members of staff in the Cabinet Office to around 15,000 or more in Departments such as the Ministry of Defence or the Home Office, and can be categorized as large bureaucracies. However, it must be remembered that no part of the government is independent of the curbs of public expenditure.

The Secretary of State or a minister is at the head of each department; below are the ministers, or Ministers of State and Under-Secretaries of State. The chief administrator is the Permanent Under-Secretary of State, which is the highest grade within the administrative civil service. The Permanent Secretary is responsible to the minister or Secretary of State and given all the responsibilities which politicians have to carry. The Permanent Secretary has the duty of managing and controlling most of the department's day-to-day work. The Permanent

Secretary also has the additional role of adviser to the minister on issues such as major departmental decisions, and the formulation of new policies and speeches, questions and legislation.

The Treasury

The Treasury is dominant in terms of economic matters and is responsible for the raising of revenue, the control of public expenditure and the management of the national economy. The 1961 Plowden Report on the Control of Public Expenditure (Treasury, 1961) introduced two new procedures, the Public Expenditure Survey Committee (PESC) and the Programme Analysis and Review (PAR).

PAR was intended to help the Treasury and non-departmental ministers to investigate any programme within any department to identify clear objectives. It was not continued after the 1980s as priority had shifted to the reduction of costs.

PESC consists of an inter-departmental group of finance officers and Treasury officials and is chaired by a Treasury Deputy Secretary. PESC seeks to ensure that there are realistic projected costings of programmes which look five years ahead and takes into consideration likely ramifications and repercussions. The committee reports to the Cabinet in May of each year and a Cabinet tier will make the final decisions. The Chief Secretary negotiates with the spending ministers using the support of senior non-departmental ministers (also known as the Star Chamber) when required. The final stage of appeal for a spending minister is the Cabinet or the Prime Minister. The report is not published but the government publishes a White Paper on public expenditure. Parliament then holds a two-day debate, but at this stage it is too late to change the outcome of the White Paper.

Despite the fact that the Treasury is a major department which has the support of the Prime Minister and Chancellor of the Exchequer, it is still not able to reduce public expenditure. This is due in part to the fact that most of the expenditure is long term; expensive services cannot be discarded and programmes which are developing may need to be allowed to continue. In addition, the Treasury does not always achieve its aim of reducing the expenditure of spending ministers and can sometimes lose out in negotiations. During the 1980s and 1990s the Treasury has managed to change the attitude to public expenditure and to engender an atmosphere of questioning expenditure levels but it still does not have total control over public expenditure.

The civil service

The civil service consists of approximately 600,000 civil servants, of whom around 1,000 are involved in policy-making processes. The policy-level civil servants are engaged in a number of major areas of work, which include administrative work, assisting in the development and implementation of policy and preparing legislation, answering parliamentary questions and letters from Members of Parliament, and briefing the minister (Stoker, 1991).

Madgwick (1984) identified the main distinctive features of the British civil service: first, the formulation of policy which is intended to ensure that policies are practicable; second, the anonymity of the civil servant – a peculiar trait as he or she advises the minister but is not answerable to Parliament. The minister who makes the final decision is responsible to the Prime Minister and the Cabinet. A third feature is that the civil service is expected to be non-political and the civil servant is expected to deal loyally with the government they are serving.

The civil service has been the subject of a number of criticisms, including suggestions that it is too vast, too expensive, is lacking initiative, a powerful elite, not able to identify the will of the public, and far too bureaucratic. In addition, the recruitment and training of staff have been criticised for mostly selecting staff from a narrow social, racial and educational origin. There is also the previously mentioned problem with the influence over the development of policy which civil servants have in relation to their ministers and the public and the secrecy which surrounds the officials at the higher levels of the civil service.

The legislature

The Member of Parliament (MP)

MPs serve on average approximately sixteen years and those who leave do so for a variety of reasons such as resignation, removal by the electors and death. The majority of MPs are middle-class by occupation and have been educated to at least undergraduate degree level. However, women and racial minorities continue to make up a minority of MPs. Thus, it can be seen that Parliament does not represent a true reflection of the population.

It is often debated whether an MP is expected to put forward their own views or the views of their constituents. The difficulty with presenting their own views is that the MP may not be able to reach a reasoned conclusion on every issue which may arise. In addition, the MP is not able to represent the views of all constituents as some may not have voted for him/her and s/he would have to consult them on any additional issues which might arise after the election. As a consequence, the MP would attempt to represent the view of the majority of their constituents, where possible.

Once or twice a week, party meetings are held to allow MPs to discuss issues on which there is no general agreement within the party and, more generally, the business of the House. This is an opportunity for ministers to meet with backbenchers. Backbenchers can sometimes be seen to 'flex their muscles' but must not be seen to direct the Cabinet, as this would be regarded as unconstitutional and incompetent as the Cabinet is responsible to the House and not the parliamentary party.

The officers of the parliamentary party are known as whips. They are expected to organise the party, ensure that as many MPs attend the House as possible, encouraging the backing of government policies and ensuring that ministers are aware of the views of the backbenchers. The Chief Whip is usually a member of the Cabinet and maintains communication with the Prime Minister. A document containing the business for the week in the House is prepared and forwarded to all members. The most important business is indicated by one, two or three underlines to indicate how important it is for the MP to attend. The ultimate sanction which the whip has is the withdrawal of the whip, which results effectively in expulsion from the parliamentary party and, if it is not revoked, the possibility of the MP losing their seat at the next election.

Parliamentary business

There are a large number of speeches, debates, committees, minutes and reports generated within the House. The Speaker of the House is responsible for controlling these functions although he or she does not have any governmental responsibility. The Speaker is able to act impartially when chairing in the House and has to ensure that parliamentary procedures and behaviour are safeguarded. In order to ensure that the rules of procedure in the House are not abused the Speaker has the power to call for order, call for the withdrawal of unparliamentary language or for the withdrawal of the member from the chamber, and to suspend the sitting.

The time of the House is controlled by the Leader of the House on behalf of the government and the remaining days (usually amounting to about twenty-nine days) are available for the opposition to raise issues which they consider important. Another twelve to twenty days (usually no more than half-day sessions) are available for private members to introduce Bills or motions. The Leader of the House will consult to a limited extent with the opposition whips when planning the timetable of the House.

Legislation

When governments enter the House they are expecting to make changes, as the election campaign will have made some attempt to exaggerate the difference between the political parties (although the 1997 election campaign has seen for the first time a possible narrowing of the differences between the two major political parties, the Conservative and Labour Parties). Some of the policies of the government can be achieved without resorting to legislation, such as the shifting of financial priorities, or a new foreign policy by introducing a new emphasis. Both of these examples may be debated within the House but do not involve the need to introduce legislation.

When legislation is introduced it is usually the result of major features of the party which have developed from debates and criticisms raised by a combination which includes the parliamentary leaders, backbenchers and the rest of the party nationally.

The legislative process

Bills are based on ideas which are developed through discussions with departments of state, interest or pressure groups, the parliamentary party and a Cabinet committee. The process mainly takes place outside Parliament, so that by the time a bill first appears in Parliament it is in draft form but is very close to what the final legislation will include.

There are public and private bills. Private bills are concerned with private matters such as a local authority seeking new powers. Public bills are promoted by the government or by a private member of Parliament; the latter are called private member's bills.

All bills have to go through an introduction and first reading where the title of the bill is read out to the House and the minister names a day for the second reading. This is merely an indicator from the government that it intends to bring a bill to the House and this allows for the bill to be published in order to give the opposition the opportunity to study it. There is therefore no debate at this stage.

The second reading is when the principal debate on the rationale of the bill takes place and at this stage the only amendment which can be accepted is a rejection of the bill. If the bill is considered to be non-contentious, it may be heard in a committee.

If the bill involves expenditure, it must be authorised by the House through the financial resolution stage, when again the House debates the financial implications.

The next stage, the standing committee, is when the bill is studied in detail and amendments are carried out. The House appoints standing committees as bills arise for consideration and between twenty-five and forty-five members are appointed to the committees. The members are selected by an all-party Committee of Selection which is managed by the whips. The representation on the committee always allows the government to have a majority (if it has a majority in the House), and a chair is selected by the Speaker from an all-party panel.

The members of a standing committee will often have a special interest in the bill under consideration or special knowledge of its theme. Most bills, if non-controversial, will be ratified by the committee after one or two sittings. However, other more contentious bills may be argued over clause by clause, especially where opposition amendments are involved.

Once the bill has been amended, it passes to the report stage, where it is formally reported to the whole House. At this stage the government may insert additional amendments.

The third reading can involve an extensive debate of the whole House similar to that which would have occurred at the second reading, and so often this stage is accepted without further debate.

The bill then passes to the House of Lords, where it goes through virtually the same process with the exception that there is no financial resolution stage, as the House of Lords cannot discuss financial matters, and the standing committee stage becomes a committee of the whole House of Lords. The bill may be amended by the Lords, but if amendments are not acceptable to the House of Commons then they may be overturned after giving the Lords the opportunity to change the amendments.

The bill usually passes through the House of Commons quickly and is then returned to the House of Lords where it receives the royal assent and so becomes an Act of Parliament.

Throughout the process the government has the means to accelerate bills through the parliamentary procedure through the closure of debate. This motion can be moved at any time during a debate and the Speaker has the right to accept or refuse such motions if he or she thinks that there has been insufficient debate. If the Speaker accepts the motion, it is then decided by a vote. If the debate concerns an amendment the Speaker can ask 'if the question has now been put' and so end the debate.

The government also has other remedies to accelerate bills, especially where the opposition are determined to argue over every clause. These include a restrictive timetable for the bill (also known as the 'guillotine'), which is managed by the grouping of clauses and amendments in the bill for debate, or by 'kangaroo', where the government jumps from one chosen amendment and clause to another.

Private member's bills

The time allowed for the introduction of bills and motions by private members (approximately twenty days with some half days) is shared out by a ballot. There is also the 'ten minute' rule whereby a member can introduce a bill to the House with a ten-minute speech. A ten-minute reply may be made and the House then votes or accepts the bill unopposed. No bill is allowed to propose the spending of money and there is no official help in the drafting of the bill. It must be supported by a minimum of one hundred members, the committee investigating the bill must have a quorum, and if it is not completed within one session of Parliament it then lapses. There are occasions when such a bill has widespread support and on these occasions the government may step in to assist its passage through Parliament.

Scrutiny of government by committees

Standing committees are legislative and are concerned with generating a bill in the final form. The minister who chairs the committee controls the work of the committee and uses his/her majority to ensure that his/her purpose is achieved.

Select committees are much more varied and range from committees overseeing expenditure to those overseeing domestic matters such as privileges.

The Committee on Public Accounts consists of fifteen backbenchers who are appointed in line with the balance of power in the House. The chair is usually a member with some financial experience from the opposition party and an official from the Treasury also attends. The committee can obtain the assistance and advice of the staff from the Comptroller and Auditor-General, who specialise in the accounts of government. This department audits the accounts of government departments and presents its findings to the committee.

The Committee on Public Accounts considers expenditure which has already been incurred to ensure that it has been spent as was originally intended, to monitor the virement by the Treasury from one department to another, and to control the means of presenting the national accounts. The reports which are generated by this committee are forwarded to the Treasury and departments. The Treasury then responds on the recommendations of the committee by way of a formal annual Treasury minute. In addition, the House debates the reports over one or two days and questions can be asked at this time.

A new system of select committees was introduced in 1979 in which a select committee is linked to a major department. These committees deal with areas such as expenditure, administration and policy of the main government departments. Each committee consists of between nine and thirteen members, specialist advisers and a chair who is chosen by the Committee of Selection.

Questions in the House

Question Time is often seen as an opportunity for backbenchers to query government policies. Questions are put down at least forty-eight hours prior to Question Time, and members are limited to two oral questions per session. Ministers attend the House on a rota basis while the Prime Minister attends once a week. If the oral questions are not answered, then a written reply is provided.

The answers to the parliamentary questions are prepared in the departments by civil servants. The minister will also receive information to help him/her answer supplementary questions which may be asked after the minister has answered the first question raised by a member, especially if the member is not satisfied with the reply to their first question. Some questions are arranged by the government in order to announce an important issue such as a new policy. The government can refuse to answer a question if there would be disproportionate cost in doing so.

During the period of Conservative government, 1979–97, Question Time became increasingly confrontational, as members increasingly used the opportunity to ask questions about the Prime Minister's engagements in order to ask supplementary questions about matters of concern.

Interest and pressure groups

Madgwick (1984) states that pressure groups try to alter government policy by the use of persuasion. While interest groups do keep a watch on government to ensure their own particular interests are not ignored, they also seek to promote their attitudes and views when needed. The changes are sought as the group has an interest in protecting or developing a concept which could be basic government policy or an administrative detail. Interest and pressure groups are an accepted part of the political system.

In order to influence government the most powerful pressure groups will seek to work within the departments and with the ministers. They may, for example, have their corporation headed by a political notable. Less powerful pressure groups use other means, such as public campaigns and contacts with ministers and members.

There are some instances where a minister is sponsored by a pressure group with regard to receiving financial assistance for an election campaign and for looking after the particular interests of the pressure group. Some may take a part-time position as an adviser or consultant to a group, or be asked to take on an honorary position such as vice-president.

Scotland and Wales

In Scotland the Scottish Office is held by the Secretary of State with the support of ministers and is based in Edinburgh. The Welsh Office is based in Cardiff with a similar structure to the Scottish Office. They represent the interests of their territories and are expected to manage areas of domestic policies such as housing and education.

The ministers in the Scottish and Welsh Offices are seen as being a part of Whitehall first and a part of their respective territories second. The minister is expected to be the spokesperson for their territory, while at the same time regulating, supporting and promoting central government policies.

Central and local government relations

As stated above, higher-level civil servants and ministers are mainly 'concerned with allocating funds, regulating standards, processing legislation and identifying new problems and issues but they are not directly responsible for day-to-day service delivery and policy implementation' (Stoker, 1991: 141). These functions are carried out by local authorities and other governmental and quasi-governmental organisations.

Rhodes (1988) states that central government is in fact a mixture of disparate departments and sections. As a consequence, there are a number of conflicting concerns, such as in local authority expenditure, where there are at least three departments involved: first, the Treasury; second, spending departments, such as the Department of Defence, with no interest in local government and so perhaps wishing to see it take the impact of any expenditure restraint; and lastly those spending departments, such as the Department of the Environment, with an interest in local government, which may attempt to preserve local authorities' expenditure levels.

In addition to the example of inter-departmental conflict described above, there is also the situation where a department, such as the Department of the

Environment, has conflicting roles to perform. Thus, the department has an advocacy role as described above, a monitoring function, and one as a controller of local government expenditure levels.

There also exists the situation whereby various elements of central government will join together with the same aim of providing a focus for co-ordination. There are also a shared culture and a commitment to common values which set central government apart from local government (Heclo and Wildavsky, 1974).

Sub-central structures

Rhodes (1988) stated that the central government departments have sub-central structures, or 'intermediate institutions', which include regional differences between and within departments. The intermediate institutions were identified as having three functions. First, they administer the duties which are handed down to them by central government and so co-ordinate the actions of a variety of agencies. Second, the intermediate institutions are responsible for regulating and promoting the implementation of central government policies. Third, they represent territorial interest to the centre.

Thus, local authorities need to develop good relations with regional offices of government departments such as the Department of the Environment to be in a better position to manage the range of policies in their areas.

Inter-governmental Relations

Dunleavy (1980) states that local government not only has to relate with central government departments but also has to function alongside non-elected local government bodies. These can range from those which may have to contact and work with the local authority, such as the nuclear power industry looking for suitable sites to dispose of waste, to those which have no contact with the local authority. There are also local authority associations, professional organisations, party institutions and trade unions. All have some impact on how local authorities are able to perform their tasks within a local framework.

Central and local government relations during the 1980s

Stoker (1991) indicates that there has been a major shift in the relations which existed between local authorities and central government during the 1970s and 1980s. This is seen as central government attempting to strengthen its power and influence over local government, especially during the period when Margaret Thatcher was Prime Minister.

The relations between central and local government have become concerned with the behaviour and performance of local government. As a result the amount of legislation, circulars and advice has increased dramatically to the extent that the local authority's discretion in certain local issues such as housing has been reduced and individual authorities that do not carry out their duties as directed by central government can now be targeted, for example by reductions in funding allocated by central government.

Whereas in the past legislation was passed to enable local authorities to function with some degree of discretion, this possibility has now been removed to a large extent and now local authorities are faced with goals and targets which are set and monitored centrally. Elcock (1986) states that, while legislation exists to allow individual members of the public to do anything which is not forbidden by law, local authorities may only do what the law permits them to do. Thus, if local authorities spend outside of the rules laid down by legislation, then they are acting outside the law (*ultra vires*) and the members of the council which supported the illegal action may be surcharged and required to repay the funds from their private finances. The members can also be barred from holding public office for a period of time which is prescribed by the courts.

Nicholas Ridley (1988), during his period as Secretary of State for the Environment, asserted that the role of local authorities would be to stop attempting to be a universal provider but to adopt instead the role of enabler. This would ensure that their functions were carried out instead by other agencies within the private and voluntary sectors.

Ridley (1988) saw this being achieved through, first, the disposal of large amounts of local authorities' assets such as houses, buildings and land; second, the need to introduce private sector principles into local authorities, as the introduction of competition would lead to better value for money and improved service delivery to the customer; finally, local authorities should question the need to adopt the role of universal provider and look to other agencies to carry out as many of their functions as possible.

Despite the fact that many local authorities did not share the same ideals as central government, they were required, under the Thatcher governments, to comply as they had to perform the day-to-day implementation of the policies. However, despite the attempts of central government to bring local authorities into line there have been some that have still managed to avoid central government demands. Stoker (1991) argues that, while central government may have increased its control over local authorities, it is over a narrow range of issues, while resistance from local authorities has become politicised.

During the Thatcher governments, fifty Acts which directly affected local government were passed and in fact legislation has become the basis on which the relations between local and central government have developed (Loughlin, 1986). Local government legislation has become more detailed and complicated and has given central government more discretionary powers.

Both the local authorities and central government have been willing to resort to the courts in order to advance or protect their concerns. This has not always been a successful strategy, as often the courts are unable in terms of design and culture to deal with these demands (Grant, 1986).

In fact, central government has attempted to overcome cases where local authorities have won victories by the passing of new legislation to revoke the courts' decision and to ensure that there is no further recourse to the courts. An example of this was the Local Government Act 1987 Section 4(1), which states that:

> Anything done by the Secretary of State before the passing of this Act for the purposes of the relevant provisions in relation to any of the initial years or intermediate years shall be deemed to have been done in compliance with those provisions.

Section 4(6) also states that:

> Subsection (1) above shall have effect notwithstanding any decision of a court (whether before or after the passing of this Act) purporting to have a contrary effect.

A feature of central–local relations has been the introduction of minimal consultation with local authorities. Major legislation affecting the local authorities, such as the abolition of the Greater London Council (GLC) and six Metropolitan Counties, was passed without any reference to the public, local authorities or other relevant bodies.

Central government has had increasingly to turn to the use of specific and supplementary grants, as mentioned above, in order to wield greater control over local government. The specific grants are targeted to special areas such as the police and cover a portion of the total cost, so leaving the local authority to make up the difference. If, therefore, central government decides to reduce its grants to local authorities or to restrict the rate of growth of the grant, it can have a major effect on the ability of local authorities to develop and improve their services (Elcock, 1986).

Central–local relations have also been affected by the willingness of central government to bypass local authorities in the carrying out of new local initiatives such as the introduction of the Urban Development Corporations.

LOCAL POLICY-MAKING PROCESSES

Elected members

Malpass and Murie (1990) state that the policy process within local authorities often lacks any clear start or finish and the passage from one stage to the next is not clearly defined. While the elected members of the local authority may be the decision-takers, the actual decision-makers are often council officers, who formulate the policies on behalf of the members. These policies are formalised through the voting of elected members in committees and the full council meetings.

Stoker (1991: 34) states that 'the traditional councillor stereotype is a white, middle-aged, white-collar male'. The survey conducted in 1985 by Widdicombe (1986) found that 81 per cent of councillors were male, 74 per cent were over forty-five years old, 69 per cent were in or had been in non-manual work and 85 per cent were owner-occupiers (see Table 10.1). The researchers also found that the number of female councillors was small across all the political parties.

Within the local authority, not all councillors are equally influential and usually it is the party with the majority within the town hall which has the

Table 10.1 Councillors by political party membership, age, gender, activity status, socio-economic group and income, 1985

	Con %	*Lab* %	*Lib* %	*Ind* %	*Other* %	*All* %
Age						
60+	37	32	22	52	30	36
45–59	42	33	30	36	33	37
18–44	19	33	49	11	33	26
Gender						
Male	78	83	79	81	81	81
Female	21	17	21	19	19	19
Activity status						
Employed full or part-time	64	59	74	47	67	60
Unemployed	2	8	3	1	2	4
Retired	24	22	13	40	30	25
Permanently sick/disabled	—	2	—	1	—	1
Looking after a home	9	6	9	9	7	8
Other	—	1	—	—	—	1
Socio-economic group						
Professional	11	6	15	8	5	9
Employers/managers	42	20	28	37	26	32
Intermediate non-manual	14	22	23	11	30	18
Junior non-manual	8	11	15	6	2	10
Skilled manual/own account non-professional	10	25	8	14	14	16
Semi-skilled manual	1	8	4	5	—	4
Unskilled manual	—	2	1	1	—	1
Armed forces/NA	12	8	8	16	5	11
Income						
£15,000+	25	7	18	14	33	16
£10,000–£14,999	22	23	29	17	2	22
£6,000–£9,000	23	28	23	27	25	26
up to £5,999	22	38	23	29	23	28
None	1	1	2	2	—	1
Refused/NA	8	4	3	10	2	7
Base	(595)	(496)	(133)	(224)	(43)	(1,557)

Source: Widdicombe, D. (1986)

most collective influence. In addition, each local councillor interprets their own role as an elected member differently. As the member is elected by their ward, they may acquire a caseload of queries from their ward which can take up a great deal of their time (Malpass and Murie, 1990). However, there are other members who see their role as being one of participating in policy issues. Again, these roles are not exclusive to each other and some members may attempt to combine both elements of their responsibilities, while often newer members may concentrate more on case-work and the emphasis then changes gradually as they become more experienced ward members.

In order to become active in housing policy a councillor would need to procure a place on the housing committee. If a councillor is unable to obtain a place on the committee then they can still contribute to policy through the private meetings of their party group. Their party groups meet on a regular basis to agree how the group will vote at the full council meeting and to consider other policy issues.

Senior council officers are involved in policy-making while the more junior staff are responsible for carrying out the implementation of the policies. However, the trade unions can influence the policy-making process and many local authorities have now introduced research and policy sections in an attempt to formalise the process further.

Local interest groups

The public are also able to have some influence on policy, depending on the relations that an interest group may be able to develop with a ward member. Those who have good working relations with members are able to use the informal and formal means to influence policy. However, those interest groups which are not so fortunate have to resort to actions such as demonstrations as they are not able to gain access through formal or informal methods (Malpass and Murie, 1990).

Stoker (1991) identifies four types of interest groups which operate within local government: first, producer or economic groups, which include trade unions and businesses; second, community groups such as residents' associations and groups representing racial and ethnic minorities; third, those groups which are established to promote a particular set of ideals and beliefs, such as local branches of the Campaign for Nuclear Disarmament; finally, the voluntary sector, established to meet a perceived need within the community on a non-commercial and non-statutory basis, such as local churches and Age Concern. However, the line between the different interest groups is not always clear.

The interpretation of group politics draws on two theories, namely pluralist and elitist. The pluralist theory states that interest groups help to vocalise and so improve local democracy. However, the elitist theory states that power is concentrated in the hands of a few and local authorities are closed organisations with only a select range of external interests drawn into the decision-making process, with producer or economic interests predominating.

However, there has been an opening up of local authorities in an attempt to remove the elitist label. First, there has been a shift in the attitudes of councillors, who now work more closely with a wide range of interest groups. There has also been a change in terms of the decentralisation of the decision-making framework, with the introduction of local area committees and the use of decentralised service delivery. This process has helped to increase participation by the public. In addition, council officers have been forced to develop a more community-based outlook through the introduction of compulsory competitive tendering (CCT) and performance information which is now published for the public.

Individuals within the policy-making process

In order to explain the position of the individuals in the policy-making process at a local level, Malpass and Murie (1990) refer to a corporate-competitive model. Councillors represent the electorate and so are able to confront the council officers as elected representatives. The council officers can claim professional knowledge and, with the growing level of specialisation in local government, can base their authority on expertise in their work.

Malpass and Murie (1990) refer to two types of rationality at work in local authorities, the political rationality of the elected members and the bureaucratic/technical rationality of the council officers. Thus, a decision which may appear to be correct politically may be a disaster technically and vice versa.

The corporate-competitive model referred to above is based on the concept that council officers and councillors negotiate their relations and so the distribution of power between the two parties is more or less fixed. The process of negotiation allows for some variation between each decision and each policy.

However, it must be remembered that the council officers have the advantage of being full-time professionals, while councillors are usually part-time with another mode of employment elsewhere and in some instances do not have the expertise in important areas such as local government finance. In addition, they also have to take care of their constituents and be aware that they must face re-election every four years. There are, however, senior councillors who do over a period of time become experts in issues such as housing. Often the local authority political leaders are full-time politicians as they often represent safe seats.

The power the officers have rests in the fact that they can decide the issues which will be presented to the committee for decision and also in the way the information is presented to the committee, which can result in the committee being directed to a positive or negative outcome. The Widdicombe survey (1986) found that councillors were becoming much more willing to challenge officers in areas such as agenda-setting for committees, the redrafting of reports, the appointment of senior officers and the expectation from councillors that they should have free access to any officer and to intervene in internal management matters.

Malpass and Murie (1990) identified five categories of decisions that comprise housing policy:

1. *Strategy* Political parties usually have some sort of broadly distinctive values approach to housing. This provides the basis for the development of a housing policy in which the overall framework of objectives and aspirations can be set out. Strategic decisions are important because they establish the values and principles from which other sorts of policy decisions can be derived.
2. *Investment* Within the overall framework of values guiding housing policy generally in a local authority, a specific category of decisions is constituted by those that concern capital-investment projects. Annual estimates of investment in housing projects are now required by the Department of the Environment. In addition, each local authority has to make a series of policy decisions about each continuing investment programme, ranging from the decision to proceed and the acceptance of a contractor's tender.
3. *Management* Policy decisions are also required for the management of the housing service after the dwellings have been built. This is a continuing management process which requires continuing policy decisions.
4. *Procedural* This policy refers to decisions about how decisions are to be made, such as the best way to include residents in the policy-making process.
5. *Administrative* This refers to the style and structure of the housing service, such as the issues regarding decentralisation of area housing offices.

THEORIES ON LOCAL GOVERNMENT POLITICS

Localist theory

The localist theory supports the need for 'autonomous, elected local authorities' (Stoker, 1991: 234). The case for local government in the localist theory includes four elements: first, the need to have as many local decision-makers as possible involved in the process; second, local government allows different views from different localities to be represented and so rather than using the same policies regardless of local needs, local authorities are able to develop and learn from the experiences of other local authorities.

A third element of the localist theory is that local government is accessible and responsive to local needs through the election of local councillors and the appointment of local government officers. These individuals live close to the people they serve and so people within the locality are able to influence the decisions of the local authority for which they work as an officer or represent as a councillor. Finally, local government is able to match local needs to the resources available locally and so is able to win local loyalty.

This theory has been supported by many, including the Widdicombe committee, which stressed the need of local authorities to encourage change, public choice and participation. However, these views are based on a pluralist approach which is against centralisation and instead supports local service delivery.

The localist theory accepts that access to local authorities' policy-making process is not the same for all, especially with regard to some local interest groups, as stated above. In addition, this theory acknowledges that local government does not always encourage changing local needs and the structure which exists within local government may not allow change to take place. The dominance of the committees in reaching decisions, the bureaucratic composition of local government and the professional influence from council officers can all control the amount of local interests and factors which impact on the local policy-making process.

The need for major changes within local government is an important element of the localist theory. These reforms would include the introduction of a charter which would clearly identify the roles of central and local governments; local income tax; proportional representation in local elections; unitary local authorities; a new management style for council officers; and more concern for the needs of their customers.

This theory has been widely accepted within local government; however, it has also been widely criticised. These criticisms include the questioning of the need for local government at all. It is accepted that there is a need for a central government which develops new policies but many would argue that the need for a local government is debatable.

Another criticism of the localist theory is that local authorities do not effectively represent the views of their residents. The market is regarded, therefore, as the best means of representing the views and desires of the population as it is less liable to the distortions which political processes encounter. Furthermore, other critics of this theory have pointed to the fact that the local political mechanism is subject to issues such as racism and sexism and so cannot effectively represent the local population.

It has been argued that local government has been at the forefront of attempting to implement effective equal opportunity policies in order to change the orientation of local government. However, critics of localist theory state that the local government structure is difficult to reform and that reform can only be achieved where local authorities make a conscious choice to bring it about.

Public choice theory

The public choice theory (also known as the New Right theory) states that the most effective means of allocating resources and decision-making is the market, and so the public sector and elected democracy are not constructive methods. The public sector is seen as driven to excessive growth and consequent expenditure and is always in excess of what the true market level of growth and expenditure would be.

Thus, local authorities are seen as much more interested in their own bureaucratic welfare than that of their residents. Departments within local authorities will therefore be more inclined to be 'pushing continuously for budgetary growth, which increases their numbers, improves promotion prospects, creates discretionary patronage and generally builds up organisational slack and improves job security' (Stoker, 1991: 240).

It is believed by the public choice theorists that there are no adequate checks in place to curb expenditure as the level of taxation can be raised if needed to meet increased spending levels. Politicians are also inadequate in this role as they are heavily influenced by the information provided to them by the bureaucrats regarding the costs of introducing, providing or improving services. The committee structure which exists in local authorities is also seen as inadequate as councillors soon come to share the interests of council officers and/or members of the public.

Thus, local government is seen as inherently wasteful and inefficient. Self-interest groups such as trade unions and professionals tend to dominate and the control which managers and politicians attempt to wield is relatively weak. The views of their customers, who see the services provided by local authorities as poor quality and relatively expensive, are not seen as an important issue.

The party political process helps to build up the public to expect more than they really need and once elected the politicians attempt to meet their promises by the use of deficit funding, which spreads the costs over many years.

In addition, interest groups are seen according to this theory as constantly pressing for more expenditure to improve the services which meet their own particular interest. The silent majority are seen as the losers, especially with regard to local authorities, where it is felt that pressure groups are much more effective in achieving their aims.

The public choice theorists regard elections as a poor reflection of the views of the population, as people are forced to vote on a range of issues. However, the market mechanism would be much more effective in offering a range of choices and options which represented the views of the public.

A number of reforms are advocated by the public choice theorists in order to balance the detrimental effect of local government. The first revision includes the concept of contracting out many of the functions which are performed by local authorities, such as housing management. This philosophy is preferred by the theorists, as competition between bureaucrats and the private sector is intended to force local government to compare the costs of service provision and so become more cost-effective as well as more efficient. It is assumed by the public choice theorists that the private sector is much more efficient and cost-effective as it is competing in the market and would not survive unless it was able to meet the demands of the market mechanism. The contracting out of services is also intended to reduce the power of the trade unions and the professionals within local government.

The second reform which the public choice theorists advocate is the breaking up of the large bureaucratic local authorities. The replacement of the large bureaucracies with smaller local authorities should take place as it is considered that local people would be better represented. In addition, the customer would be able to make choices and move to those smaller local authorities which offer the best value for money instead of having to rely on large monopoly bureaucracies.

Further changes recommended by the public choice theorists include the need to link the income and other rewards which council officers receive to their ability to achieve savings and budgeting targets such as performance-related pay. It is suggested that councillors should use consultants who are more able to challenge the information provided by council officers and are also much more objective.

However, there have been a number of criticisms of this theory: first, the supposed over-supply of services by local government overlooks the fact that the role of local authorities is to meet the needs of the public and often the market has to supplement the supply in order to meet demands. One example of this is with regard to the supply of temporary accommodation for the homeless, which has had to be supplemented by bed-and-breakfast accommodation or the renting of properties in the private sector.

There has also been criticism raised with regard to the use of the market mechanism as the barometer of public needs. The market is seen as inherently defective as it does not represent the full range of views from customers in terms of preferences and quality.

Markets are seen as creating insecurity and instability among both producers and consumers.

Those members of the public who are on low incomes or claiming welfare benefits and who therefore have to rely heavily on the services provided by the local authorities may find that it is not so easy for them to make choices or to go elsewhere in order to achieve value for money.

Finally, although it can be discounted that local authority officers are concerned with budget maximisation, commentators argue that local government is inclined to staff maximisation since officers are motivated to fulfil their policies and serve the public.

The dual state thesis

The dual state (or dual politics) thesis was developed by Cawson (1978) and Saunders (1986). This thesis is based on the difference between social investment and social consumption as functions of the state. Social investment is concerned with maintaining the production of goods and services in the economy by supporting the profitability of firms in the private sector. Thus, the state would produce raw materials and services and provide the physical infrastructure and financial support, all of which help the private sector to maintain its profits. In contrast, social consumption is aimed at supporting the consumption of various groups in the population who cannot fulfil their needs through the private sector, such as housing.

It is suggested that different types of political policies have developed as a direct result of provision of services through production or consumption orientation. The former is seen as a much more closed sector which works with a few chosen groups, such as successful producers and professional bodies, whereas the latter case involves a much wider group of bodies and people with competing and diverse needs.

This thesis supports the view that investment in social consumption is managed nationally by central government while the consumption is organised and managed by local authorities. Saunders (1981) states that the provision of social consumption through local authorities has resulted in various consumption groups competing for services on the grounds of need. While local government is seen as effective in providing the needs of the local people, Saunders (1981) considers that the local councils are hampered by the policies from central government.

The dual state thesis has been criticised on a number of points. The division of production and consumption services between central and local government has been discredited as central government is also responsible for social consumption areas such as welfare benefits. Furthermore, many local authorities are increasingly involved in social investment with economic implications as they become more involved in partnerships with the private and quasi-private sector. There has also been some criticism for not addressing the officials who are involved in the policy-making process and the effect they can have on the outcome.

Local state theory

Duncan and Goodwin (1988) suggest that the local state provides an understanding of how society functions and why. It also represents both the differing local interests of local authorities and the national policies which affect the way that these local interests are implemented. This can result in conflict between central–local control especially if the policies of central government clash with the policies which the local authority may wish to implement in order to reflect its differing local structure in terms of class, ethnicity and/or culture.

Criticisms of this theory include the suggestion that the model is too simplistic and does not reflect the fact that local politics is often dominated by national policies and so local politics is not as dominant in the minds of the public as may be suggested. There is also concern that this thesis does not portray a true picture of the role that central government has with regard to local needs. Finally, the local state is criticised for being too ideological and not concerned with turning ideology into practice.

HOUSING ASSOCIATIONS

Structure

The National Housing Federation (1995) states that an effective control framework within housing associations is extremely important and is possible through the use of four key aspects. First, with regard to the planning framework a mission statement will be produced which will state what the organisation is trying to achieve and a corporate plan developed as a medium-term account of the housing association's goals. The central framework is further broken down into budgets and annual plans which can also be departmental in large organisations (see Figure 10.1).

Second, the delegation framework is another aspect of control which has been identified by the NHF (1995). In this case the board has to delegate work to the staff employed within the organisation. Thus, the staff would be responsible for deciding the staff structure which the organisation will use, the completion of job descriptions for staff and statements which define the limits of authority which senior managers may have.

The third area of control relates to the policy framework, which sets out the standards which the housing association is expected to achieve. The policies will identify the legal, ethical and regulatory standards in areas such as equal opportunities, tenant participation and borrowing strategies. There will be external factors which may attempt to influence what the standards should be, such as tenants, local authorities, the Housing Corporation, NHF and legislation. However, it is up to the housing association to decide how best to achieve the standards which have been set.

Finally, the procedural framework has been identified by the NHF (1995) as an aspect of control in housing associations as it sets out how the tasks should be carried out in order to meet the organisation's objectives. The key to this framework is that the housing association decides the rules which the board/committee members will follow, and the board/committee then decides on the overall procedures or standing orders which staff should adhere

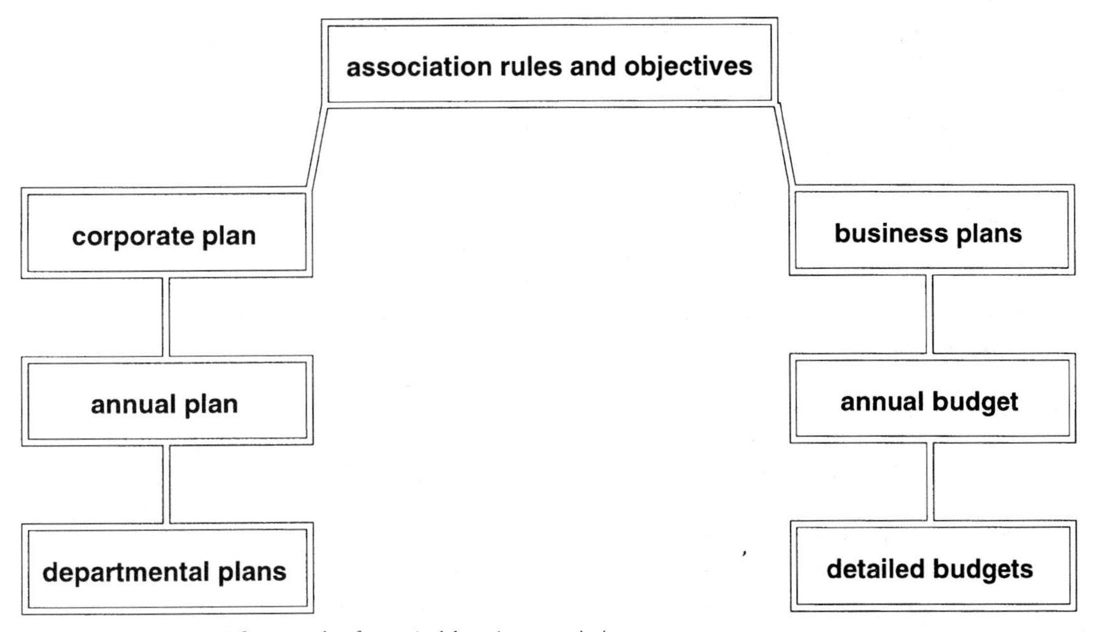

Figure 10.1 The control framework of a typical housing association

to, and issues financial regulations and procedure manuals for the use of staff (see Figure 10.2).

Board/committee members

Members of the board or committee of a housing association are legally responsible for the work carried out by the organisation. Board members have much more responsibility for the housing association than do committee members. Members need to have an effective control system in place to ensure that the housing association functions well. The members are responsible for the quality and direction of the work of the housing association and how best to achieve their aims.

It is essential that the members regularly review the strategy and direction of the organisation to ensure that best practice is in place. It is also important that the members monitor the work of the housing association to ensure that the control systems in place are actually being implemented by staff. There may be reasons for the control systems not functioning such as fraud, failure of the systems to work well in practice and lack of training.

The NHF (1995) has identified nine important areas which a successful committee or board needs to address to ensure that it is functioning effectively. They are as follows:

1 a strong committee;
2 a relevant committee structure;
3 well-conducted meetings;
4 high-calibre staff;
5 an effective staff structure;
6 clear objectives and plans;
7 clear policies and procedures;
8 clear delegation;
9 effective monitoring and compliance checks.

The NHF (1995) states that it is just as important for members to establish strategy and guidance for staff as it is for staff to identify aims and direction for the members to endorse. The system encourages the need to work as a partnership with both staff and members seeking to take responsibility for the goals which are set and agreed.

The members may identify areas such as policy which they would like to evaluate, review or adopt. The members are effectively involved in risk management and are concerned to ensure that risks are identified and monitored closely through the use of systems such as contingency plans (NFHA, 1992). The senior staff within the

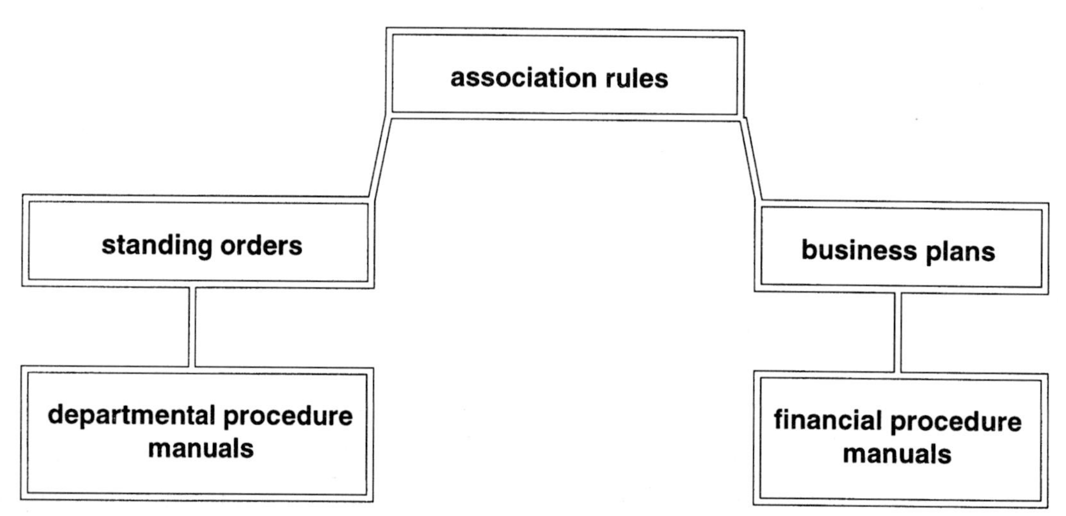

Figure 10.2 The procedural framework of a typical housing association

organisation will therefore prepare reports as requested and may also initiate reports of their own which will include the possible range of choices available to the members.

The business plan of the housing association is of particular importance to members as it will cover a three- to five-year period. In addition, there is the annual budget which is also scrutinised by the members. It is important that the resources which are required to meet the objectives of the organisation are available and managed in an effective manner. Thus, it is important that income generating is effective and the delegated powers to staff to manage the budgets are monitored closely by the members who have overall responsibility for the organisation's business and financial plans.

CONCLUSIONS

This chapter initially considers the executive, legislative and judicial functions of central government and looks at the local policy-making process and the policy-making process in housing associations.

Clearly, political activity is a necessity and our society cannot manage without it. Politics is about our society's conflicts and disagreements and the various means of attempting to resolve them. The continuing cohesion of our society is based on how well such disagreements are contained, modified, postponed or settled.

The 1980s and 1990s have seen significant changes in the relations between local government and the Housing Corporation with central government. While central government has remained the same structurally throughout the period, the way that disagreements between central government and local government have been dealt with has left many local authorities and the Housing Corporation implementing policies which may be contrary to the political ideology by which they work.

Since this chapter provides the governmental context in which housing policy and strategy is developed and applied, it is of particular relevance to Chapter 3 on housing markets and policy, Chapter 6 on planning and housing developments, Chapter 7 on environmental health and housing and Chapter 9 on management.

QUESTIONS FOR DISCUSSION

1 Discuss whether there has been an increase or a decrease in the importance of party politics in local government during the last ten years.

2 Discuss the methods by which central government controls and influences local government, giving examples of how they have been used since 1979.

3 'Since 1979 experience has shown that determined ministers are not controlled by their civil servants.' Discuss.

4 Critically examine the stages through which a public bill passes in both Houses of Parliament, with reference to recent housing legislation.

5 Discuss the changing role of the housing association movement since 1979.

RECOMMENDED READING

Balchin, P. (1995) *Housing Policy: An Introduction*, Routledge, London.

Cope, H. (1990) *Housing Associations: Policy and Practice*, Macmillan, London.

Elcock, H. (1986) *Local Government: Politicians, Professionals and the Public in Local Authorities*, 2nd edn, Methuen, London.

Madgwick, P.J. (1984) *Introduction to British Politics*, 3rd edn, Hutchinson, London.

Stoker, G. (1991) *The Politics of Local Government*, 2nd edn, Macmillan, London.

REFERENCES

Balchin, P. (1995) *Housing Policy: An Introduction*, Routledge, London.

Cawson, A. (1978) 'Pluralism, Corporation and the Role of the State', in *Government and Opposition*, vol. 13.

Duncan, S. and Goodwin, M. (1988) *The Local State and Uneven Development*, Polity, Cambridge.

Dunleavy, P. (1980) *Urban Political Analysis*, Macmillan, London.
Elcock, H. (1986) *Local Government: Politicians, Professionals and the Public in Local Authorities*, 2nd edn, Methuen, London.
Gilroy, R. and Woods, R. (eds) (1994) *Housing Women*, Routledge, London.
Grant, M. (1986) 'The Role of the Courts in Central–Local Relations', in M. Goldsmith (ed.) *New Research in Central–Local Relations*, Gower, Aldershot.
Heclo, H. and Wildavsky, A. (1974) *The Private Government of Public Money*, Macmillan, London.
Loughlin, M. (1986) *Local Government in the Modern State*, Sweet & Maxwell, London.
Madgwick, P.J. (1984) *Introduction to British Politics*, 3rd edn, Hutchinson, London.
Malpass, P. and Murie, A. (1990) *Housing Policy and Practice*, 3rd edn, Macmillan, Basingstoke.
National Federation of Housing Associations (NFHA) (1992) *Risk Management for Committee Members*, National Federation of Housing Associations, London.
National Housing Federation (NHF) (1995) *Internal Control: A Guide for Voluntary Board Members*, National Housing Federation, London.
Rhodes, R. (1988) *Beyond Westminster and Whitehall: The Sub-Central Government of Britain*, Allen & Unwin, London.
Ridley, N. (1988) *The Local Right: Enabling Not Providing*, Centre for Policy Studies, London.
Saunders, P. (1981) *Social Theory and the Urban Question*, Hutchinson, London.
—— (1986) 'Reflections on the Dual Politics Thesis: The Argument, Its Origins and Its Critics', in M. Goldsmith and S. Villadsen (eds) *Urban Political Theory and the Management of Fiscal Stress*, Gower, Aldershot.
Stoker, G. (1991) *The Politics of Local Government*, 2nd edn, Macmillan, London.
Treasury (1961) *The Report of the Committee on the Control of Public Expenditure* (under the Chairmanship of Lord Plowden), HMSO, London.
Widdicombe, D. (1986) *Research Volume II: The Local Government Councillor*, Cmnd 9799, HMSO, London.

11

CONCLUSION

Paul Balchin, Maureen Rhoden and John O'Leary

Although good-quality, secure housing is essential to the well-being of every household, housing professionals witnessed a substantial increase in the housing problems faced by individuals and families during the last two decades of the twentieth century. House-building by local authorities and housing associations plummeted from over 107,000 starts in 1978 to 34,000 in 1996, and an unacceptable number of houses (within the total housing stock) remained unfit or in need of urgent repairs. In the owner-occupied sector, the homes of more than 1 million people were repossessed during the slump in the housing market (1990–95) and nearly 2 million households were affected by negative equity. Throughout the 1990s, because of the shortage of affordable housing, homelessness was at least twice as high as it had been in the late 1970s. These problems were, in large part, the result of Conservative governments reducing investment in social housing by half in real terms (1979–96), decreasing the availability of renovation grants in the private sector, and terminating the duty of local authorities to provide permanent housing for homeless families.

As to the future, within the broad context of housing markets and finance, the housing needs of disadvantaged segments of society, town planning, construction, environmental health, property and housing law, management and local government, there will be both opportunities to solve the housing problems inherited from the latter years of the twentieth century and continuing causes for concern.

Clearly, well into the first decade of the twenty-first century, the relationship between housing supply and housing need will be substantially determined by policies employed by the Labour Party following its return to office after the general election of 1 May 1997.

Under the Local Authority (Finance) Bill 1997, the incoming government proposed that £5 billion of capital receipts from the sale of council houses would be gradually released and re-invested in building new houses and rehabilitating new ones. Since most receipts are held by suburban authorities with little housing need, while areas of greatest need (the large metropolitan authorities) often have little or no money at their disposal (having used receipts to pay off debts), the government proposed to redistribute funds through the established credit approval system. Capital-starved authorities would thus be permitted to increase their spending, initially by £900 million, 1997–8 to 1998–9, while cash-rich authorities would be unable to spend their receipts above the formerly imposed 25 per cent limit.

The impact of this approach, however, might be little more than marginal; the Labour government, in accepting the plans of its predecessors, committed itself to a £1.3 billion cut in housing investment, 1997–8 to 1998–9; many inner city authorities could use the redistributed capital to rehabilitate their stock rather than add to it, and at most only 40,000 homes could be built each year instead of the 100,000 widely accepted as necessary.

If metropolitan authorities, in particular, were encouraged to divert their extra capital resources to housing associations (which could then top up these funds with private finance), or if local authorities were permitted to borrow privately, one would expect to see an appreciably greater number of houses being built or rehabilitated.

This would be consistent with Labour's declaration that (subject to the support of tenants concerned) it would support a three-way partnership between public, private and housing association sectors to promote good social housing; for example, through the deployment of private finance to improve the condition of stock and to provide greater diversity and choice (Labour Party, 1997).

Recognising, however, that most families wish to own their own homes, the Labour Party (wary of the boom-and-bust policies which had caused the collapse of the housing market in the early 1990s) pledged to work with mortgage lenders to encourage the provision of more flexible mortgages to protect families at times of job insecurity, to safeguard borrowers from the sale of disadvantageous mortgage packages, and to tackle the problem of gazumping (Labour Party, 1997). Similarly, to counter house price inflation, mortgage interest relief at source (MIRAS) was reduced from 15 to 10 per cent in Labour's budget of 2 July 1997 and stamp duty was increased from 1 to 1.5 per sent on purchases above £250,000 and to 2 per cent on transactions above £500,000.

Labour also aimed to support efficiently run private rented housing and, by means of a licensing scheme, to provide protection for the most vulnerable tenants – those in houses in multiple occupation. It also favoured the simplification of rules restricting the purchase of freeholds by leaseholders, and the introduction of 'commonhold' to enable people living in flats to own their own homes individually and to own and manage the common parts collectively.

Despite the above policy initiatives, there will still be people homeless through no fault of their own. Thus for those in 'priority need', the incoming Labour government recognised that it was important to restore the duty of local authorities to provide permanent rather than temporary housing – a 'safety net' for the most disadvantaged.

All of the above measures, if effectively implemented, should ensure that there is an increase in the degree to which housing needs are satisfied. There are, however, a number of causes for concern relating to equality and efficiency with regard to housing investment and finance.

First, investment in the social housing sector is still severely constrained by limitations on public expenditure – house-building in this sector remaining at an unsatisfactorily low level. An increased reliance on private finance by housing associations and other social sector landlords inevitably pushes up rents and necessitates an escalation of housing benefit payments. A shift of emphasis from benefits to bricks-and-mortar subsidies should be therefore seriously considered as a more efficient and less inflationary means of satisfying the need for affordable social rented housing.

Second, owner-occupiers and particularly first-time (and relatively low-income) house-buyers might be better protected by a form of mortgage benefit than by the provision of more flexible mortgages or the continued existence of MIRAS.

Third, in line with many other European countries, to ensure an adequate supply of efficiently run and profitable private rented housing, landlords in the United Kingdom might be willing to trade their right to charge market rents (under shorthold) for depreciation allowances and a return to a system of 'fair rents' – not least in respect of houses in multiple occupation. But in the United Kingdom (at the time of writing) they have been denied this form of support.

Fourth, for the homeless not in 'priority need', a greater quantity and higher quality of temporary accommodation should be made available under an improved 'Rough Sleepers Initiative', and full income support should be available from the age of eighteen rather than twenty-five.

Finally, the very specific housing problems of the

elderly, women, black and ethnic minority households and those suffering from ill-health need to be addressed and solutions found. With regard to the elderly, it can only be hoped that the Royal Commission promised by the Labour Party in its 1997 manifesto satisfactorily works out a fair system for funding long-term care and that a 'long-term care charter' is introduced defining the standard of services which people are entitled to expect from housing, health and social services.

There are also a number of important issues to be addressed by policy-makers at the interface between housing and planning as we enter the new millennium. There is, for example, a view held in parts of the planning movement that planning-based mechanisms should be used to help achieve the provision of affordable housing. The Department of the Environment was lobbied for changes to *Circular 13/9* (DoE, 1996), which in its present form limits the ability of local authorities to negotiate for the inclusion of social housing in smaller and medium-sized housing developments on a threshold basis. The threshold approach might be considered unnecessarily restrictive, given the overall bias towards the provision of market housing (see Chapters 2, 3 and 6). A market-orientated view of the situation is that market forces will seek out and provide niches of housing provision in any particular geographical area. For example, the market should, following this view, bring forward 'affordable housing' where incomes are lower than average (for whatever reason). A 'market-sceptic' view would be that the housing market does respond quite well to market signals from the upper reaches of the market, but that market forces either falter or do not work in income-depressed localities. In the latter areas demand is depressed, leaving housing needs largely un-met. Differences of opinion will remain at the political level about the efficacy of markets to resolve the problem and indeed the degree to which regulation is needed, or desirable, to steer or bend market forces in the direction of social housing provision.

Housing design is another issue. During the mid-1990s, the Department of the Environment championed the cause of creating quality environments through research, publicity, ministerial speeches, sponsorship of design competitions and ultimately in policy (DoE, 1997). The 'discovery' by the DoE and sponsorship of the concept of urban design by former Secretary of State John Gummer looks set to continue under the present administration. Governmental enthusiasm for quality environments has coincided (fortuitously or deliberately) with a recognition that there may have to be significant land releases to provide new homes to meet projected household formation. The professional bodies and pressure groups have been supportive of the quality initiative on the part of government, although the difficulties of implementation at the grass-roots level lie ahead. One of the difficulties in the housing field is to persuade some of the volume house-builders to adopt and support subtle qualitative concepts embodied in urban design. Volume house-builders and local authorities have not historically experienced cordial relationships, the house-builders viewing the involvement of local authorities in design matters as unnecessary interference in the market mechanism. The history of the implementation of design guidance is perhaps an indicator as to how the new quality initiative might fare in the real world of implementation. The historical results have been patchy, resting largely on the political will of council members and officer expertise at the local level.

The concept of sustainability, a third planning issue, has gained considerable momentum and support in the world of planning and development over the last decade and Local Agenda 21 strategies are beginning to surface at the local authority level. Sustainability has begun to shape ideas about settlement design and density, and the indications are that this process will permeate the debate about where to provide for the new households projected for the year 2016 (see Town and Country Planning Association, 1996). One school of thought is that the solution lies not just in developing the available brownfield sites (for example, the Royal Docks or Greenwich Peninsula) but in developing them at high density along with other windfall sites within urban envelopes. Another school of thought, led by the Town and Country Planning Association, is that the solution lies in developing stand-alone and self-sustaining new settlements on carefully selected sites beyond existing conurbations. Such settlements could achieve the broad principles of sustainability if carefully designed and would have the benefit, supporters claim, of taking the development pressure off the existing urban areas where 'town cramming' is thought to have become an issue. There are obvious parallels here with the long-abandoned new towns programme, although it is unlikely in the present political and economic climate that such settlements could be solely driven and financed by the public sector. Partnership-driven new settlements are the more likely scenario, if indeed a political consensus develops around the principle of new settlement building.

With regard to environmental health, it is of considerable concern whether the present regime of grants available for rehabilitation in the private sector and governmental funding of housing renewal in the social sectors will be sufficient to make serious inroads into the problem of poor housing (at the most, only £5 billion of capital receipts will be available from the sale of council houses). Leather and Morrison (1997) reported that there were 1.6 million homes unfit for habitation in the United Kingdom (or 1 in 14 of the total housing stock), and a large minority of the remainder needed urgent repairs. A sum of £20 billion needs to be spent on repairing council houses alone. Demolition of unfit housing, moreover, plummeted so dramatically over the period 1970–95 that in England and Wales new housing built at the end of the twentieth century would have to stand for 5,600 years before being replaced at current rates of demolition.

A significant part of housing policy will, of course, need to be implemented through changes in property and housing law; for example, the intended protection of tenants in houses in multiple occupation; the introduction of 'commonhold' to enable people living in flats to own their homes individually but the whole property collectively; and the re-imposition of a duty on local authorities to protect households in priority need who are homeless through no fault of their own.

Finally, if Labour's 1997 manifesto is implemented, housing management will need to respond positively to the challenges resulting from local authorities no longer having to put their services out to tender. With the end of compulsory competitive tendering, local authorities (including their housing managers) will be required to publish a local performance plan with targets for service improvement, and be expected to achieve them. Increasingly this will be within the arena of public–private partnership arrangements, for example the setting up of non-profit-making local housing companies (in which local authorities will have seats on the board) to take over the ownership and management of council estates under the provisions of the Housing Act 1996.

REFERENCES

Department of the Environment (DoE) (1996) *Circular 13/9. Planning and Affordable Housing*, DoE, London.

—— (1997) *General Policy and Principles* (superseding 1992 version), DoE, London.

Labour Party (1997) *New Labour Because Britain Deserves Better*, Labour Party, London.

Leather, P. and Morrison, T. (1997) *The Stae of UK Housing: A Factfile on Dwelling Conditions*, Policy Press, Bristol.

Town and Country Planning Association (1996) *The People: Where Will They Go?*, TCPA, London.

INDEX

Entries in bold denote major section/chapter devoted to subject. t denotes table, f denotes figure and p denotes plate.